Animal Nutrition

NIPA® GENX ELECTRONIC RESOURCES & SOLUTIONS P. LTD.
New Delhi-110 034

About the Authors

Dr. Anand Kumar Pathak was born on 5th January, 1974 and did graduation and post graduation from Veterinary College, Durg and PhD from IVRI, Izatnagar. He is currently working as Associate Professor in the Division of Animal Nutrition, Faculty of Veterinary Sciences and AH, SKUAST-Jammu. He started his carrier as Junior Research Fellow in the Department of Medicine and Animal Nutrition, Veterinary College, Durg followed by Scientist Cum General Manager (Poshak Feeds India Pvt Ltd), Karnal and Research Associate in PN Laboratory AN Division IVRI Izatnagar, Assistant Professor and then Associate Professor Division of Animal Nutrition SKUAST-Jammu. He has 23 years of research experience in the field of Clinical Nutrition (Ruminants and Poultry). He was instrumental and pioneer worker in designing, fabricating and developing manually operated multi nutrient block, densified complete feed block and densified functional feed block making machines and commercialized condensed tannins enriched multi nutrient block (MNB-CT) technology. He has published more than 80 research and review papers in the journals of national and international repute, 11 books, 62 book chapters, 41 popular articles, presented papers/ delivered invited lectures (more than 203) in conferences, workshops and trainings. He has successfully guided 10 MVSc students in the field of Clinical Nutrition (Ruminants & Poultry). Dr. Pathak is recipient of Awards like Distinguished Scientist Award, Award for Excellence in Research, Dr. Verghese Kurien Best Scientist Award, Best Innovator Award for multi nutrient and densified complete feed block making machines, Publons Peer Reviewer Award, Best Reviewer Award, NAVS Associate, Outstanding Scientist Award, Presidential Appreciation Award and ANA Dr. M S Rahal ANFT Best Paper Published Award, more than 22 Best Oral and Poster Presentation Awards and his 2 MVSc students got Sushila Pathak memorial Award for Best Thesis for master in Veterinary Sciences. He is also active member of many scientific societies and bodies as life member and portfolio holder and Fellow of Bose Science Society, Tamil Nadu, India, Fellow of International Society for Research and Development, UK, Fellow of National Academy of Veterinary Nutrition and Animal Welfare, Bareilly and Fellow of Animal Nutrition Association and recipient of CLFMA Award-2023. He was associated with many projects as PI and Co-PIs; in trainings as Coordinator and Co-coordinators, and village visit and re-visit programmes of the University. He has an expertise in feed formulation and preparation of least cost, eco-friendly functional feeds for livestock and poultry from locally available conventional and un-conventional feed resources. He has developed 3 entrepreneurs and established 3 DFFBs making units in 3 villages of Jammu as small scale business enterprises for self employment generation for rural unemployed youth under JKST&IC, DST project as PI.

Prof. (Dr.) Nitya Nand Pathak, Birth 1st July, 1943. V&P.: Nagwa, District Ballia, Utta Pradesh, Bharat.

Education: Upto 8th Village School Nagwa and Akbar; High School-Government High School, Ballia; Intermediate- S.C. College, Ballia; B.Sc. K.K. College, Lucknow University; B.V.Sc. & A.H. 1966 & M.V.Sc. 1970 Veterinary College, Mathura; Ph.D. 1973 Indian Veterinary Research Institute, Izatnagar - Agra University.

Teaching: Post graduate 28 years.

Research Subjects: Cattle, Buffalo, Sheep, Goats, Horse, Ass, Pigs, Rabbits, Fowl, Guinea Fowl, Laboratory Animals, New forages.

Research Papers in National Journals 275+ and International Journals 80+.

Text and Reference Books on Cattle, buffalo, Yak, Mithun, Animal Nutrition, Animal Production, Forage, Avian Nutrition, Nutritional Microbiology, Disaster Management, Nutrients requirements, Analytical Techniques-45.

Advisory Services: Project Leader, Buffalo Development Project, S.R. Vietnam 1982-84; Programme Group Small Holder Animal Production, International Livestock Research Institute, Addis Ababa, Ethiopia.

Prizes/Awards: 1st Prize Essay Competition 1961-62. K.K. College, Lucknow; 1st Prize, Essay Competition 1962-63., Veterinary College, Mathura; 1st prize, Cattle Judging 1965-66, ICAR, New Delhi. Jawaharlal Nehru Award, 1974, ICAR, New Delhi. Institute level Best Teacher Award 1996, Indian Veterinary Research Institute, Izatnagar. National Research Development Corporation Award 2005 and World Intellectual Property Organization, Geneva award certificate 2006.

Before starting of Centre of Advanced Studies in Animal Nutrition, conducted one Summer Institute, ICAR as Co-Director; and two on Clinical and Wildlife Nutrition as Director. In these courses teachers, scientists and research scholars participated from all corners of Bharat. Founder President, Animal Nutrition Association, ICAR-IVRI, Izatnagar. Founder President, Pashu Poshan Kalyan Samiti and National Academy of Veterinary Nutrition and Animal Welfare, Bareilly. Founder President - National Association of Wildlife and Zoo Veterinarians, Bareilly. Former President, Animal Nutrition Society of India,

ICAR-NDRI, Karnal. Former Vice President, Indian Society for Buffalo Development, ICAR- CIRB, Hissar.

Students under guidance and co-guidance-Achievements Dr. S.V. Deshmukh, Dean & Registrar, Veterinary University, Nagpur; Dr. A. Mandal, Director, ICAR-CARI, Izatnagar; Dr. S.S. Giri, Expert Animal Production, SARC Centre, Dhaka, Bangladesh; Dr. A. Sahoo, Director, ICAR-CSWRI, Malpura, Avikanagar and NRC on Camel, Bikaner; and Now Director, NIANP, Bangalore; Dr. R. Bhar, Incharge, IVRI Regional Centre, Palampur; Dr. B.C. Mondal, Controller-Examinations and Head, Animal Nutrition, GB Pant University, Pantnagar; Prof. Dr. S.P. Tiwari, Dean, Veterinary College, Anjora and Vice Chancellor, Nanaji Deshmukh Veterinary University, Jabalpur; Dr. K.K. Baruah, Director, ICAR-NRC Yak and Dean, College of Fisheries, Agricultural University, Assam; Dr. Hyder Ali, Joint Director, Animal Husbandry and Veterinary Services, Andhra Pradesh.

Animal Nutrition
Nutrients and Their Role in Animal Health

Anand Kumar Pathak
BVSC & AH, MVSc (Hons), PhD (IVRI),
FBSS, FISRD, FNAVNAW, FANA
CLFMA Awardi
Associate Professor
Division of Animal Nutrition
Faculty of Veterinary Sciences & Animal Husbandry
Sher-e-Kashmir University of Agricultural Sciences & Technology of Jammu
R.S. Pura-181102, Jammu (UT of J & K), India

Nitya Nand Pathak
BVSc & AH, MVSc, PhD, Ex-Director, CIRB
Hisar & Founder President of
National Academy of Veterinary Nutrition and Animal Welfare
Bareilly (UP), India

NIPA® GENX ELECTRONIC RESOURCES & SOLUTIONS P. LTD.
New Delhi-110 034

NIPA® GENX ELECTRONIC
RESOURCES & SOLUTIONS P. LTD.
101,103, Vikas Surya Plaza, CU Block
L.S.C. Market, Pitam Pura, New Delhi-110 034
Ph : +91-11-43860225, Mob.: +91 9717133558, 9540816132
E-mail: newindiapublishingagency@gmail.com
Website: www.nipaersources.com

Print ISBN: 978-93-58873-83-2
ebook ISBN: 978-93-58871-88-3

Composed and Designed by NIPA®.

Preface

The text book of Animal Nutrition: Nutrients and their role in Animal Health is a vital resource that delves into the complex relationship between animal nutrition and metabolic health. It begins with an overview of the basic principles of nutrition, detailing the essential nutrients such as carbohydrates, proteins, fats, vitamins and minerals. It highlights their roles in the physiological processes of various species of livestock and poultry. It provides a detailed exploration of nutritional and metabolic disorders, emphasizing their diagnosis, management and prevention. This makes it an essential guide for veterinary practitioners, students and researchers. Each chapter is structured to provide clarity on nutrient metabolism, mechanism of action, pathophysiology, control and prevention of disease/ disorder through nutritional management while discussing species-specific dietary nutrients and their requirements.

The book highlights the critical role of proper nutrition in preventing metabolic disorders such as ketosis, hypocalcemia and lactation tetany, etc. It examines the consequences of nutritional imbalances on acid-base balance and metabolic stability, providing insights into identifying and addressing these issues through tailored feeding strategies and dietary interventions.

One of the key features of this book is its focus on formulating balanced diets. It provides step-by-step guidelines for developing diets tailored to the unique needs of animals, factoring in age, breed, health status and production goals. It also delves into specialized areas such as nutritional management of diseases, weight control and feeding practices during different life stages like growth, reproduction and lactation.

Dehydration and its metabolic implications are discussed in depth, including its impact on electrolyte levels and acid-base equilibrium. The text outlines diagnostic methods and treatment protocols, such as rehydration therapy and fluid management. It also emphasizes the importance of nutritional support during recovery from dehydration. Another significant aspect is the discussion of emerging trends in veterinary nutrition. Topics like impact of nutrition on immune function and the role of omega-3 fatty acids, probiotics and prebiotics in gut health are explored in detail. These insights enable veterinarians to adopt evidence-based practices and stay updated on advancements in the field.

Blood transfusions are another focus area, with detailed guidelines on indications, techniques, and compatibility testing. The book also explores laboratory tests for assessing haemostasis, including coagulation profiles and platelet function tests, which are critical for managing bleeding disorders in animals.

Through the book is a combination of theoretical knowledge and practical approaches and a vital resource for understanding the interplay between diet and animal health. The textbook of veterinary nutrition equips veterinarians with the tools to address complex nutritional and metabolic challenges, the knowledge to make informed nutritional decisions, ultimately improving animal health and welfare. Its integration of nutrition with clinical pathology and therapeutic strategies sets it apart as a comprehensive guide of veterinary nutrition for advancing the field of veterinary science.

We hope that this textbook will inspire readers to appreciate the significance of veterinary nutrition in the health and well-being of animals. By fostering a deeper understanding of the subject, we aim to contribute to the development of skilled professionals who are committed to advancing animal health, productivity and welfare in a sustainable and ethical manner. It is not merely a repository of information but a tool designed to empower its readers. It is our sincere hope that it will serve as a valuable companion for all those dedicated to the study and practice of veterinary nutrition, veterinary sciences, animal care and welfare.

Anand Kumar Pathak
FVSc & AH, SKUAST-J, R S Pura, Jammu
Nitya Nand Pathak
National Academy of Veterinary Nutrition & Animal Welfare
Bareilly (U.P.), India

Contents

1

Introduction

Nutrition plays a vital role in the health, well-being and longevity of animals. The science behind nutrition encompasses the dietary needs of animals and how these needs are met through various forms of food/ feed, supplements and feed additives. A well-balanced diet is fundamental to the health of both domesticated and wild animals. Veterinary nutrition takes into account the physiology, biology and pathology of animals, aiming to optimize their health, performance and longevity through the right nutritional approaches.

Proper nutritional management is crucial not only for growth and reproduction but also for disease prevention and recovery. Veterinary nutrition encompasses understanding metabolic and nutritional disorders, gastrointestinal disturbances, metabolic imbalances and systemic conditions. It focuses on providing animals with balanced diets to meet their physiological needs. The right blend of carbohydrates, proteins, fats, vitamins, and minerals supports growth, immune function and production. Deficiencies or excesses in nutrition can lead to disorders such as metabolic imbalances, gastrointestinal disturbances, and organ dysfunction. Understanding nutritional science enables veterinarians to address and prevent conditions like ketosis, fatty liver, and metabolic acidosis. Veterinary nutrition and the understanding of associated metabolic and systemic disorders are essential for ensuring animal health and productivity. From preventing nutritional deficiencies to managing complex metabolic conditions, veterinarians play a critical role in maintaining the well-being of animals and poultry. By addressing these challenges through research, innovation, and education, we can ensure sustainable animal husbandry and improved welfare.

Nutritional and metabolic disorders are a significant concern in animal and poultry management. These disorders occur when there is an imbalance between an animal's nutrient intake and its physiological requirements, leading to impaired health, productivity, and, in severe cases, mortality. Nutrition is a cornerstone of animal and poultry health, determining growth, reproduction, production, and immunity. Balanced diets provide essential nutrients, including carbohydrates, proteins, fats, vitamins, minerals, and water, which collectively sustain metabolic processes and maintain overall health. Any deficiency, excess, or imbalance in these nutrients can disrupt physiological functions, resulting in nutritional and metabolic disorders.

In commercial animal and poultry production, where the goal is often to optimize productivity, the margin for error in nutritional management is narrow. Intensive farming systems and genetic selection for high production (e.g., milk, meat, eggs) have increased the metabolic demands of animals and birds, making them more susceptible to dietary imbalances and metabolic diseases.

Protein and energy deficiencies are prevalent in poorly managed systems where feed quality is inadequate. Symptoms include reduced growth, poor body condition, low productivity, and increased susceptibility to diseases. Overfeeding or nutrient imbalance in certain animals and birds can lead to obesity and associated complications, including reduced fertility, fatty liver disease, and cardiovascular stress.

Improper feed formulation, whether due to a lack of knowledge or poor-quality feed ingredients, can lead to nutrient imbalances. High-density animal and poultry production often places immense metabolic demands on individuals, increasing their susceptibility to disorders. Temperature extremes, humidity, and other environmental factors can affect feed intake and nutrient metabolism. While genetic improvement has enhanced production traits, it has also increased the risk of metabolic stress and disorders. Overfeeding, underfeeding, or inconsistent feeding routines can disrupt metabolic balance. Diagnosis of these disorders requires a combination of clinical signs, laboratory analyses, and nutritional assessments. Blood biochemistry, urine tests, and feed analysis are commonly employed to identify nutrient deficiencies or metabolic imbalances.

Nutritional imbalances and metabolic disorders often arise from inadequate diets or poor feed quality. Disorders such as ketosis, caused by negative energy balance, and fatty liver syndrome, due to lipid accumulation, are common in ruminants. In poultry, metabolic disorders like calcium deficiencies can lead to rickets, osteomalacia, parturient paresis and reduced productivity. Addressing these issues requires careful dietary planning and monitoring of metabolic markers.

Traumatic reticulitis, also known as hardware disease, occurs when sharp objects ingested by cattle penetrate the reticulum, causing inflammation and infection. Proper feed management and magnetic boluses are preventive measures. Environmental extremes can cause hypothermia or hyperthermia in animals. Hypothermia results from prolonged exposure to cold, leading to slowed metabolism, while hyperthermia, often seen during heat stress, can cause organ dysfunction. Proper housing and ventilation are essential in preventing these conditions. Grain engorgement syndrome results from the overconsumption of fermentable carbohydrates, leading to ruminal acidosis. Symptoms include dehydration, bloating, and systemic acidosis. Preventive measures include gradual dietary transitions and limiting grain intake.

Displacement or torsion of digestive compartments, such as the abomasum, often results from dietary imbalances or metabolic stress. These conditions require surgical intervention and preventive dietary management. Electrolyte imbalances, such as metabolic acidosis or alkalosis, and respiratory-related acid-base disturbances can severely impact animal health. Proper hydration, balanced feed, and monitoring of electrolyte levels are key preventive measures. Bovine pulmonary emphysema is a respiratory condition linked to specific dietary compounds, such as high levels of tryptophan. Pica, or abnormal eating behaviour, often results from mineral deficiencies or psychological stress, indicating the need for dietary supplementation.

Ketosis, a metabolic disorder due to negative energy balance, is prevalent in high-producing dairy cattle. In contrast, spontaneous hyperglycemia and diabetes mellitus are less common in animals but can result from hormonal imbalances or pancreatic dysfunction. Fatty liver occurs in over-conditioned animals or those under stress, such as during lactation. The accumulation of fat in the liver impairs metabolic function, necessitating dietary adjustments to prevent excessive lipid mobilization.

The overuse of antibiotics as growth promoters or additives in animal feeds can lead to resistance, disrupted gut microbiota, and reduced feed efficiency. Rational antibiotic use and alternative growth promoters are essential to mitigate these risks.

Torsion or impaction of segments like the rumen and reticulum is a life-threatening condition caused by dietary errors or physical obstructions. Management involves immediate veterinary intervention and prevention through proper feed formulation. Radiation exposure can damage cellular DNA, affecting immunity and reproduction. Similarly, heat stress impacts metabolic and reproductive performance, highlighting the importance of proper housing and temperature regulation. Congenital or acquired coagulation defects can cause hemorrhagic conditions in animals. Laboratory tests for blood groups, transfusion compatibility, and hemostatic markers help diagnose and manage these disorders effectively.

Pituitary abnormalities, such as over-secretion or deficiencies, can cause metabolic disorders, impacting growth and reproduction. Similarly, thyroid imbalances, including hypothyroidism and hyperthyroidism, affect metabolic rate and calcium metabolism. Dehydration is a common issue in sick animals, impacting electrolyte balance and blood pressure. Calcinosis, caused by excessive calcium deposition, often results from imbalanced diets in grazing animals.

Animal health and productivity depend significantly on balanced mineral metabolism, proper electrolyte regulation, and efficient circulatory and cardiac function. Nutritional deficiencies or imbalances, metabolic disturbances, and congenital or acquired conditions can lead to severe physiological challenges in animals. This manuscript containing very important chapters explore the key concepts of role of nutrients, their interrelationship, mineral disorders, electrolyte-related metabolic issues, haemorrhagic disorders and the role of laboratory tests in assessing haemostasis and cardiovascular health, etc.

Minerals such as calcium, phosphorus, magnesium, sodium, potassium, sulfur, and trace elements (e.g., cobalt, copper, zinc) are vital for metabolic, structural, and enzymatic functions. Deficiencies or toxicities of these minerals can lead to specific diseases. Calcium imbalances, such as hypocalcemia, are critical in dairy cattle during lactation. Conditions like rickets, osteomalacia, and osteoporosis result from prolonged calcium or vitamin D deficiencies. Addressing these involves dietary supplementation and mineral balancing. Parakeratosis is a condition of the rumen lining caused by excessive grain feeding, leading to impaired absorption and inflammation. Rumenitis often accompanies parakeratosis and can progress to systemic infections if untreated. Ruminants are prone to ingesting toxic materials, pesticides or molds, leading to conditions like rumenitis or acute tympany. Deficiencies like cobalt (leading to anemia) and sulfur (affecting microbial fermentation) can disrupt normal rumen function, emphasizing the need for balanced mineral intake.

Phosphorus is essential for various biological processes, including energy metabolism, bone formation, and cellular signalling. Disorders of phosphorus metabolism in animals often manifest as hypophosphatemia characterized by reduced phosphorus levels, this condition causes poor growth, muscle weakness, and reproductive failure. In cattle, it may result in creeper syndrome, where animals exhibit stiffness and recumbency. Though, hyperphosphatemia is the excess phosphorus, often due to kidney dysfunction or dietary imbalances, can lead to soft tissue mineralization and secondary calcium imbalances. Proper phosphorus supplementation and dietary management are critical in preventing these disorders. The calcium-to-phosphorus ratio is especially significant in ruminants and poultry, where deviations can result in severe health consequences. Hypomagnesemia known as grass tetany, often occurs in grazing animals, leading to neuromuscular excitability and convulsions. Copper deficiency causes anemia, reduced growth, and coat discoloration. Zinc deficiency leads to parakeratosis, poor wound healing, and immune suppression. Mineral toxicities due to excessive intake of minerals like selenium can lead to poisoning, manifesting as lameness, hair loss, and reproductive failure. Balanced feeding programs and mineral supplements are crucial for maintaining animal health.

Electrolytes (sodium, potassium, chloride, bicarbonate) play a central role in maintaining acid-base balance, fluid homeostasis, and cellular function. Disturbances in these balances can lead to metabolic or respiratory acidosis and alkalosis. Metabolic acidosis caused by increased acid production (e.g., lactic acid in grain overload) or bicarbonate loss (e.g., diarrhea). Symptoms include rapid breathing, lethargy, and decreased feed intake. However, metabolic alkalosis results from excessive bicarbonate or loss of hydrogen ions, often due to prolonged vomiting or abomasal displacement leads to muscle weakness, slowed respiration, and anorexia. Furthermore, respiratory acidosis occurs when carbon dioxide builds up due to respiratory failure or hypoventilation. Symptoms include disorientation and elevated blood pressure, while, respiratory alkalosis caused by excessive carbon dioxide exhalation due to hyperventilation or heat stress causes tachypnea, weakness, and dizziness. Management of these conditions involves correcting the underlying cause, fluid therapy, and electrolyte balancing.

Haemorrhagic disorders are characterized by excessive bleeding due to coagulation abnormalities. Congenital coagulation defects often hereditary, such as Hemophilia A (factor VIII deficiency) or von Willebrand's disease, resulting in prolonged bleeding and poor clot formation, whereas, acquired coagulation defects may result from vitamin K deficiency (due to warfarin poisoning or liver dysfunction) or consumption of moldy sweet clover containing dicumarol. Symptoms include spontaneous bleeding, hematomas, and prolonged bleeding after injury. Early diagnosis and intervention, such as vitamin K supplementation, are crucial.

Laboratory tests for haemostasis required for effective diagnosis of haemorrhagic disorders requires laboratory tests to assess clotting mechanisms and blood parameters. Key tests include prothrombin time (PT) and activated partial thromboplastin time (APTT), which evaluate the extrinsic and intrinsic pathways of the coagulation cascade. Platelet count detects thrombocytopenia, a common cause of bleeding. Fibrinogen levels detect low levels may indicate disseminated intravascular coagulation (DIC) or liver dysfunction.

Blood group typing is crucial for safe transfusions, as mismatched blood groups can cause fatal reactions. Cross-matching ensures donor and recipient blood compatibility, preventing hemolytic reactions. Coagulation factor assays used to identify specific deficiencies, such as factor VIII in Hemophilia A. Blood groups vary among species, with cattle, horses, dogs, and cats having unique systems. For example in cattle 11 blood group systems (e.g., A, B, C), in horses seven blood groups (e.g., A, C, Q), in dogs (dog erythrocyte antigen) system with eight groups. Blood transfusion is a life-saving procedure, often required in cases of severe anemia or hemorrhage. Compatibility testing is essential to avoid hemolytic reactions.

Blood volume, pressure, and heart muscle disorders indicate cardiovascular health which is integral to overall animal well-being. Disorders affecting blood volume, pressure, and heart muscles include hypovolemia, caused by dehydration or hemorrhage, leads to shock. Hypertension, though rare in animals, can result from kidney disease or endocrine disorders. Moreover, hypertrophy is the thickening of the heart walls due to chronic high blood pressure or valve defects. Though, atrophy indicates weakening of the heart muscle from disuse or malnutrition, whereas, pace-making abnormalities such as arrhythmias result from disruptions in the heart's electrical conduction system, often due to electrolyte imbalances.

The interplay between mineral metabolism, nutritional and metabolic disorders, electrolyte balance, coagulation pathways, and cardiovascular health is complex and critical to animal survival. Nutritional deficiencies/ excess, metabolic disorders, and haemorrhagic conditions can severely impact productivity and welfare if not addressed promptly. Laboratory diagnostics, combined with preventive measures such as balanced diets, mineral supplementation, and proper electrolyte management, play an essential role in maintaining animal health. These comprehensive understanding enables veterinarians to prevent, diagnose and treat these conditions effectively, ensuring the well-being of animals under their care.

2

Role of Nutrients in Animal Health, Production and Welfare

Nutrients play a fundamental role in ensuring the health, productivity, reproduction, immunity, and overall welfare of animals. The balance and availability of macronutrients (carbohydrates, proteins and fats), micronutrients (vitamins and minerals) and water are crucial for optimal physiological and metabolic processes. This chapter explores the role of nutrients in different aspects of animal health and well-being.

Macronutrients

Carbohydrates

Carbohydrates are the primary energy source for most animals, playing a critical role in sustaining metabolic processes, maintaining body temperature, and supporting physical activity. In ruminants, carbohydrates are fermented in the rumen to produce volatile fatty acids (VFAs) such as acetate, propionate and butyrate, which serve as energy substrates.

Health: Adequate carbohydrate intake prevents energy deficits, which can lead to ketosis, fatty liver, or hypoglycemia, particularly in lactating animals.

Production: Energy from carbohydrates drives milk production in dairy animals and growth in meat-producing species.

Reproduction: Insufficient carbohydrate availability can delay puberty, impair estrus cycles, and reduce fertility.

Immunity: Energy deficits weaken immune responses, increasing susceptibility to infections.

Welfare: Optimal carbohydrate nutrition minimizes stress and supports natural behaviours like grazing or rumination.

Proteins

Proteins are essential for growth, tissue repair, and the synthesis of enzymes, hormones, and antibodies. Protein quality is determined by its amino acid composition and digestibility.

Health: Proteins support muscle development, organ function, and wound healing.

Production: High-quality protein is critical for meat, milk, and egg production. For instance, lysine and methionine are vital in poultry for optimal growth and egg yield.

Reproduction: Protein deficiencies impair ovarian and testicular functions, reducing reproductive efficiency.

Immunity: Amino acids like glutamine and arginine enhance immune responses and promote the healing of damaged tissues.

Welfare: Protein adequacy supports vitality and minimizes diseases associated with deficiency, such as anaemia or poor coat condition.

Fats

Fats are concentrated energy sources and are important for the absorption of fat-soluble vitamins (A, D, E, K).

Health: Essential fatty acids (e.g., omega-3 and omega-6) play roles in maintaining skin integrity, brain function, and anti-inflammatory pathways.

Production: In dairy cattle, fat supplementation can improve milk fat percentage, while in poultry, it enhances feed efficiency.

Reproduction: Fats are precursors for steroid hormones like estrogen and progesterone, influencing reproductive cycles.

Immunity: Omega-3 fatty acids boost immune function and reduce inflammation.

Welfare: Dietary fats improve coat shine, energy levels, and body condition, reflecting better welfare.

Micronutrients

Vitamins

Vitamins are organic compounds required in small quantities for enzymatic reactions and overall health. They further classified as fat soluble vitamins (such as Vitamin A, D, E and K) and water soluble vitamins (Vitamin B complex and vitamin C).

Vitamin A: Supports vision, epithelial integrity, and reproductive health. Deficiency can lead to night blindness and reproductive failure in animals and nutritional roup inpoultry.

Vitamin D: Regulates calcium and phosphorus metabolism, crucial for bone health and eggshell formation in poultry.

Vitamin E: Acts as an antioxidant, protecting cells from oxidative damage and enhancing immunity.

Vitamin K: Essential for blood clotting and bone metabolism.

B-complex vitamins: Play a role in energy metabolism, red blood cell production, and nervous system function.

Vitamin C: Act as antioxidants, required for collagen synthesis and enhance immunity.

Health and Welfare: Adequate vitamin intake ensures physiological stability and reduces stress-related disorders.

Production: Vitamins enhance feed efficiency, growth rates, and productivity in livestock.

Minerals

Minerals are inorganic elements critical for structural, metabolic, and physiological functions.

Macro minerals: Calcium, phosphorus, magnesium, sodium, potassium, sulphur, and chloride are needed in larger amounts.

Calcium and Phosphorus: Vital for bone strength, milk production, and eggshell quality.

Magnesium: Prevents grass tetany in grazing animals.

Sodium and Potassium: Maintain electrolyte balance and nerve function.

Trace minerals: Zinc, copper, iron, manganese, selenium, and iodine are required in trace amounts.

Zinc and Copper: Support skin health, immune function, and reproduction.

Selenium: Acts as an antioxidant and prevents white muscle disease.

Iodine: Essential for thyroid hormone synthesis, regulating metabolism.

Health: Minerals prevent deficiency-related disorders, such as rickets or anemia.

Production: Adequate mineral levels optimize productivity, including wool, milk, and meat yields.

Reproduction: Minerals influence fertility, embryonic development, and parturition.

Immunity: Zinc and selenium play pivotal roles in immune modulation and oxidative stress reduction.

Welfare: Balanced mineral nutrition promotes vitality and reduces disease susceptibility.

Water

Water is often overlooked but is the most critical nutrient. It facilitates digestion, thermoregulation, nutrient transport, and waste excretion.

Health: Dehydration leads to electrolyte imbalances, reduced feed intake, and impaired physiological functions.

Production: Water intake directly correlates with milk yield, growth rates, and feed conversion efficiency.

Reproduction: Water is vital for maintaining uterine health and supporting embryonic development.

Immunity: Adequate hydration supports lymphatic circulation and detoxification.

Welfare: Clean and accessible water improves animal comfort and reduces stress.

Energy balance

The balance between energy intake and expenditure is crucial for optimal performance.

Health: Negative energy balance predisposes animals to metabolic disorders like ketosis in dairy cows or hepatic lipidosis in cats.

Production: High-energy diets maximize growth, milk production, and reproductive performance.

Reproduction: Energy deficits delay puberty and impair ovulation.

Immunity: Energy supports immune cell proliferation and response to infections.

Welfare: Balanced energy intake promotes natural behaviours and reduces stress.

Immunity and disease resistance

Nutrients such as proteins, vitamins, and minerals directly influence the immune system.

Proteins: Required for the synthesis of immunoglobulins and cytokines.

Vitamins: Vitamin E and C act as antioxidants, reducing oxidative damage during infections.

Minerals: Zinc and selenium enhance immune cell function and protect against oxidative stress.

Omega-3 fatty acids: Reduce inflammation and support tissue repair

Nutritional deficiencies weaken the immune response, increasing susceptibility to diseases like mastitis, pneumonia, or gastrointestinal infections.

Reproduction

Nutrition significantly impacts reproductive performance in both males and females.

Females: Adequate energy, protein, and micronutrients ensure regular estrus cycles, conception, and healthy offspring development.

Males: Nutrients like zinc and selenium improve sperm quality and libido.

Deficiencies or imbalances can lead to reduced fertility, embryonic mortality, and postpartum complications.

Welfare and behaviour

Proper nutrition enhances animal welfare by supporting physical health, reducing stress, and promoting natural behaviours.

Stress reduction: Adequate energy and nutrient intake minimize physiological stress, improving behaviour and productivity.

Comfort and longevity: Balanced nutrition reduces disease incidence, ensuring a longer and more productive lifespan.

Nutrients are integral to all aspects of animal health, production, reproduction, immunity, and welfare. A balanced diet tailored to the species, physiological stage, and production goals ensures optimal performance and well-being. Continuous monitoring and adjustment of nutritional strategies are essential for sustaining animal health and productivity in modern livestock management.

a) Role of carbohydrates

Carbohydrates are the primary source of energy in animal diets and play an essential role in various physiological and metabolic processes. They are critical for maintaining health, enhancing production, supporting reproduction, and promoting overall welfare.

Carbohydrates are organic compounds composed of carbon, hydrogen, and oxygen. They include sugars, starches, and fiber, which vary in complexity and digestibility. In animals, carbohydrates provide the energy required for bodily functions, including movement, thermoregulation, and metabolic processes.

Types of carbohydrates

Simple carbohydrates: Monosaccharides (e.g., glucose) and disaccharides (e.g., sucrose), which are rapidly digestible.

Complex carbohydrates: Polysaccharides (e.g., starch) that provide sustained energy.

Fiber: Indigestible carbohydrates (e.g., cellulose) that play a role in gut health and fermentation in ruminants.

Carbohydrates and animal health

Energy provision

Carbohydrates are the primary energy source for animals. Energy is necessary for maintaining body temperature, muscle function, and metabolic activities. Insufficient carbohydrate intake leads to energy deficits, resulting in conditions such as:

Ketosis in lactating dairy cows,

Hypoglycemia in neonatal animals,

Fatty liver syndrome caused due to excessive mobilization of body fat to compensate for energy deficits.

Gut health

In ruminants, carbohydrates are fermented by rumen microbes to produce volatile fatty acids (VFAs), such as acetate, propionate and butyrate. These VFAs serve as energy substrates and are critical for:

Maintaining optimal rumen pH,

Preventing conditions like ruminal acidosis,

Supporting epithelial health in the gut,

In monogastric animals, dietary fiber promotes gut motility and the growth of beneficial gut microbiota, improving overall digestive health.

Prevention of disease

Adequate carbohydrate intake prevents stress-related disorders and enhances the animal's ability to recover from illnesses. For instance, glucose is vital for healing wounds and producing immune cells during infections.

Carbohydrates and production

Dairy production

Carbohydrates are critical for milk production in dairy animals. Propionate, a VFA derived from carbohydrate fermentation, is a key precursor for gluconeogenesis in the liver, which supplies glucose for lactose synthesis in milk. Insufficient carbohydrates can lead to:

Reduced milk yield and quality,

Low milk fat percentage, especially in high-producing dairy cows,

Meat production

In meat-producing animals, carbohydrates ensure efficient growth by providing energy for protein synthesis and muscle deposition. Adequate carbohydrate feeding improves feed conversion ratios, leading to higher weight gains and better carcass quality.

Egg production

In poultry, carbohydrates are crucial for meeting the energy demands of egg production. Insufficient carbohydrate intake results in lower egg production rates, smaller egg size, and poor shell quality.

Carbohydrates and reproduction

Female reproduction

Carbohydrates are essential for reproductive health in female animals. Energy derived from carbohydrates supports:

Ovulation: Adequate energy reserves are necessary for follicular development and ovulation.

Gestation: Carbohydrates provide the energy needed to support fetal growth and development.

Lactation: Energy derived from carbohydrates ensures sufficient milk production during the postpartum period.

Carbohydrate deficiencies can lead to:

Delayed puberty

Irregular estrous cycles

Increased risk of abortion or stillbirths

Male reproduction

In males, carbohydrates provide the energy required for spermatogenesis and mating behaviours. Glucose is particularly important for maintaining testicular function and sperm motility.

Carbohydrates and nutritional status

Carbohydrates play a pivotal role in maintaining an animal's overall nutritional balance. They contribute to:

Energy density: Carbohydrates make up a significant portion of most animal diets, ensuring adequate energy intake.

Protein sparing: Adequate carbohydrate intake prevents protein from being used as an energy source, allowing it to be utilized for tissue repair and growth.

Body condition: Animals with sufficient carbohydrate intake maintain optimal body weight and condition, which is crucial for productivity and reproduction.

Deficiency in carbohydrates can result in:

Poor growth rates

Muscle wasting due to protein catabolism

Reduced fat reserves, leading to poor thermoreigulation

Carbohydrates and immunity

Immune function

Carbohydrates are directly and indirectly involved in enhancing immune responses:

Direct role: Glucose is a critical energy source for immune cells like lymphocytes and macrophages, enabling effective pathogen elimination.

Indirect role: VFAs produced in ruminants support gut immunity by promoting the growth of beneficial gut microbes.

Stress and recovery

Stressful conditions, such as transportation or weaning, increase animals' energy demands. Adequate carbohydrate nutrition ensures sufficient energy availability, reducing stress-related immunosuppression and enhancing recovery from illnesses.

Carbohydrate and antioxidant defence

Carbohydrates, particularly fibre, help modulate oxidative stress by promoting the growth of beneficial gut bacteria, which produce short-chain fatty acids with anti-inflammatory properties.

Carbohydrates and welfare

Animal welfare is closely linked to their nutritional status. Adequate carbohydrate intake promotes:

Energy for activity: Animals are more active and exhibit natural behaviors when their energy needs are met.

Thermoregulation: Energy derived from carbohydrates helps maintain body temperature during extreme weather conditions.

Reduction in stress: Sufficient energy intake reduces the physiological and psychological stress associated with hunger or malnutrition.

Conversely, carbohydrate deficiencies can lead to:

Behavioural changes, such as lethargy or aggression

Poor coat condition and reduced vitality

Increased susceptibility to diseases, reducing overall quality of life

Challenges and strategies in carbohydrate feeding

Balancing carbohydrates

In ruminants, feeding highly fermentable carbohydrates (e.g., grains) in excess can lead to acidosis, while inadequate fibre reduces rumen function. Proper balancing of forage and concentrate is essential.

In monogastrics, ensuring a balance of simple and complex carbohydrates prevents blood sugar fluctuations and promotes sustained energy release.

Feed quality

The quality of carbohydrate sources affects digestibility and energy availability. High-quality forage, grains, and by-products like molasses can enhance the nutritional value of diets.

Meeting specific needs

Different species and physiological stages (e.g., growth, lactation, gestation) require tailored carbohydrate levels to optimize health and productivity.

Carbohydrates are indispensable for animal health, production, reproduction, immunity, and welfare. As the primary energy source, they drive metabolic processes, support physiological functions, and enhance productivity. Proper carbohydrate nutrition not only prevents deficiencies but also promotes optimal growth, reproductive performance, and overall well-being. By ensuring balanced and species-specific carbohydrate intake, producers can maximize productivity while safeguarding animal welfare.

Role of soluble and insoluble (fibrous) carbohydrates

Carbohydrates are a vital component of animal diets, broadly classified into soluble carbohydrates (e.g., sugars and starches) and insoluble carbohydrates (fibres like cellulose and hemicellulose). These forms play complementary roles in supporting animal health, productivity, reproduction, immunity, and welfare. This essay explores their specific contributions.

Soluble carbohydrates

Soluble carbohydrates, including simple sugars and starches, are easily digestible and provide a quick source of energy. They are particularly important in monogastric animals and ruminants during periods of high energy demand.

Health

Soluble carbohydrates supply glucose, essential for brain function, muscle activity, and thermoregulation.

In ruminants, soluble carbohydrates are fermented in the rumen to produce volatile fatty acids (VFAs), particularly propionate, a precursor for glucose synthesis in the liver. This prevents energy-deficiency disorders like ketosis.

Production

In dairy cattle, energy from soluble carbohydrates supports milk production, while in poultry; it enhances growth rates and egg production. In meat-producing animals, soluble carbohydrates improve feed conversion efficiency and growth performance.

Reproduction

Glucose from soluble carbohydrates is critical for ovulation, pregnancy maintenance, and foetal development. Energy shortages can delay puberty, reduce conception rates, and lead to pregnancy losses.

Nutritional status

Soluble carbohydrates provide high-energy density, preventing protein from being used as an energy source (protein sparing). Adequate intake ensures optimal body condition and performance.

Immunity

Soluble carbohydrates are crucial for energy-intensive immune processes, supporting lymphocyte proliferation and antibody production. They reduce stress during critical periods like weaning and transportation, enhancing immune resilience.

Welfare

Sufficient energy from soluble carbohydrates promotes vitality and reduces the risk of hunger-induced stress, contributing to better behavioural and physical welfare.

Insoluble (fibrous) carbohydrates

Insoluble carbohydrates, or fibre, include structural components of plants like cellulose, hemicellulose, and lignin. They play a crucial role in maintaining gut health, particularly in ruminants and hindgut fermenters.

Health

Fibrous carbohydrates promote rumen health in ruminants by stimulating saliva production, which buffers rumen pH and prevents acidosis. In monogastrics, fibre enhances gut motility and supports the growth of beneficial gut microbiota, reducing the risk of gastrointestinal disorders.

Production

For ruminants, fibrous carbohydrates ensure efficient rumen fermentation and nutrient absorption, directly influencing milk and meat production. In poultry, dietary fibre improves gut health, leading to better nutrient utilization and egg production.

Reproduction

Adequate fibre intake helps maintain stable body condition scores, which are critical for reproductive performance. Fibre promotes satiety, reducing over-conditioning that can impair fertility in breeding animals.

Nutritional status

Insoluble carbohydrates contribute bulk to the diet, preventing overconsumption of high-energy feeds and supporting optimal nutrient balance. They enhance digestion and nutrient absorption through healthy gut function.

Immunity

Fibre fermentation produces short-chain fatty acids (SCFAs) like butyrate, which strengthen gut integrity and modulate immune responses. A healthy gut microbiome, supported by fibre, reduces pathogen colonization and enhances overall immunity.

Welfare

Fibre promotes natural feeding behaviours like chewing and rumination, reducing stress and promoting psychological well-being. It minimizes the risk of digestive issues like bloating and colic, contributing to better overall welfare.

Balancing soluble and insoluble carbohydrates

A balance between soluble and insoluble carbohydrates is critical for optimizing animal health and productivity:

In ruminants, excessive soluble carbohydrates can lead to rumen acidosis, while insufficient fibre disrupts rumen function. In monogastrics, too much soluble carbohydrate may cause digestive upset, while adequate fibre promotes gut health.

Soluble and insoluble carbohydrates play distinct but complementary roles in animal health, production, reproduction, immunity, and welfare. Soluble carbohydrates provide immediate energy for metabolic functions, while insoluble carbohydrates maintain gut health and support long-term productivity and well-being. A balanced inclusion of both forms is essential for optimizing performance and promoting sustainable animal welfare.

b) Role of fats and lipids

Fats and lipids are essential macronutrients in animal diets, providing concentrated energy, supporting cellular functions, and serving as carriers for fat-soluble vitamins (A, D, E and K). Their roles extend beyond energy supply to influencing health, production, reproduction, immunity, and overall welfare. This essay explores their significance in various aspects of animal physiology and productivity.

Fats are composed of triglycerides, which include three fatty acid molecules attached to a glycerol backbone. Fatty acids are categorized based on their chemical structure:

Saturated fats: No double bonds in the carbon chain.

Unsaturated fats: Contain one or more double bonds, including monounsaturated (MUFA) and polyunsaturated fatty acids (PUFA).

Essential fatty acids (EFAs): Omega-3 and omega-6 fatty acids, which animals cannot synthesize and must obtain from their diet.

Lipids also include phospholipids (components of cell membranes) and sterols (e.g., cholesterol).

Fats and lipids in animal health

Energy provision

Fats are the most energy-dense nutrient, providing approximately 2.25 times more energy than carbohydrates or proteins. They are vital for meeting the high energy demands of production, growth, and reproduction.

Ruminants: In ruminants, fats bypass the rumen and are digested in the small intestine. Excessive fat in the diet can interfere with rumen microbial activity, so careful management is necessary.

Monogastrics: Fats are digested and absorbed efficiently in monogastrics, contributing significantly to their energy requirements.

Cellular functions

Lipids are integral to cell membranes, maintaining their fluidity and functionality. They are also involved in signal transduction and the formation of lipid-based signaling molecules, such as prostaglandins.

Thermoregulation

In cold environments, fats stored in adipose tissue serve as insulation and energy reserves, helping animals maintain body temperature.

Skin and coat health

Fats contribute to skin integrity and a shiny coat, reflecting good health. Essential fatty acids (EFAs) like omega-3 and omega-6 play a critical role in preventing dermatological disorders.

Fats and lipids in animal production

Dairy production

Fat supplementation in dairy diets can increase the energy density of feed, improving milk yield and fat content. Unsaturated fatty acids enhance milk quality, influencing its nutritional and processing properties.

Meat production

In meat animals, dietary fats improve feed efficiency and weight gain by providing concentrated energy. Omega-3 and omega-6 fatty acids influence the quality of meat, including its marbling, flavour, and omega fatty acid profile, which are critical for consumer preference and health benefits.

Egg production

In poultry, fats in the diet are crucial for optimizing egg production, egg size, and yolk quality. Linoleic acid (an omega-6 fatty acid) is particularly important for egg formation.

Fats and lipids in reproduction

Female reproduction

Fats are precursors for steroid hormones like estrogen and progesterone, which are essential for ovarian function, estrous cycles, and pregnancy maintenance. Omega-3 fatty acids improve uterine health, embryo development, and reduce inflammation during gestation. Energy from fats supports lactation, ensuring milk production to meet the offspring's nutritional needs.

Male reproduction

Dietary fats influence testicular function and sperm quality. Omega-3 fatty acids, in particular, enhance sperm motility and reduce oxidative stress in reproductive tissues.

Deficiencies in fats and essential fatty acids can lead to reproductive disorders, including delayed puberty, poor conception rates, and embryonic loss.

Fats and lipids in nutritional status

Fats improve the palatability and energy density of diets, encouraging feed intake, particularly in high-producing animals or those under stress.

Protein sparing

Dietary fats reduce the need for protein to be used as an energy source, allowing proteins to be utilized for growth and repair.

Fat-soluble vitamin absorption

Fats are necessary for the absorption and transport of fat-soluble vitamins:

Vitamin A: Supports vision and immune function.

Vitamin D: Regulates calcium and phosphorus metabolism, promoting bone health.

Vitamin E: Acts as an antioxidant, protecting cells from oxidative damage.

Vitamin K: Essential for blood clotting and bone metabolism.

Deficiency in dietary fat can impair vitamin absorption, leading to deficiencies and associated health issues.

Fats and lipids in immunity

Fats and fatty acids play a direct role in modulating immune responses:

Omega-3 fatty acids: These have anti-inflammatory properties, which are beneficial in controlling chronic inflammation and improving recovery from infections.

Omega-6 fatty acids: Essential for the synthesis of eicosanoids, which are signaling molecules involved in immune function and inflammation.

Vitamin E and antioxidants: Fat-associated vitamin E strengthens the immune system by neutralizing free radicals, reducing oxidative stress in immune cells.

Fats and lipids in welfare

The inclusion of fats in animal diets improves welfare by supporting physical and psychological health:

Energy reserve: Adequate fat reserves allow animals to cope with periods of feed scarcity or high energy demand, reducing stress.

Behavioural comfort: Animals with sufficient dietary fat exhibit better energy levels and natural behaviours, such as grazing, foraging, or mating.

Stress reduction: Omega-3 fatty acids improve stress resilience by modulating hormonal and inflammatory responses to stressors like transportation, environmental changes, or social challenges.

Managing fats and lipids in diets

Balanced inclusion

Excessive fat intake can lead to digestive disturbances (e.g., rumen dysfunction in ruminants) or obesity. Conversely, inadequate fat intake results in energy deficits, poor growth, and compromised reproductive and immune functions.

Quality of fats

High-quality fats and oils (e.g., fish oil, flaxseed, or soybean oil) ensure adequate intake of essential fatty acids while minimizing the risk of rancidity or contamination.

Species-specific needs

Different species and production stages require tailored fat inclusion. For instance: High-yielding dairy cows benefit from energy-dense fats during lactation. Growing poultry need dietary fats for rapid weight gain and optimal egg production.

Fats and lipids play a multifaceted role in animal health, production, reproduction, immunity, and welfare. As a concentrated energy source, they support growth, lactation, and reproduction while enhancing immunity and reducing stress. Essential fatty acids, in particular, have profound effects on inflammation, cell integrity, and reproductive success. Proper management of dietary fats, considering quality, quantity, and species-specific requirements, is critical for optimizing animal performance and ensuring welfare. By incorporating balanced fat sources, producers can achieve sustainable, productive, and welfare-friendly animal systems.

c) Role of proteins, amino acids and protein quality

Proteins are macronutrients essential for growth, maintenance, and various physiological processes in animals. Composed of amino acids, proteins are vital for cellular function, enzyme activity, and immune responses. Protein quality determined by amino acid composition, digestibility and bioavailability plays a crucial role in ensuring optimal health, production, reproduction, immunity, and welfare.

Proteins are composed of 20 amino acids, categorized into:

Essential amino acids (EAAs): Cannot be synthesized by the body (e.g., lysine, methionine, and tryptophan, etc.) and must be obtained through diet.

Non-essential amino acids: Synthesized in the body (e.g., alanine, glutamine, etc.).

Conditionally essential amino acids: Become essential under specific conditions, such as stress or illness (e.g., arginine, glutamine).

High-quality proteins contain balanced EAAs that match the animal's physiological needs, ensuring efficient utilization.

Proteins and amino acids in animal health

1. ***Tissue development and maintenance:*** Proteins are the primary structural components of muscles, skin, hair, and internal organs. Adequate protein intake supports tissue repair, wound healing, growth, ensuring the overall health and resilience of animals.
2. ***Enzyme and hormone production:*** Proteins serve as precursors for enzymes and hormones critical for metabolic regulation. For instance: Insulin, a protein hormone, regulates glucose metabolism. Enzymes like proteases and lipases facilitate digestion and nutrient absorption.
3. ***Energy supply:*** Under conditions of energy scarcity, proteins can be catabolized to provide energy. However, this is inefficient and detrimental to muscle and tissue health, emphasizing the importance of adequate dietary protein.

Proteins and amino acids in animal production

1. ***Growth and weight gain:*** Proteins are crucial for muscle development and overall growth in production animals. Adequate dietary protein ensures higher feed conversion efficiency and optimal weight gain, directly impacting meat yield.

   High-quality proteins: Improve growth performance in broilers and swine. *Limiting amino acids*: Deficiency in critical amino acids (e.g., lysine, methionine) slows growth and reduces feed efficiency.
2. ***Milk production:*** In lactating animals, proteins provide the building blocks for milk proteins, essential for calf development. High-quality protein sources enhance milk yield and composition, including casein and whey proteins critical for offspring growth.
3. ***Egg production:*** In poultry, dietary proteins and specific amino acids like methionine and lysine directly influence egg size, production rate, and shell quality.

Proteins and amino acids in reproduction

Female reproduction

Foetal growth: Amino acids such as arginine and methionine are crucial for placental development and foetal growth.

Reproductive hormones: Proteins are necessary for the synthesis of hormones like estrogen and progesterone, essential for ovulation, implantation, and pregnancy maintenance.

Lactation: Adequate protein levels support milk production, ensuring the nutritional needs of offspring are met.

Male reproduction

Proteins and amino acids are required for spermatogenesis and the production of seminal plasma. Arginine and glutamine improve sperm motility and quality, enhancing fertility.

Proteins and amino acids in nutritional status

Protein balance

Positive protein balance: Achieved when protein intake meets or exceeds the body's requirements, supporting growth and productivity.

Negative protein balance: Occurs during protein deficiency, leading to muscle wasting, poor growth, and reduced productivity.

Protein quality

High-quality proteins contain all essential amino acids in proportions that align with an animal's needs. Examples include:

Animal-derived proteins: Fishmeal, meat meal, meat cum bone meal and milk proteins.

Plant-based proteins: Soybean meal, groundnut cake and canola meal often supplemented with synthetic amino acids to improve quality.

Proteins and amino acids in immunity

Immune cell development

Amino acids like glutamine, arginine, and cysteine are crucial for the proliferation and activation of immune cells (e.g., lymphocytes and macrophages). Proteins support the synthesis of antibodies, cytokines, and acute-phase proteins, strengthening immune defences.

Stress response

During stress (e.g., disease, transportation), protein and amino acid requirements increase due to heightened metabolic demands. Providing high-quality proteins ensures animals maintain their immune competence under such conditions.

Gut immunity

Amino acids like threonine are vital for mucin production, protecting the intestinal lining from pathogens. Proteins promote gut microbiota health, indirectly enhancing immunity.

Proteins and amino acids in welfare

Behaviour and stress reduction

Adequate protein intake reduces behavioural issues like aggression and lethargy associated with malnutrition. Tryptophan, a precursor of serotonin, plays a role in mood regulation and stress resilience.

Physical health

Protein deficiency leads to poor coat condition, weak hooves, and reduced vitality. High-quality proteins support robust musculoskeletal health, enabling natural behaviours like movement and foraging.

Sustainability

Sustainable protein nutrition enhances productivity and reduces the environmental footprint of livestock systems, promoting long-term welfare.

Protein deficiency and excess

Deficiency

Protein deficiency results in:

Poor growth rates and weight loss

Reproductive failures such as delayed puberty and infertility

Weak immune responses, increasing susceptibility to infections

Behavioural changes, such as reduced activity and aggression

Excess

Excessive protein intake can lead to:

Increased nitrogen excretion contributes to environmental pollution.

Higher feed costs with no additional benefits if protein exceeds the animal's requirements.

Strategies for optimizing protein nutrition

Balancing amino acids

Formulating diets with adequate levels of essential amino acids minimizes protein waste and improves efficiency. Supplemental amino acids like lysine, methionine, and threonine are used to enhance diet quality.

Stage-specific needs

Protein requirements vary by life stage and production phase. For instance: Growing animals require more protein for muscle development. Lactating or pregnant animals need additional protein to support offspring.

Improving protein quality

Processing methods like heat treatment and enzyme supplementation can enhance protein digestibility and availability. Combining plant and animal protein sources provides a more balanced amino acid profile.

Proteins and amino acids are indispensable for animal health, production, reproduction, immunity, and welfare. As structural and functional molecules, they ensure tissue development, metabolic regulation, and immune competence. Protein quality, determined by amino acid composition and digestibility, is critical for optimizing performance while minimizing environmental impacts. By providing balanced, high-quality protein sources tailored to species-specific needs, producers can enhance productivity, ensure animal well-being, and promote sustainable livestock management.

d) Role of minerals

Minerals are essential nutrients required in varying amounts for optimal physiological function in animals. They are broadly classified into:

Macro-minerals: Required in larger amounts (e.g., calcium, phosphorus, magnesium, sodium, potassium, chlorine and sulphur).

Micro-minerals (trace minerals): Needed in smaller amounts (e.g., zinc, iron, copper, selenium, iodine and manganese, etc.).

Macro-minerals: Roles and dietary requirements

Calcium (Ca)

Role: Calcium is crucial for bone and teeth formation, muscle contraction, blood clotting, and nerve transmission.

Dietary requirement

Growing animals: 0.5%-1% of the diet.

Lactating cows: 1.2%-1.5% of the diet.

Deficiency: Leads to rickets, osteomalacia, and milk fever in lactating animals.

Excess: Can interfere with the absorption of other minerals like phosphorus and zinc.

Phosphorus (P)

Role: Essential for bone development, energy metabolism (ATP), and DNA synthesis.

Dietary requirement

Growing animals: 0.3%-0.5% of the diet.

Lactating cows: 0.35%-0.45%.

Deficiency: Causes reduced growth, poor reproduction, and pica (abnormal eating behaviour).

Excess: Environmental pollution due to phosphorus excretion.

Magnesium (Mg)

Role: Involved in enzyme activation, muscle function, and nerve transmission.

Dietary requirement: 0.1%-0.3% of the diet, depending on species.

Deficiency: Leads to grass tetany in ruminants, characterized by muscle spasms and convulsions.

Excess: Rare but can interfere with calcium absorption.

Sodium (Na)

Role: Maintains osmotic balance, acid-base regulation, and nerve function.

Dietary requirement: 0.1%-0.2% of the diet.

Deficiency: Causes reduced feed intake, dehydration, and poor growth.

Excess: Results in salt toxicity if water is limited.

Potassium (K)

Role: Important for muscle function, nerve transmission, and maintaining acid-base balance.

Dietary requirement: 0.5%-1% of the diet.

Deficiency: Causes weakness, reduced feed intake, and decreased milk yield in lactating animals.

Excess: May interfere with magnesium absorption, leading to grass tetany.

Sulphur (S)

Role: Component of sulphur-containing amino acids (e.g., methionine, cysteine) and vitamins (e.g., biotin, thiamine).

Dietary requirement: 0.15%-0.2% of the diet in ruminants.

Deficiency: Results in poor wool production, reduced feed efficiency, and impaired growth.

Excess: Causes sulphur toxicity, leading to polioencephalomalacia in ruminants.

Micro-minerals: Roles and dietary requirements

Zinc (Zn)

Role: Supports enzyme activity, skin health, wound healing, and immune function.

Dietary requirement: 30-60 ppm (parts per million), depending on species.

Deficiency: Causes poor growth, skin lesions, and impaired immunity.

Excess: May interfere with copper and iron absorption.

Iron (Fe)

Role: Critical for hemoglobin synthesis, oxygen transport, and enzyme function.

Dietary requirement: 50-100 ppm.

Deficiency: Results in anaemia, weakness, and reduced growth.

Excess: Can cause oxidative stress and interfere with zinc and copper metabolism.

Copper (Cu)

Role: Involved in enzyme activation, hemoglobin synthesis, and connective tissue formation.

Dietary requirement: 5-20 ppm, depending on species.

Deficiency: Causes anaemia, de-pigmentation of hair or wool, and poor reproduction.

Excess: Leads to copper toxicity, especially in sheep.

Selenium (Se)

Role: Acts as an antioxidant (component of glutathione peroxidase) and supports immune function.

Dietary requirement: 0.1-0.3 ppm.

Deficiency: Causes white muscle disease, poor growth, and impaired immunity.

Excess: Results in selenium toxicity (alkali disease), characterized by hair loss and hoof deformities.

Iodine (I)

Role: Essential for thyroid hormone synthesis, which regulates metabolism.

Dietary requirement: 0.2-0.5 ppm.

Deficiency: Leads to goiter, reduced growth, and reproductive failure.

Excess: Rare but can cause iodine toxicity, affecting thyroid function.

Manganese (Mn)

Role: Important for bone formation, enzyme activation, and reproduction.

Dietary requirement: 20-40 ppm.

Deficiency: Causes skeletal deformities, poor growth, and reproductive issues.

Excess: Rare but may interfere with iron absorption.

Cobalt (Co)

Role: Required for vitamin B_{12} synthesis by rumen microbes.

Dietary requirement: 0.1-0.2 ppm in ruminants.

Deficiency: Leads to poor appetite, weight loss, and anemia.

Excess: Rare and typically non-toxic.

Role of minerals in animal health

Bone health: Calcium, phosphorus, and magnesium are vital for strong bones and teeth.

Nerve and muscle function: Sodium, potassium, and magnesium support nerve impulse transmission and muscle contraction.

Antioxidant defence: Selenium and zinc protect cells from oxidative damage.

Role of minerals in animal production

Dairy production: Calcium and phosphorus are essential for milk production, while potassium and magnesium support lactation.

Meat production: Zinc, iron, and copper enhance muscle growth and meat quality.

Egg production: Calcium and phosphorus are critical for eggshell formation and overall productivity in poultry.

Role of minerals in reproduction

Female reproduction

Calcium and phosphorus are essential for foetal bone development.

Selenium and iodine support hormonal balance and pregnancy maintenance.

Male reproduction

Zinc and selenium improve sperm quality and fertility.

Role of minerals in nutritional status

Minerals maintain electrolyte balance, enzyme activation, and metabolic regulation. Proper mineral supplementation prevents deficiencies and optimizes nutrient utilization.

Role of minerals in immunity

Immune cell function: Zinc, selenium, and iron are vital for the development and activation of immune cells.

Antioxidant support: Selenium and copper reduce oxidative stress during infections and stress.

Role of minerals in welfare

Behaviour: Adequate mineral intake prevents behavioural issues linked to deficiencies, such as pica or lethargy.

Stress resilience: Minerals like magnesium and selenium help animals cope with stress, improving overall welfare.

Challenges in mineral nutrition

Deficiency: Results from poor-quality feed, imbalanced diets, or inadequate supplementation.

Excess: Can lead to toxicity, reduced feed intake, or interference with other nutrients.

Strategies for optimizing mineral nutrition

Balanced diets: Ensure proper inclusion of minerals to meet species-specific requirements.

Supplementation: Use mineral blocks, premixes, or fortified feeds to address deficiencies.

Monitoring: Regularly assess mineral status through feed analysis and animal health monitoring.

Macro and micro minerals are indispensable for animal health, production, reproduction, immunity, and welfare. Proper mineral nutrition ensures optimal growth, productivity, and resilience to stress, contributing to sustainable livestock management. Balancing mineral intake according to species-specific needs is critical for preventing deficiencies and excesses, ensuring long-term animal welfare and productivity.

e) Role of vitamins (fat-soluble and water-soluble)

Vitamins are organic compounds essential for various biochemical and physiological functions in animals. They are categorized into:

Fat-soluble vitamins: Stored in body fat and liver (e.g., vitamins A, D, E, and K).

Water-soluble vitamins: Not stored in significant amounts and require daily intake (e.g., B-complex vitamins and vitamin C).

Fat-soluble vitamins: Roles and requirements

Vitamin A (Retinol)

Roles

Vision: Essential for forming rhodopsin in the retina.

Growth: Supports epithelial cell development and differentiation.

Immunity: Enhances mucosal immunity and antibody production.

Reproduction: Promotes foetal growth and placental development.

Dietary requirement: 2200–5000 IU/kg dry matter (DM) for most livestock.

Deficiency: Causes night blindness, poor growth, infertility, and increased susceptibility to infections.

Excess: Leads to toxicity, causing skeletal deformities and reduced feed intake.

Vitamin D (Calciferol)

Roles

Calcium and phosphorus metabolism: Facilitates absorption and utilization for bone health.

Immune support: Modulates immune cell activity.

Dietary requirement: 300–1000 IU/kg DM.

Deficiency: Causes rickets in young animals and osteomalacia in adults.

Excess: Results in hypercalcemia, kidney damage, and soft tissue calcification.

Vitamin E (Tocopherol)

Roles

Antioxidant: Protects cell membranes from oxidative damage.

Immunity: Enhances disease resistance by supporting immune cells.

Reproduction: Improves fertility and reduces embryonic mortality.

Dietary requirement: 10–50 IU/kg DM.

Deficiency: Leads to white muscle disease, poor fertility, and immune dysfunction.

Excess: Generally non-toxic but may interfere with other fat-soluble vitamins.

Vitamin K

Roles

Blood clotting: Essential for synthesizing clotting factors.

Bone health: Aids in calcium deposition in bones.

Dietary requirement: 0.5–1.5 mg/kg DM.

Deficiency: Causes prolonged bleeding, hemorrhages, and poor bone development.

Excess: Rare and not well-documented.

Water-soluble vitamins: Roles and requirements

B-complex vitamins

Vitamin B-complex group includes thiamine (B_1), riboflavin (B_2), niacin (B_3), pantothenic acid (B_5), pyridoxine (B_6), biotin (B_7), folic acid (B_9), and cobalamin (B_{12}).

General roles

Energy metabolism: Act as coenzymes in carbohydrate, protein, and fat metabolism.

Nervous system: Support neural function and neurotransmitter synthesis.

Red blood cell production: Promote hemoglobin synthesis and oxygen transport.

Specific B-vitamin roles and requirements

Thiamine (B_1): Prevents nervous disorders; required at 1–2 mg/kg DM.

Riboflavin (B_2): Supports energy production; required at 4–6 mg/kg DM.

Niacin (B_3): Prevents pellagra-like symptoms; required at 10–40 mg/kg DM.

Pantothenic acid (B_5): Aids in fat metabolism; required at 8–15 mg/kg DM.

Biotin (B_7): Maintains hoof and skin health; required at 0.1–0.3 mg/kg DM.

Folic acid (B_9): Promotes DNA synthesis; required at 0.5–1.0 mg/kg DM.

Cobalamin (B_{12}): Essential for red blood cell production; required at 10–20 µg/kg DM.

Deficiency: Causes poor growth, anemia, nervous disorders, and metabolic disturbances.

Excess: Rare due to water solubility and excretion.

Vitamin C (Ascorbic acid)

Roles

Antioxidant: Neutralizes free radicals and protects tissues.

Collagen synthesis: Promotes wound healing and tissue repair.

Immunity: Enhances disease resistance by supporting immune cells.

Dietary requirement: Not required in most livestock as they synthesize it endogenously; supplemental levels of 100–200 mg/day may benefit stressed or sick animals.

Deficiency: Causes scurvy, poor wound healing, and weak immunity in species that cannot synthesize it (e.g., guinea pigs).

Excess: Generally safe but may cause gastrointestinal upset at very high doses.

Role of vitamins in animal health

Growth and development

Fat-soluble vitamins (e.g., A, D) promote bone growth and epithelial health.

B-complex vitamins aid in energy metabolism, ensuring proper growth rates.

Tissue repair

Vitamin C and biotin support collagen formation, aiding in wound healing and skin health.

Disease prevention

Antioxidants like vitamins E and C protect tissues from oxidative stress, reducing the risk of chronic diseases.

Role of vitamins in production

Dairy production

Vitamin D ensures calcium metabolism, critical for milk production.

Vitamin E improves milk quality by reducing oxidative spoilage.

Meat production

B-complex vitamins enhance feed efficiency and muscle growth.

Antioxidants prevent oxidative rancidity in meat, improving shelf life.

Egg production

Vitamin A and biotin promote eggshell integrity and yolk quality.

Vitamin D ensures adequate calcium deposition in eggshells.

Role of vitamins in reproduction

Fertility

Vitamin A supports ovarian function and embryonic development.

Vitamin E reduces oxidative stress in reproductive tissues, improving fertility.

B-complex vitamins (e.g., folic acid and B_{12}) aid in DNA synthesis and cell division.

Pregnancy and lactation

Vitamin D ensures foetal bone development.

Vitamin A supports foetal organogenesis.

Role of vitamins in nutritional status

Vitamins optimize nutrient metabolism, ensuring efficient utilization of carbohydrates, proteins, and fats. Deficiencies disrupt metabolic pathways, leading to poor growth and production.

Role of vitamins in immunity

Immune cell function

Vitamin A enhances mucosal immunity by maintaining epithelial integrity.

Vitamin E boosts disease resistance by protecting immune cells from oxidative damage.

Stress management

Vitamin C supports adrenal function, helping animals cope with stress during transportation or illness.

Role of vitamins in welfare

Behaviour: Adequate vitamin intake prevents lethargy and abnormal behaviours caused by deficiencies.

Stress resilience: Vitamins like C and E enhance stress tolerance, improving overall welfare.

Challenges in vitamin nutrition

Deficiency: Caused by poor-quality feed, imbalanced diets, or increased requirements during stress or illness.

Excess: Over-supplementation, especially of fat-soluble vitamins, can lead to toxicity.

Strategies for optimizing vitamin nutrition

Balanced diets: Formulate diets that meet species-specific requirements for all vitamins.

Supplementation: Use premixes or fortified feeds to address deficiencies.

Monitoring: Regularly assess vitamin status through health monitoring and feed analysis.

Fat and water-soluble vitamins are essential for animal health, production, reproduction, immunity, and welfare. By supporting metabolic processes, enhancing immunity, and preventing diseases, vitamins ensure optimal growth, productivity, and well-being. Providing balanced diets and addressing specific vitamin needs at different life stages are critical for achieving sustainable and welfare-oriented livestock management.

f) Role of water

Water is the most essential nutrient for animals, constituting up to 70-80% of their body mass. It plays a central role in physiological processes, from metabolism to temperature regulation. Despite its significance, water often receives less attention compared to other nutrients. This essay explores the roles and dietary requirements of water in various aspects of animal health, production, reproduction, immunity, and welfare.

Roles of water in animal health

Maintaining homeostasis

Water is vital for regulating body temperature, maintaining blood volume, and ensuring the balance of electrolytes.

Metabolic functions

Water is a solvent for biochemical reactions, enabling nutrient transport, waste excretion, and energy metabolism.

Hydration and cellular function

Adequate water intake ensures cellular integrity, aids in nutrient absorption, and facilitates enzymatic activities.

Dietary requirements for water

Factors influencing water needs

Water requirements vary based on:

Species: Ruminants require more water than monogastric animals.

Body weight: Larger animals need more water.

Physiological stage: Lactating or pregnant animals have higher water demands.

Environmental conditions: Heat stress increases water intake.

Diet composition: High-salt or high-protein diets elevate water needs.

General guidelines for water intake

Cattle: 30-50 liters/day.

Sheep and Goats: 4-10 liters/day.

Horses: 20-45 liters/day.

Poultry: 0.2-0.4 liters/day per bird.

Swine: 8-12 liters/day.

Role of water in animal production

Dairy production

Water is a major component of milk, constituting 87-88% of its volume. Insufficient water reduces milk yield and quality.

Meat production

Adequate hydration ensures optimal feed conversion, muscle growth, and meat quality. Dehydrated animals experience reduced growth rates and poor carcass composition.

Egg production

In poultry, water consumption directly influences egg production and shell quality. Dehydration can lead to smaller, malformed eggs.

Role of water in reproduction

Female reproduction

Adequate amount of water ensures proper hydration of reproductive tissues.

Supports placental fluid production, which is vital for foetal development.

Male reproduction

Adequate water maintains hydration for normal semen production and sperm motility.

Pregnancy and lactation

Increased water requirements during pregnancy ensure foetal growth.

Lactating animals require 2-3 times more water to support milk production.

Role of water in nutritional status

Feed utilization

Water aids in digesting feed, dissolving nutrients, and transporting them to cells.

Appetite and feed intake

Dehydrated animals often reduce feed intake, leading to malnutrition and weight loss.

Metabolic efficiency

Proper hydration ensures optimal metabolism, preventing energy wastage and nutrient deficiencies.

Role of water in immunity

Toxin elimination

Water supports kidney function, helping to excrete metabolic waste and toxins, reducing the risk of illness.

Immune cell function

Adequate hydration is critical for the transport of immune cells and antibodies to infection sites.

Stress and recovery

Water reduces heat stress and aids recovery during disease or vaccination periods.

Role of water in welfare

Behaviour and comfort

Animals with unrestricted access to clean water exhibit natural behaviors and reduced signs of stress.

Stress management

During transport or heat waves, adequate hydration is essential to prevent dehydration-related welfare issues. Providing clean, cool water prevents conditions like heatstroke, which compromise animal welfare.

g) Critical nutrients for dairy animals

High-producing dairy cattle and buffaloes have specific nutritional requirements to support milk production, reproduction, immunity, and overall health. Among these, critical amino acids, fatty acids, minerals, and vitamins play a central role in ensuring optimal physiological function and productivity.

Critical amino acids

Amino acids are the building blocks of proteins essential for milk synthesis, growth, reproduction, and immune function. While dairy cattle and buffaloes can synthesize some amino acids, others must be provided through the diet (essential amino acids).

Lysine

Role

Essential for milk protein synthesis,

Promotes growth and improves feed efficiency,

Supports immune function

Sources: Soybean meal, canola meal, and fishmeal.

Methionine

Role

Acts as a methyl donor in metabolic processes

Enhances milk fat synthesis.,

Protects liver function and prevents fatty liver syndrome.

Sources: Corn, soybean meal, and rumen-protected methionine supplements.

Threonine

Role

Supports gut health and mucin production.

Plays a role in milk protein synthesis,

Sources: Blood meal, soybean meal, and rumen-protected supplements.

Histidine

Role

Crucial for hemoglobin formation

Improves milk protein concentration

Sources: Fishmeal, alfalfa, and bacterial protein from the rumen.

Critical fatty acids

Fatty acids are key components of cell membranes, energy metabolism, and milk fat synthesis. They are categorized into essential fatty acids, which must be supplied through the diet.

Linoleic acid (C18:2, Omega-6)

Role

Precursor for prostaglandin synthesis

Supports immune function and reproduction

Sources: Corn, sunflower oil, and cottonseed oil.

Alpha-Linolenic acid (C18:3, Omega-3)

Role

Reduces inflammation and enhances immune response

Supports ovarian function and embryo development

Sources: Flaxseed, canola oil, and fresh forage.

Eicosapentaenoic acid (EPA) and Docosahexaenoic acid (DHA)

Role

Improve fertility by enhancing uterine health.

Support milk fat synthesis and reduce ketosis risk.

Sources: Fish oil and marine algae supplements.

Critical minerals

Minerals are divided into macro-minerals (required in larger amounts) and micro-minerals (required in trace amounts). Both categories are vital for various physiological functions.

Critical macro-minerals

Calcium (Ca)

Role

Essential for bone health, muscle contraction, and milk production

Prevents milk fever and hypocalcemia

Sources: Limestone, di-calcium phosphate, and legume forages.

Phosphorus (P)

Role

Crucial for energy metabolism (ATP synthesis)

Supports skeletal development and milk production

Sources: Di-calcium phosphate and cereals.

Magnesium (Mg)

Role

Involved in nerve function and enzyme activation

Prevents grass tetany

Sources: Magnesium oxide and dolomitic limestone.

Potassium (K)

Role

Maintains acid-base balance and hydration

Supports milk production during heat stress.

Sources: Potassium chloride and forage.

Sodium (Na)

Role

Regulates osmotic pressure, acid-base balance, and nerve impulses

Sources: Salt blocks and mineral mixes.

Critical micro-minerals

Zinc (Zn)

Role

Promotes skin and hoof health

Enhances immune function and reproduction

Sources: Zinc sulphate and zinc methionine.

Copper (Cu)

Role

Supports enzyme function and iron metabolism

Enhances fertility and immune response

Sources: Copper sulphate and copper oxide.

Selenium (Se)

Role

Functions as an antioxidant with vitamin E

Prevents retained placenta and white muscle disease

Sources: Sodium selenite and selenium-enriched yeast.

Iodine (I)

Role

Crucial for thyroid hormone synthesis

Supports growth and lactation

Sources: Potassium iodide and iodized salt.

Iron (Fe)

Role

Required for hemoglobin and oxygen transport

Prevents anaemia, especially in young animals

Sources: Ferrous sulfate and forages.

Critical fat-soluble vitamins

Vitamin A

Role

Maintains epithelial integrity

Enhances reproduction and vision

Sources: Green forages, alfalfa, and vitamin A supplements.

Vitamin D

Role

Facilitates calcium and phosphorus absorption

Prevents rickets and milk fever

Sources: Sunlight, sun-cured hay, and vitamin D supplements.

Vitamin E

Role

Acts as an antioxidant

Prevents mastitis and supports immune function

Sources: Fresh forage and vitamin E supplements.

Vitamin K

Role

Essential for blood clotting

Sources: Green leafy forages and gut microbial synthesis.

Water-soluble vitamins

Vitamin B-complex includes thiamine (B_1), riboflavin (B_2), niacin (B_3), and cobalamin (B_{12}).

Role

Support energy metabolism, nerve function, and red blood cell production.

Sources: Bacterial synthesis in the rumen and dietary supplements.

Vitamin C

Role

Acts as an antioxidant and supports stress resilience.

Sources: Endogenously synthesized in most ruminants, but supplementation may help during stress.

Critical nutrients such as amino acids, fatty acids, minerals, and vitamins are essential for sustaining the health, productivity, and welfare of dairy cattle and buffaloes. A balanced diet that meets these nutritional requirements not only optimizes milk production and reproductive efficiency but also improves immunity and reduces the risk of metabolic diseases. Strategic supplementation, high-quality feed and proper management practices ensure that dairy animals thrive and contribute to sustainable dairy farming.

h) Dietary requirements and dose of nutrients

The nutrition of animals is a complex interplay of various nutrients. Among these, carbohydrates, fats, proteins, and amino acids are macronutrients critical to health, production, reproduction, immunity, and overall welfare.

Soluble and insoluble carbohydrates

Carbohydrates are the primary energy source for animals, divided into soluble (sugars and starches) and insoluble (fibrous) carbohydrates.

Soluble carbohydrates

Soluble carbohydrates, like sugars and starches, are rapidly digestible and provide immediate energy.

Functions

Quick energy for metabolic processes

Drives microbial protein synthesis in the rumen

Supports milk production and body maintenance

Dietary requirements and doses

Ruminants: 18-24% of the dry matter (DM) intake as non-fibre carbohydrates (NFC).

Monogastrics: 40-50% of the diet, depending on production stage.

Sources: Corn, barley, wheat, molasses.

Insoluble carbohydrates (Fibre)

Insoluble carbohydrates, primarily fibre, include cellulose, hemicellulose, and lignin. These are fermented slowly in the rumen to produce volatile fatty acids (VFAs), which are vital for energy.

Functions

Maintains rumen health and prevents acidosis.

Supports milk fat synthesis.

Promotes gut motility and faecal consistency.

Dietary requirements and doses

Ruminants: 28-34% neutral detergent fibre (NDF) of the DM.

Monogastrics: 5-10% crude fibre in the diet.

Sources: Forages, silage, hay, and by-products like soybean hulls.

Fats and lipids

Fats and lipids are dense energy sources, containing 2.25 times more energy than carbohydrates or proteins.

Functions

Provides long-term energy storage

Essential for the absorption of fat-soluble vitamins (A, D, E, K)

Enhances reproductive performance by supporting ovarian and hormonal function

Contributes to milk fat synthesis and immune resilience

Dietary requirements and doses

Ruminants: 5-6% of the DM as fats (exceeding 7% may depress fibre digestion).

Monogastrics: 5-10% of the diet.

Poultry: 4-7% for broilers and layers.

Sources

Vegetable oils (soybean, canola and flaxseed).

Animal fats (tallow, lard).

Bypass fats for ruminants (calcium salts of fatty acids).

Proteins and amino acids

Proteins are critical for structural development, enzymatic functions, and immune response. Amino acids, the building blocks of proteins, are categorized as essential (must be supplied through diet) and non-essential (synthesized by the animal).

Proteins

Functions

Tissue repair and growth

Milk and egg protein synthesis

Production of enzymes, hormones, and antibodies

Dietary requirements and doses

Ruminants: Crude protein (CP) should constitute 12-18% of the DM depending on the production stage.

Metabolizable protein (MP): 8-14% of DM.

Monogastrics

Swine: 14-20% of the diet.

Poultry: 16-24% depending on growth or production.

Sources: Soybean meal, groundnut cake, canola meal, alfalfa, fishmeal.

Critical amino acids

Lysine

Supports growth, milk protein synthesis, and reproduction

Dose: 7.0-7.2% of digestible protein (ruminants) and 1.1-1.2% of the diet (monogastrics).

Methionine

Enhances milk fat synthesis and prevents fatty liver syndrome

Dose: 2.2-2.4% of digestible protein (ruminants) and 0.35-0.45% of the diet (monogastrics).

Threonine

Aids in gut health and immune function

Dose: 5.0-5.3% of digestible protein (ruminants) and 0.7-1.0% of the diet (monogastrics).

Histidine

Crucial for milk protein and hemoglobin production

Dose: 2.5-3.0% of metabolizable protein for high-yielding dairy cows.

Non-ruminant requirements

Amino acid-balanced diets are critical for poultry and swine. Synthetic amino acids (e.g., lysine and methionine) are often added to achieve balance.

Nutritional roles in animal systems

Animal health

Balanced carbohydrates ensure a stable energy supply, preventing metabolic disorders like ketosis or acidosis. Proteins and amino acids support tissue repair and immune cell production. Essential fatty acids reduce inflammation and enhance cellular integrity.

Production

Carbohydrates and fats provide the energy required for milk, meat, and egg production. Proteins and amino acids are directly used in milk protein and meat synthesis. High-fat diets improve milk fat percentage and energy-corrected milk.

Reproduction

Soluble carbohydrates improve ovarian activity and conception rates. Fats support hormonal function and improving fertility. Methionine and lysine enhance embryo development.

Nutritional status

Insoluble carbohydrates promote gut health and nutrient absorption. Balanced protein intake prevents weight loss and supports overall nutritional adequacy.

Immunity

Proteins and amino acids enhance antibody and immune cell production. Omega-3 fatty acids modulate inflammatory responses, reducing disease risk. Adequate carbohydrate intake prevents immunosuppression during stress.

Welfare

Proper nutrition reduces the risk of metabolic disorders, ensuring optimal growth and health. Balanced diets support behavioural well-being by promoting comfort and productivity.

Soluble and insoluble carbohydrates, fats, proteins, and amino acids are indispensable for animal health, productivity, reproduction, immunity, and welfare. Understanding their roles and providing optimal dietary doses ensures sustainable and efficient livestock production. Balanced diets not only maximize production but also enhance animal welfare, contributing to the overall success of animal husbandry practices.

3

Significance of Nutrient Interrelationship in Livestock and Poultry

Livestock nutrition is a complex field that involves the interplay of various nutrients to ensure optimal health, growth, reproduction and productivity. Nutrients do not function in isolation; rather, they interact in multiple ways synergistically or antagonistically affecting digestion, absorption, metabolism, and overall animal performance. Understanding these interrelationships is crucial for formulating balanced diets that meet the specific needs of different livestock species at various physiological stages.

Nutrients in livestock diets can be broadly classified into macronutrients (carbohydrates, proteins, and fats) and micronutrients (vitamins and minerals). Each nutrient has a specific role, but their interactions influence bioavailability and utilization. For instance, dietary proteins and carbohydrates have a direct impact on energy metabolism, affecting the animal's ability to utilize other nutrients efficiently. Excess protein can increase energy expenditure due to deamination, while a deficiency in energy can limit protein synthesis. Similarly, the balance between dietary fat and carbohydrate affects energy density and digestion, influencing growth performance and body composition.

Among minerals, calcium (Ca) and phosphorus (P) exhibit a well-known interrelationship. Both are essential for bone development, but their absorption and utilization depend on the Ca:P ratio in the diet. An imbalance whether an excess or deficiency of one can result in metabolic disorders such as rickets or osteoporosis. Additionally, excess phosphorus reduces calcium absorption, leading to skeletal deformities, while excessive calcium can interfere with magnesium and zinc absorption, affecting overall metabolic functions. The interaction between iron, copper, and zinc is another critical aspect of mineral nutrition. High dietary iron can interfere with copper absorption, leading to anaemia, while excessive zinc can inhibit copper metabolism, causing growth retardation and immune suppression.

Vitamins also have intricate interactions with minerals and other nutrients. Vitamin D plays a key role in calcium and phosphorus metabolism, influencing bone health and immune function. A deficiency of vitamin D can lead to improper calcium absorption, even if dietary calcium levels are adequate. Vitamin A is involved in vision, immune response, and reproduction, but its excessive intake can interfere with vitamin D and calcium metabolism, leading to bone abnormalities. The balance of vitamin E and selenium is essential for antioxidant defence mechanisms, preventing oxidative stress and muscle degeneration in livestock.

In addition to direct nutrient interactions, factors such as anti-nutritional components in feed, feed processing methods, and gut microbiota composition further influence nutrient utilization. Certain plant-based feedstuffs contain phytates, tannins, and oxalates that bind to minerals and reduce their bioavailability. Fermentation and enzymatic treatments can help mitigate these anti-nutritional factors, improving nutrient digestibility and absorption. Moreover, the gut microbiota plays a crucial role in modulating nutrient metabolism, particularly in ruminants, where microbial fermentation determines the availability of volatile fatty acids, amino acids, and vitamins.

A thorough understanding of nutrient interrelationships is essential for optimizing livestock nutrition programs. It helps prevent nutritional deficiencies, toxicities, and metabolic disorders while maximizing feed efficiency, animal health, and productivity. Nutritionists and animal scientists must consider these complex interactions when formulating diets, ensuring that animals receive not just the required amounts of nutrients but also the right balance for optimal physiological function.

a) Calcium (Ca), phosphorus (P) and vitamin D interrelationship

The interrelationship between calcium (Ca), phosphorus (P), and vitamin D is crucial in maintaining bone health, metabolic functions, and overall physiological well-being in animals. These nutrients work together to regulate bone mineralization, enzyme activity, muscle contraction, nerve transmission, and cellular signalling. Any imbalance can lead to severe metabolic disorders such as rickets, osteoporosis, and hypocalcaemia. Understanding their mechanism of interaction helps in formulating balanced diets and preventing nutritional deficiencies in livestock.

Significance of the Ca-P-vitamin D interaction

1. ***Bone formation and maintenance:*** Ca and P are the primary minerals in hydroxyapatite, which provides structural strength to bones and teeth.

   The Ca:P ratio in the diet should be optimal (typically 1.2:1 to 2:1 in most species) to ensure proper mineralization.
2. ***Metabolic regulation:*** Ca is essential for muscle function, nerve transmission, blood clotting, and enzyme activation.

P plays a role in energy metabolism (ATP formation), nucleic acid synthesis, and cell signalling.

3. ***Role of vitamin D in Ca and P homeostasis:*** Vitamin D (mainly as 1,25-dihydroxycholecalciferol or calcitriol) regulates Ca and P absorption from the intestine, their mobilization from bones, and renal reabsorption.

   Deficiency in vitamin D leads to poor Ca absorption, resulting in hypocalcaemia and weak bone structure.

Mechanism of action of Ca, P and vitamin D interrelationship

1. ***Dietary absorption (intestine):*** Dietary Ca and P are absorbed in the small intestine.

   Vitamin D enhances the expression of Ca-binding proteins, increasing Ca absorption.

2. ***Bone mobilization (skeletal system):*** When blood Ca levels drop, the parathyroid gland releases parathyroid hormone (PTH). Thus, PTH stimulates bone resorption, releasing Ca and P into the bloodstream.

   Vitamin D further enhances Ca mobilization from bones.

3. ***Renal re-absorption (Kidneys):*** PTH and vitamin D regulate Ca re-absorption in the kidneys to prevent excessive urinary loss.

   P levels are controlled by fibroblast growth factor-23 (FGF-23) and PTH, which increase P excretion when levels are high.

4. ***Homeostasis regulation:*** When Ca and P levels are adequate, calcitonin (secreted by the thyroid) inhibits bone resorption, maintaining balance

   Any imbalance in vitamin D, Ca, or P disrupts this cycle, leading to metabolic disorders.

This intricate regulatory system ensures that animals maintain proper Ca and P levels for optimal physiological function. Proper dietary balance and vitamin D supplementation are essential for preventing metabolic diseases and improving livestock health and productivity.

Flow Chart of Ca, P and vitamin D Mechan

Dietary intake of Ca and P

↓

Absorption in the small intestine

↓

(Vitamin D enhances Ca and P absorption)

↓

Blood Ca and P levels

↓ ↓

If low: If high:

↓ ↓

PTH released Calcitonin released

↓ ↓

Bone resorption Bone mineralization
(Ca and P mobilized) (Ca and P deposited

↓ ↓

Renal reabsorption Increased urinary excreti

↓ ↓

Vitamin D activation Homeostasis maintain
(Liver → Kidney)

b) Vitamin E and selenium interrelationship

The interrelationship between vitamin E and selenium (Se) plays a crucial role in antioxidant defence, immune function, reproduction, and overall health in animals. Both act as key components of the body's antioxidant system, protecting cells from oxidative damage caused by free radicals. Deficiencies in either nutrient can lead to serious health problems, including muscle degeneration, impaired immune response, reproductive failures, and increased susceptibility to diseases. Understanding their interaction helps in formulating balanced diets to enhance animal health, productivity, and disease resistance.

Significance of vitamin E and Selenium interaction

1. **Antioxidant protection:** Vitamin E (α-tocopherol) is a fat-soluble antioxidant that protects cell membranes from lipid peroxidation. The Se is an essential component of glutathione peroxidase (GSH-Px), an enzyme that neutralizes harmful peroxides, preventing oxidative stress. Together, they prevent oxidative damage to cell membranes, especially in muscles, the liver, and immune cells.
2. **Muscle health and prevention of white muscle disease (WMD):** Se and vitamin E deficiency can lead to WMD, a condition characterized by muscle degeneration, stiffness, and weakness in young animals, particularly in calves, lambs, and poultry.

3. **Immune function:** Both nutrients enhance the immune response by protecting immune cells from oxidative damage. Se supports the activity of neutrophils and macrophages, while vitamin E modulates cytokine production, improving disease resistance.
4. **Reproductive health:** The Se is essential for sperm motility and overall fertility in both male and female animals. Vitamin E supports embryonic development and prevents reproductive disorders such as retained placenta and low fertility rates.
5. **Growth and productivity:** Adequate levels of vitamin E and Se improve growth performance, feed efficiency, and overall productivity in livestock.

Mechanism of action of vitamin E and Se

1. **Lipid peroxidation prevention:** Free radicals generated from metabolic processes attack cell membranes, leading to lipid peroxidation. Vitamin E donates electrons to stabilize free radicals, preventing cell membrane damage.
2. **Glutathione Peroxidase (GSH-Px) activity:** Selenium, as a component of GSH-Px, converts harmful peroxides (H_2O_2) and lipid peroxides) into non-toxic compounds (water and alcohol). This action reduces oxidative stress, protecting muscle cells and organs.
3. **Regeneration of vitamin E:** Once vitamin E neutralizes a free radical, it becomes oxidized. Se-containing enzymes help regenerate active vitamin E, ensuring continued antioxidant protection.
4. **Immune modulation and cellular protection:** Vitamin E supports immune cell function by maintaining membrane integrity. Se aids in the production of selenoproteins, which regulate cellular processes and protect against oxidative damage.

Flow chart of vitamin E and Se mechanism

Metabolic processes → Free radical production

↓

Lipid peroxidation (Cell membrane damage)

↓

Vitamin E neutralizes free radicals

↓

Vitamin E becomes oxidized (inactive)

↓

Se (as glutathione peroxidase) reduces peroxides

↓

Vitamin E is regenerated and antioxidant protection continues

↓

Cell membrane integrity maintained → Healthy muscle, immunity, and reproduction

This interaction ensures optimal antioxidant defence, protecting animals from oxidative stress, muscle degeneration, immune suppression, and reproductive disorders. Proper dietary supplementation of vitamin E and Se is essential for maintaining health, productivity, and disease resistance in livestock.

c) Vitamin A and zinc interrelationship

Vitamin A and zinc (Zn) have a crucial interrelationship in various physiological functions, including vision, immune response, reproduction, growth, and cellular differentiation. Their combined role is particularly significant in the formation of visual purple (rhodopsin), a light-sensitive pigment in the retina necessary for night vision. Zinc is essential for vitamin A metabolism, transport, and activation, making their interaction vital for maintaining normal vision and preventing deficiencies.

Significance of vitamin A and zinc interaction

1. **Vision and rhodopsin formation: Vitamin A (retinol)** is a precursor of **retinal,** which forms rhodopsin in the rod cells of the retina.

   Zinc is required for the enzyme **retinol dehydrogenase,** which converts retinol into retinal, a critical step in the visual cycle.

   Deficiencies in either nutrient lead to night blindness and impaired vision.

2 **Immune function:** Vitamin A supports epithelial integrity and immune responses, while zinc enhances immune cell function and enzyme activity.

   Both nutrients work together to protect against infections and oxidative stress.

3. **Reproduction and growth:** Vitamin A is essential for embryonic development, while zinc supports DNA synthesis and cell division.

   Deficiencies lead to reproductive failures and stunted growth in livestock.

4. **Cell differentiation and skin health:** Vitamin A regulates skin and mucosal tissue development, while zinc is necessary for wound healing and keratinization.

   Their combined action prevents skin disorders and promotes healthy epithelial tissues.

Mechanism of action in vision (Formation of visual purple - Rhodopsin)

1. ***Vitamin A metabolism and transport:*** Dietary vitamin A (retinyl esters) is absorbed in the intestine and converted into retinol.

   Zinc-dependent enzyme retinol dehydrogenase converts retinol into retinal (the active form in vision).

2. **Rhodopsin formation in the retina:** Retinal binds with opsin (a protein in rod cells) to form rhodopsin (visual purple).

   Rhodopsin is essential for low-light vision and enables animals to detect dim light.

3. **Light exposure and visual cycle:** When light strikes rhodopsin, it triggers a chemical change, converting 11-cis-retinal to all-trans-retinal.

   This change generates an electrical signal sent to the brain for vision.

4. **Regeneration of Rhodopsin:** All-trans-retinal is recycled back to 11-cis-retinal via the enzyme retinal isomerase.

   Zinc is required for this enzymatic process, ensuring a continuous supply of rhodopsin.

Flow chart of visual cycle (visual purple; Rhodopsin) formation

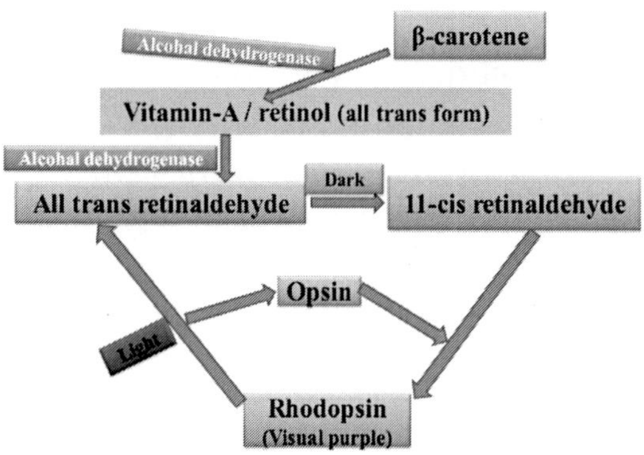

or

Dietary vitamin A (Retinyl esters)

↓

Intestinal absorption → Retinol formation

↓

Zinc-dependent enzyme (Retinol dehydrogenase)

↓

Retinol → Retinal (Active form for vision)

↓

Retinal + Opsin → Rhodopsin (Visual purple)

↓

Light exposure → Rhodopsin breaks down

↓

11-cis-Retinal → All-trans-retinal

↓

Enzyme: Retinal isomerase (Zinc-dependent) converts back

↓

Rhodopsin regenerated → Continuous night vision

The interrelationship between vitamin A and zinc is essential for vision, particularly in rhodopsin synthesis and the regeneration of the visual cycle. Zinc ensures the proper metabolism of vitamin A, enabling its role in maintaining normal eyesight, immune function, and cellular growth. Deficiencies in either nutrient can lead to night blindness, impaired immunity, poor growth, and reproductive failures in animals. Proper dietary management, ensuring adequate vitamin A and zinc intake, is crucial for livestock health and productivity.

d) B vitamins and trace minerals interrelationship

B vitamins and trace minerals play essential roles in metabolism, enzyme activation, energy production, immunity, and overall physiological functions in animals. Their interrelationship is based on their complementary roles in enzymatic processes, coenzyme formation, and cellular functions. Deficiencies in either can impair metabolism, leading to poor growth, reduced immunity, and metabolic disorders.

Significance of B vitamins and trace minerals interaction

1. **Energy metabolism: B-complex vitamins (B_1, B_2, B_3, B_5, B_6, B_7, B9 and B_{12})** are involved in carbohydrate, protein and fat metabolism.

   Trace minerals (iron, copper, zinc, selenium, manganese, cobalt, iodine) serve as cofactors for enzymes in metabolic pathways.

2. **Enzyme activation and co-enzyme formation: Vitamin B_1 (Thiamine)** works with magnesium and manganese in energy metabolism.

   Vitamin B_2 (Riboflavin) and **Niacin (B_3)** assist **iron and copper** in electron transport chain enzymes.

Vitamin B_{12} (Cobalamin) requires **cobalt** for its synthesis and function in red blood cell production.

3. **Red blood cell formation and oxygen transport: Iron** and **copper** are essential for hemoglobin synthesis, while B_{12} and folic acid (B_9) are needed for red blood cell formation.

   Deficiencies lead to anaemia and poor oxygen transport in animals.

4. **Antioxidant and immune function: Selenium (Se)** is a key component of glutathione peroxidase, which protects cells from oxidative damage, supported by **B-complex vitamins**.

   Pyridoxine (B_6) and **zinc** enhance immune responses and protein metabolism.

5. **Nervous system function and growth: B vitamins** (B_1, B_6 and B_{12}), magnesium, manganese and zinc are essential for nerve transmission and muscle coordination. Deficiencies can lead to neurological disorders, poor growth, and reproductive failures.

Mechanism of action of B vitamins and trace minerals

1. **Dietary absorption:** B vitamins and trace minerals are absorbed in the intestine and transported to target tissues.
2. **Enzyme and denzyme activation:** B vitamins act as coenzymes, while trace minerals serve as enzyme cofactors in metabolic reactions.
3. **Energy and red blood cell production:** B vitamins and iron/copper support ATP synthesis and hemoglobin formation.
4. **Antioxidant and immune support:** Selenium and B vitamins protect against oxidative stress and boost immunity.

Flow chart of B vitamins and trace minerals mechanism

Dietary intake of B vitamins and trace minerals

↓

Intestinal absorption

↓

Coenzyme formation (B vitamins) & enzyme activation (trace minerals)

↓

Energy metabolism (ATP production)

↓

Red blood cell synthesis (B_{12}, iron, Copper)

↓

Antioxidant defence (B vitamins + selenium)

↓

Immune function, growth, and nervous system health

The interaction between B vitamins and trace minerals is vital for animal health, influencing metabolism, immunity, and overall productivity. Ensuring their adequate dietary supply prevents metabolic disorders, anemia, oxidative stress, and growth deficiencies in livestock.

e) Vitamin k, calcium (Ca), and manganese (Mn) interrelationship

Vitamin K, calcium (Ca), and manganese (Mn) play crucial roles in blood clotting, bone formation, and metabolic functions in animals. Their interrelationship is primarily centered on bone mineralization and coagulation, where vitamin K activates calcium-binding proteins, and manganese contributes to cartilage and bone development. Deficiencies or imbalances in any of these nutrients can lead to impaired blood clotting, weak bones, and metabolic disorders in livestock.

Significance of vitamin K, calcium, and manganese interaction

1. **blood clotting (Coagulation cascade): Vitamin K** is essential for synthesizing clotting factors (prothrombin and other coagulation proteins).

   Calcium acts as a cofactor in blood clotting pathways, ensuring proper coagulation.

   Manganese supports enzyme function in clotting factor synthesis.

2. **Bone formation and mineralization: Vitamin K** activates **osteocalcin**, a calcium-binding protein needed for bone formation.

   Calcium is the primary mineral in bones and teeth, providing structural integrity.

   Manganese is required for cartilage formation and bone matrix development, preventing skeletal deformities.

3. **Enzyme activation and metabolic functions: Vitamin K** is involved in liver enzyme activation for blood clotting.

   Calcium and manganese regulate muscle contraction, nerve signalling and metabolic enzyme activity.

   Manganese is a cofactor for glycosyltransferases, which support bone and joint health.

4. **Deficiency consequences: Vitamin K deficiency** causes poor blood clotting, internal bleeding and weak bones.

   Calcium deficiency leads to rickets, osteoporosis and poor muscle function.

   Manganese deficiency causes skeletal abnormalities, poor growth, reproductive issues.

Mechanism of action of vitamin K, Ca and Mn

1. **Dietary absorption:** Vitamin K is absorbed in the intestine (fat-soluble), while calcium and manganese are absorbed in the small intestine.
2. **Activation of clotting factors:** Vitamin K-dependent **γ-glutamyl carboxylase** activates prothrombin, requiring calcium for clot formation.
3. **Bone matrix formation and mineralization:** Vitamin K activates osteocalcin, which binds calcium to bones.

   Manganese-dependent **glycosyltransferases** synthesize proteoglycans for cartilage and bone strength.

Flow chart of B vitamins and trace minerals mechanism

Dietary intake of B vitamins and trace minerals

↓

Intestinal absorption

↓

Coenzyme formation (B vitamins) & enzyme activation (trace minerals)

↓

Energy metabolism (ATP production)

↓

Red blood cell synthesis (B_{12}, iron, Copper)

↓

Antioxidant defence (B vitamins + selenium)

↓

Immune function, growth, and nervous system health

The interaction between vitamin K, calcium, and manganese is essential for blood clotting, bone health, and metabolism in animals. Ensuring a balanced supply of these nutrients in livestock diets helps prevent clotting disorders, skeletal deformities, and metabolic dysfunctions, promoting overall animal health and productivity.

f) Copper (Cu), Molybdenum (Mo), and Sulfur (S) interrelationship

Cu, Mo and S have a complex interrelationship that affects enzyme activity, metabolism, and overall health in animals. Their balance is crucial because Mo and S can interfere with Cu absorption, leading to Cu deficiency, which impacts growth, immunity, reproduction, and enzyme functions. Understanding their interaction helps in formulating proper livestock diets to prevent metabolic disorders.

Significance of Cu, Mo and S interaction

1. **Copper's role in animal health:** Essential for enzyme activation, hemoglobin formation, immune function, reproduction, and coat

pigmentation. It also involved in collagen formation and bone development.

2. **Molybdenum's role and its antagonistic effect on copper:** Component of enzymes like xanthine oxidase and sulfite oxidase,

   Excess Mo binds to Cu, forming insoluble thiomolybdates in the rumen, reducing Cu absorption.

3. **Sulfur's role and its interaction with Cu and Mo:** S is needed for amino acid synthesis (methionine, cysteine) and keratin formation.

   When excess S is present, it enhances Mo's ability to form thiomolybdates, further reducing Cu availability.

4. **Deficiency and toxicity concerns: Copper deficiency** (caused by excess Mo and S): Poor growth, anaemia, reproductive failures, weak bones, faded coat colour.

   Excess copper toxicity (low Mo and S): Liver damage, jaundice, hemolysis.

   Excess molybdenum: Leads to teart disease in cattle (diarrhea, poor health).

Mechanism of action of Cu, Mo and S

1. **Dietary intake:** Animals consume Cu, Mo and S from feed and water.
2. **Formation of thiomolybdates in the rumen:** Excess Mo and S combine to form thiomolybdates, which bind to Cu, making it unavailable for absorption.
3. **Reduced copper absorption in the intestine:** Cu bound to thiomolybdates becomes insoluble and excreted, leading to Cu deficiency.
4. **Impaired enzyme functions and metabolic disruptions:** Low Cu affects cytochrome oxidase (energy metabolism), superoxide dismutase (antioxidant defense), and ceruloplasmin (iron transport).

Flow chart of Cu-Mo-S mechanism

Dietary Cu, Mo and S intake

↓

Excess Mo + S → Thiomolybdate formation in rumen

↓

Thiomolybdates bind to Cu → Insoluble Cu complexes

↓

Reduced Cu absorption → Cu deficiency

↓

Enzyme dysfunction → Anemia, weak bones and poor growth

The balance between Cu, Mo, and S is critical for enzyme activity, metabolism, and overall health in animals. Excess Mo and S reduce Cu availability, leading to deficiency-related disorders. Proper dietary management is essential to prevent imbalances and ensure optimal animal health and productivity.

g) Calcium (Ca) and zinc (Zn) interrelationship

Ca and Zn interact in various physiological processes, including bone formation, enzyme function, immune response, and metabolism. However, their balance is crucial, as excessive Ca can interfere with Zn absorption, leading to deficiencies and metabolic disorders in animals.

Significance of Ca and Zn interaction

1. **Bone health and growth:** Calcium is the primary mineral in bones and teeth, providing structural strength.

   Zinc plays a role in bone mineralization and collagen synthesis.

2. **Enzyme activation and metabolism:** Zinc is a cofactor for over 300 enzymes, including those involved in protein synthesis, immune function, and reproduction.

   High dietary calcium levels reduce zinc absorption by competing for absorption sites in the intestine.

3. **Immune function and reproduction:** Zinc is essential for immune cell function and hormone production.

   Excess calcium may disrupt zinc metabolism, leading to poor immunity and reproductive issues.

Maintaining a proper Ca:Zn ratio in animal diets is essential for optimal growth, immunity, and enzyme function. Excessive calcium can impair zinc absorption, leading to poor growth, weak immunity, and metabolic imbalances in livestock.

h) Iron (Fe) and Copper (Cu) interrelationship

Iron and copper have a crucial interrelationship in hemoglobin synthesis, enzyme function, and red blood cell formation. Cu is essential for iron metabolism, and an imbalance between these minerals can lead to anemia, poor growth, and metabolic disorders in animals.

Significance of Fe and Cu interaction

1. ***Hemoglobin and red blood cell formation:*** Iron is a key component of hemoglobin, essential for oxygen transport.

   Cu is required for iron absorption and mobilization from storage sites.

2. ***Enzyme activation and iron metabolism:*** Cu-dependent enzyme ceruloplasmin converts Fe^{2+} (ferrous) to Fe^{3+} (ferric), enabling iron transport in the blood.

   Low Cu levels impair iron metabolism, leading to iron accumulation in tissues and anaemia.

3. **Deficiency consequences:** Iron deficiency: Leads to anemia, weakness, and poor growth.

   Cu deficiency: Reduces iron utilization, causing secondary iron deficiency anaemia.

The balance between iron and copper is essential for hemoglobin synthesis, enzyme function, and oxygen transport. A deficiency in Cu can impair iron metabolism, leading to anemia, while excess iron can interfere with copper absorption, affecting overall health in animals.

i) Sulfur (S) and selenium (Se) interrelationship

Sulfur (S) and selenium (Se) interact in various metabolic processes, primarily in protein synthesis, antioxidant defense, and enzyme function. Both are essential for muscle health, immune function, and growth, but an imbalance can lead to nutritional disorders in animals.

Significance of S and Se interaction

1. ***Protein and antioxidant enzyme synthesis:*** S is a component of S-containing amino acids (methionine, cysteine), which are precursors for selenoproteins.

   Se is essential for glutathione peroxidase (GSH-Px), an enzyme that protects cells from oxidative damage.

2. ***Absorption and antagonistic effects:*** High dietary S competes with Se for absorption in the intestine, potentially leading to Se deficiency.

S-rich diets (e.g., high sulfate water or feed) can reduce Se bioavailability, affecting antioxidant function and immunity.

3. ***Deficiency consequences:*** Low Se leads to white muscle disease, poor immunity and reproductive issues.

   Excess S: Can cause polioencephalomalacia (PEM) and interfere with Se metabolism.

S and Se must be balanced in livestock diets to support antioxidant defence, muscle function, and immune health. Excess S can limit Se absorption, leading to deficiency-related disorders in animals.

j) Vitamin A and vitamin E interrelationship

Vitamin A and vitamin E have a complementary relationship in growth, immune function, reproduction, and antioxidant defense in animals. Both are fat-soluble vitamins that work together to maintain cell integrity, vision, and disease resistance. However, an imbalance can lead to oxidative stress, immune suppression, and poor reproductive performance.

Significance of vitamin A and vitamin E interaction

1. ***Antioxidant protection and vitamin A stability:*** Vitamin E is a potent antioxidant that protects vitamin A from oxidation, ensuring its bioavailability.

   Inadequate vitamin E levels can lead to vitamin A degradation, reducing its effectiveness in vision and immunity.

2. ***Immune function and growth:*** Vitamin A supports epithelial tissue integrity, while vitamin E enhances immune cell function.

   Both vitamins help in disease resistance, wound healing, and growth in livestock.

3. ***Reproductive health:*** Vitamin A is essential for embryonic development and fertility, while vitamin E prevents oxidative damage to reproductive cells.

   A deficiency in either can result in infertility, poor foetal development and reduced hatchability in poultry.

Vitamin A and vitamin E work synergistically to support growth, immunity, and reproduction in animals. Proper dietary balance is essential, as low vitamin E can reduce vitamin A efficiency, while excess vitamin A can increase oxidative stress, raising vitamin E demand.

k) Cobalt (Co) and vitamin B12 interrelationship

Cobalt, a trace mineral is nutritionally essential in the diet of ruminants; although, the required level is very low i.e. 0.1 ppm. Co deficiency in ruminants is actually a vitamin B_{12} deficiency brought out by the inability of rumen microorganisms to synthesize vitamin B_{12}, as Co is an integral component/ part of vitamin B_{12}. Co is dietary essential only in ruminants, whereas in monogastric animals it is required to be given in the form of vitamin B_{12}.

Vitamin B_{12} is very important in carbohydrate metabolism of ruminants, as vitamin B_{12} is a component of the enzyme methyl melonyl CoA isomerise (mutase) which catalyses the conversion of methyl malonyl Co A to succinyl Co A. Thus for the utilization of propionic acid through TCA cycle the presence of vitamin B_{12} is must. In the deficiency of vitamin B_{12} it is observed that there is an increase in the proportion of odd-number fatty acids in lipids as a result of increase in the availability of propionic acid. Increase in the availability of later is the result of decrease in activity of vitamin B_{12} dependent enzyme methylmalonyl CoA mutase, leading to higher level of methyl malonyl CoA in blood plasma. The later is considered as an index of Co or vitamin B_{12} deficiency, and it affects the voluntary feed intake.

l) Vitamin C and minerals interrelationship

Vitamin C (ascorbic acid) and its oxidation product, dehydroascorbic acid play an important role in various oxidation-reduction mechanism in living cells. Vitamin C also plays a significant role in mineral metabolism due to their chelating behaviour with minerals. Vitamin C enhances the absorption of various dietary minerals and their distribution throughout the body. It promotes the absorption of non-haem iron from the food and acts by reducing the ferric iron at the acidic pH in the stomach and by forming complexes with iron ions that stay in the solution at alkaline condition of duodenum. It also helps in the reduction and release of ferric iron from the tight linkages with plasma protein and its incorporation into ferritin.

m) Vitamin D and magnesium (Mg) interrelationship

Vitamin D and magnesium (Mg) work together to regulate Ca metabolism, bone health and overall physiological functions of animals, whereas magnesium is essential for the activation and function of vitamin D, while vitamin D helps in Ca absorption, which indirectly influences Mg metabolism. An imbalance between these two nutrients can lead to bone disorders, muscle dysfunction and metabolic imbalances.

Significance of vitamin D and magnesium interaction

1. ***Vitamin D activation requires magnesium:*** Mg acts as a cofactor for enzymes that convert vitamin D into its active form (calcitriol) in the liver and kidneys.

   Without adequate Mg, vitamin D cannot effectively regulate Ca and P metabolism.

2. ***Bone health and calcium balance:*** Vitamin D enhances Ca absorption in the intestines, ensuring proper bone mineralization.

   Mg helps maintain Ca balance by influencing parathyroid hormone (PTH) secretion, which regulates Ca homeostasis.

3. ***Muscle and nerve function:*** Both vitamin D and Mg are essential for neuromuscular function.

   Deficiencies can lead to muscle cramps, weakness, and impaired nerve signalling in animals.

Vitamin D and Mg have a synergistic relationship, where magnesium is essential for vitamin D activation, and vitamin D supports Ca homeostasis. Ensuring adequate levels of both nutrients is vital for bone strength, muscle function and overall metabolic health in livestock.

4

Nutritional and Metabolic Disorders in Livestock and Poultry

Nutritional and metabolic disorders in livestock and poultry arise from imbalances in nutrient intake, environmental factors, or the inability of the body to metabolize nutrients properly. These disorders impact animal health, productivity, and overall economic performance. Early identification and proper management are critical to mitigating their effects.

Nutritional and metabolic disorders can be broadly categorized as:

1. **Disorders caused by deficiencies or excesses of nutrients**
2. **Disorders related to imbalances in energy, protein, or minerals**
3. **Disorders resulting from inadequate metabolic adaptations to physiological or environmental changes**

Common nutritional and metabolic disorders in livestock include:

A. Energy and protein-related disorders

1. Ketosis (Acetonemia) in dairy cattle

Cause: Negative energy balance during early lactation, leading to fat mobilization and ketone body accumulation.

Symptoms: Reduced milk production, weight loss, acetone-like breath, and decreased appetite.

Management

Increase energy density in diets with concentrates and high-quality forages.

Supplement with propylene glycol or glucose precursors.

Monitor body condition scores to prevent over-conditioning during the dry period.

2. Fatty liver syndrome

Cause: Excessive fat deposition in the liver during periods of energy deficiency.

Symptoms: Reduced feed intake, poor reproductive performance, and decreased productivity.

Management

Provide balanced diets with adequate energy and protein.

Avoid over-conditioning prepartum cows.

3. Protein deficiency

Cause: Diets low in protein or amino acids.

Symptoms: Poor growth, reduced milk production, and impaired reproduction.

Management

Include high-quality protein sources like soybean meal or fishmeal.

Supplement essential amino acids for monogastric animals.

Mineral imbalances

1. Milk Fever (Hypocalcemia)

Cause: Calcium deficiency during the transition period in dairy cows.

Symptoms: Weakness, inability to stand, and decreased milk production.

Management

Provide anionic salts prepartum to promote calcium mobilization.

Administer calcium borogluconate intravenously during clinical cases.

Grass tetany (Hypomagnesemia)

Cause: Low magnesium levels, often due to grazing on lush, potassium-rich pastures.

Symptoms: Muscle tremors, convulsions, and death in severe cases.

Management

Supplement magnesium in feed or mineral blocks.

Avoid grazing animals on high-risk pastures without supplementation.

Sodium deficiency

Cause: Inadequate salt in the diet.

Symptoms: Pica, reduced appetite, and poor growth.

Management: Provide free-choice salt or incorporate salt into diets.

Phosphorus deficiency

Cause: Low-phosphorus diets or poor soil quality.

Symptoms: Poor growth, bone abnormalities, and reproductive issues.

Management: Supplement with dicalcium phosphate or rock phosphate.

Copper and selenium deficiency

Cause: Diets low in trace minerals.

Symptoms: Weakness, poor immunity, and reproductive failures.

Management: Use mineral supplements tailored to the region's deficiency status.

Vitamin deficiencies

1. Vitamin A deficiency

Cause: Diets lacking green forages or vitamin A supplementation.

Symptoms: Poor vision, reproductive failure, and weak calves.

Management: Provide vitamin A-rich feeds or injectable supplements.

2. Vitamin D deficiency

Cause: Lack of sunlight or dietary sources of vitamin D.

Symptoms: Rickets in young animals and osteomalacia in adults.

Management: Include vitamin D-rich feeds or administer supplements.

3. Vitamin E and selenium deficiency

Cause: Deficient soils or feeds.

Symptoms: White muscle disease in calves, poor growth, and reproductive issues.

Management: Use selenium-enriched mineral supplements and vitamin E.

Metabolic disorders in poultry

1. Ascites syndrome

Cause: Rapid growth and inadequate oxygen supply in broilers.

Symptoms: Fluid accumulation in the abdomen, labored breathing, and mortality.

Management

Optimize ventilation and avoid overcrowding.

Adjust energy levels in the diet.

Fatty liver hemorrhagic syndrome (FLHS)

Cause: High-energy diets and low activity levels in laying hens.

Symptoms: Obesity, liver rupture, and reduced egg production.

Management: Reduce dietary energy density and include vitamin E and choline.

Cage layer fatigue

Cause: Calcium deficiency in laying hens.

Symptoms: Weak bones, paralysis, and decreased egg production.

Management: Provide calcium-rich diets with adequate vitamin D3.

Preventive and nutritional management strategies

1. ***Balanced diets***: Formulate rations based on species, age, production stage, and environmental conditions. Include all essential nutrients, including macronutrients, micronutrients, and vitamins.
2. ***Regular monitoring***: Analyze feed and forage quality for nutrient content and deficiencies. Regularly monitor body condition scores and production parameters.
3. ***Supplementation***: Provide mineral and vitamin supplements based on regional deficiencies. Use feed additives like rumen-protected fats, amino acids, and buffers as needed.
4. ***Transition diets***: For dairy cows, provide close-up rations rich in anionic salts during the dry period. For poultry, gradually transition diets during growth and laying periods.
5. ***Environmental management***: Ensure proper ventilation, stocking density, and water quality to reduce stress. Prevent access to toxic plants or feeds that may cause nutritional imbalances.
6. ***Education and training***: Train farmers to recognize early signs of nutritional and metabolic disorders. Promote the use of technology like feed testing kits and nutritional software.

Nutritional and metabolic disorders in livestock and poultry can significantly impact animal health, productivity, and profitability. These disorders often result from imbalances in energy, protein, minerals, or vitamins. Preventive measures, including balanced diets, regular monitoring, and environmental management, are essential to minimize the occurrence of these disorders. Collaboration among farmers, veterinarians, and nutritionists is critical for developing effective strategies to ensure optimal health and productivity in livestock and poultry systems.

5

Traumatic Reticulitis / Traumatic Reticuloperitonitis

Traumatic reticulitis/ traumatic reticulo-peritonitis (TRP) commonly referred to as hardware disease is a significant condition in ruminants such as cattle, buffalo, sheep, and goats. It results from the ingestion of foreign objects like nails, wires, or pieces of metal, which settle in the reticulum (a compartment of the stomach) and cause injury or perforation of the reticular wall.

Traumatic reticulo-peritonitis (TRP) is a serious condition in ruminants. It occurs when a foreign object, such as a piece of metal, punctures the reticulum (the second compartment of the stomach), leading to inflammation of the peritoneum (the membrane lining the abdominal cavity). This can result in pain, infection, and in severe cases, systemic complications such as abscesses, pericarditis (inflammation of the sac surrounding the heart), or sepsis. Nutritional management is a key aspect in both the prevention and recovery from TRP, as it aids in reducing the risk of ingestion of foreign objects and supports healing after the condition occur.

The TRP occurs when ruminants, particularly cattle, buffaloes and sheep ingest sharp foreign objects such as nails, wires, or pieces of metal. Since ruminants do not chew their food thoroughly during the initial stages of eating, these objects can easily pass into the rumen and then the reticulum. The reticulum's honeycomb structure makes it susceptible to the penetration of sharp objects. Once a foreign object punctures the reticulum wall, it can cause localized infection (reticulitis) or more widespread inflammation (reticulo-peritonitis), potentially leading to more severe systemic issues such as liver abscesses, pericarditis, and septicemia.

The condition is associated with substantial economic losses due to decreased productivity, veterinary costs, and in severe cases, death. Effective management of traumatic reticulitis includes prompt treatment, preventive measures, and proper nutritional strategies to reduce the risk of recurrence.

Pathophysiology of traumatic reticulitis

The reticulum's anatomy and physiology predispose ruminants to this condition:

1. Ruminants lack the ability to fully differentiate foreign materials from food while grazing or feeding.
2. Heavy foreign objects gravitate to the reticulum due to its location and mixing activity.
3. When sharp objects lodge in the reticulum, they may penetrate its walls, causing inflammation (traumatic reticulitis) or even perforation, leading to complications like:

Peritonitis: Inflammation of the peritoneum due to leakage of gastrointestinal contents.

Pleuropneumonia: If the foreign object penetrates into the thoracic cavity.

Abscesses: Formation around the reticulum or in other tissues.

Vagal indigestion: Damage to the vagus nerve impairing digestive motility.

Clinical signs of traumatic reticulitis

The symptoms of traumatic reticulitis vary based on the severity and stage of the disease:

1. **Acute phase**

   Sudden drop in appetite and milk production,

   Fever (pyrexia) and lethargy,

   Reluctance to move or lie down due to pain,

   Arched back or hunched posture indicative of abdominal discomfort,

   Positive grunt test or withers pinch test due to pain response.

2. **Chronic phase**

   Weight loss and poor body condition,

   Decreased rumen motility,

   Bloating or distension due to impaired motility,

   Gradual decline in productivity,

3. **Complicated cases**

   Symptoms of peritonitis (abdominal distension, fever, and sepsis),

   Symptoms of thoracic involvement (respiratory distress, pleuritis),

Table 1: Clinical signs of animals with traumatic reticulitis

Clinical Signs	Percent (%) affected
Decreased milk production	65
Abnormal or muffled heart sounds	65
Anorexia	38
Abdominal pain	30
Fever	28
Weight loss	21.6
Ruminal atony	21.6
Depression	18.3
Down	13.3
Dyspnea	8.3
Sudden death	8.3
Neurological abnormalities	6.6
Diarrhea	5.0

Diagnosis of traumatic reticulitis

Diagnosis is based on a combination of history, clinical signs, and diagnostic tools:

1. ***History and physical exam***: Hunched posture, positive grunt test, and reduced feed intake.
2. ***Radiography***: Identifies foreign objects in the reticulum.
3. ***Ultrasound***: Detects inflammation or abscesses around the reticulum.
4. ***Metal detectors***: Identify metallic foreign objects.
5. ***Exploratory laparotomy***: Occasionally performed to confirm diagnosis and remove foreign material.

Treatment of traumatic reticulitis

Treatment varies depending on the severity of the condition:

1. Conservative management

Oral administration of magnets: To immobilize metallic foreign objects and prevent further damage.

Antibiotics: Broad-spectrum antibiotics (e.g., penicillin or oxytetracycline) to control secondary infections.

Anti-inflammatory drugs: NSAIDs to reduce pain and inflammation.

Supportive therapy: Providing fluids, electrolytes, and energy to support recovery.

2. Surgical intervention

For severe cases or when conservative management fails, rumenotomy (surgical removal of foreign objects) may be required.

Rumenotomy should be performed under aseptic conditions in all animals, in standing position. Affected animals should be sedated with Triflupromazine hydrochloride @ 0.1 mg/Kg body weight. Local analgesia was achieved by inverted L-block using 2 per cent Lignocaine hydrochloride solution. A vertical incision was made in the left paralumbar fossa, 6-8 cm ventral to the transverse processes of lumbar vertebrae. Left flank laparo-rumenotomy should be performed employing Weingart's frame. The rumen was exteriorized and its wall was incised. The contents of rumen and reticulum were thoroughly searched and coins, nails, needles, plastic and rubber materials were retrieved. In animals if the adhesions between the reticulum and diaphragmatic wall were identified. Attempts to separate the adhesions were not successful. After the retrieval of foreign bodies, the rumen contents were removed up to one third of the existing quantity. For proper rumen function and for maintaining rumen ecosystem, fresh rumen cud obtained from the sheep at the slaughter houses was placed in the rumen. Probiotic drugs containing yeast extract and sugar were also placed in the rumen.

Intra venous crystalloids like dextrose normal saline (DNS) and electrolyte solutions were infused for 5 to 7 days in order to combat the dehydration.

Broad spectrum antibiotics like strepto penicillins were administered @ 10 mg/ Kg body weight for a period of one week.

Meloxicam @ 0.2 mg/Kg SC was given once daily for three days.

Post-mortem examination should be carried out in the dead animals.

Animal that died after two days showed signs of reticulo peritonitis like flakes of pus in the peritoneal fluid. The texture of the spleen was completely changed, with several foci of abscessation. Adhesions of reticular wall with diaphragm, hence it is diagnosed as traumatic reticulo-peritonitis.

Animal that died after nine days had a long penetrating nail/ sharp object in the reticulum that is penetrating in to the pericardial sac. Heart showed severe fibrinous pericarditis with thick fibrinous nodules. The contour of the heart altered from its normal structure. However, the interior of the heart was normal and free from any pathological lesions. The animal was diagnosed to have died of traumatic pericarditis.

3. Management of complications

Peritonitis or abscesses may require prolonged antibiotic therapy or drainage procedures.

Prevention of traumatic reticulitis

Preventive strategies focus on reducing the risk of foreign object ingestion:

1. ***Use of ruminal magnets:*** Administering magnets to adult cattle to trap metallic objects in the reticulum and prevent them from causing damage.
2. ***Improved feed quality:*** Avoiding feeding contaminated silage or hay, Using metal detectors during feed processing to remove foreign objects,
3. ***Environmental management:*** Regularly inspecting pastures and housing for stray wires, nails, and other sharp objects as well as proper disposal of farm equipment waste.

Nutritional management of traumatic reticulitis

Proper nutritional strategies play a pivotal role in both preventing and managing traumatic reticulitis. These strategies aim to ensure optimal rumen health, improve immunity, and reduce the likelihood of foreign object ingestion.

1. Enhancing forage quality

High-quality forages: Feeding high-quality silage, hay, and pasture minimizes the ingestion of contaminants.

Sorting and cleaning feed: Screening and cleaning forages to remove foreign materials is essential.

2. Balanced diets for rumen health

A balanced diet rich in fibre and adequate in energy and protein is crucial to maintain healthy rumen function.

Long-stem forage: Encouraging chewing and rumination, which helps sift out foreign particles during prehension.

3. Adequate mineral supplementation

Calcium and phosphorus: Deficiencies can lead to pica, increasing the risk of foreign object ingestion. Balanced mineral supplementation reduces this behavior.

Trace elements: Zinc and selenium improve immunity, reducing susceptibility to infections such as peritonitis.

4. Probiotics and prebiotics

Supplements containing beneficial microbes can improve digestion and support the immune response, aiding recovery in affected animals.

5. Feeding practices to prevent pica

Avoid overcrowding: Ensure adequate feeding space to reduce stress and abnormal eating behaviours.

Correcting deficiencies: Address deficiencies in salt, phosphorus, or other nutrients that may trigger pica.

6. Energy-dense diet during recovery

Provide easily digestible, energy-dense feeds to animals recovering from traumatic reticulitis. This includes:

High-quality concentrates.

Liquid energy supplements or molasses-based feeds.

7. Supportive feeding during illness

Offer palatable feeds to animals showing reduced appetite.

Provide small, frequent meals to minimize discomfort and maximize nutrient intake.

Monitoring and long-term nutritional strategies

1. ***Regular health checks:*** Monitor body condition, feed intake, and milk production to detect signs of subclinical reticulitis or related complications early.
2. ***Periodic feed testing:*** Analyzing feed quality ensures it meets the nutritional needs of the animals and is free from contaminants.
3. ***Promoting rumen function:*** Use of rumen buffers (e.g., sodium bicarbonate) to stabilize pH and maintain a healthy rumen environment.
4. ***Avoiding risky feeding practices:*** Refrain from feeding animals on bare ground where foreign objects are likely to be ingested.

Case study: Application of nutritional management

In a dairy farm with recurrent cases of traumatic reticulitis, the following measures were implemented:

1. ***Screening and magnet administration***: Metal detectors were used to identify metallic objects in feed, and ruminal magnets were administered to all adult cows.

2. ***Dietary adjustments***: The ration was reformulated to include more high-quality fiber and mineral supplementation.
3. ***Environmental cleanup***: Pastures and barns were inspected and cleared of debris.
4. ***Monitoring***: Monthly veterinary visits ensured no recurrence.

Outcome: The incidence of traumatic reticulitis dropped significantly, and milk production improved due to better health and reduced stress.

Traumatic reticulitis is a preventable yet potentially life-threatening condition in ruminants. While treatment options exist, prevention through good nutritional management and environmental control is far more effective. Ensuring high-quality feed, addressing mineral deficiencies, and maintaining optimal rumen health are critical to minimizing risks. Proper feeding practices, combined with the use of ruminal magnets and regular farm inspections, can significantly reduce the occurrence of traumatic reticulitis, enhancing animal welfare and productivity.

Causes and risk factors of TRP

Ingestion of foreign objects: Cattle can ingest foreign objects if there is contamination in their feed or if metal objects are present in the grazing area. Improperly stored feed or silage that contains wire nails, or other debris can contribute to this risk.

Poor feeding practices: Feed contaminated with hardware or other foreign materials is one of the primary causes of TRP.

Environmental factors: Pastures with scrap metal, construction debris, or other sharp objects increase the likelihood of TRP.

Nutritional strategies for prevention and management of TRP

1. ***Prevention through proper feed management:*** The primary goal of preventing TRP is to reduce the risk of ruminants ingesting foreign objects. Proper feed management practices play a crucial role in this regard.

   a) ***Magnets in feed mills:*** Fitting the magnets in the feed mills in different locations and in mixers helps to capture metallic objects before they enter into the finished feeds and in the feed supply. Many dairy farms already use magnets in their feed processing systems to prevent hardware from being distributed in the feed.

   Magnetic feeders: Some farms install magnetic devices at strategic points in the feeding system to capture metal particles that could otherwise be ingested by the animals.

b) ***Careful feed sourcing and inspection:*** Purchasing feed from reputable suppliers who follow strict quality control measures helps ensure that feed is free of contamination. Additionally, farmers should inspect feed and forage for debris before feeding it to their animals. Silage or hay should be carefully checked for any metallic or foreign objects that could pose a risk to the herd.

   Forage quality control: Using metal detectors and visual inspections of hay, silage, and other forages can reduce the risk of hardware contamination in feed.

c) ***Using reticulum magnets in cattle and buffaloes:*** Administering rumen magnets is a common preventive measure in cattle prone to hardware disease. Rumen magnets are designed to lodge in the reticulum, where they attract and hold metallic objects that could otherwise puncture the reticulum wall. This practice helps prevent TRP, especially in environments with a higher risk of metal ingestion.

2. ***Nutritional support for healing and recovery:*** Once cows/ buffaloes have developed TRP, nutritional support plays an essential role in the healing process. A proper diet can help reduce inflammation, support immune function, and promote tissue repair.

a) ***Ensuring adequate energy and protein intake:*** Cattle suffering from TRP may experience a loss of appetite due to pain and discomfort, leading to reduced feed intake and weight loss. Ensuring that the animal consumes enough energy and protein is essential to prevent further weakening and to support healing.

   High-energy feeds: Providing easily digestible, energy-dense feeds, such as high-quality forages (e.g., alfalfa hay) and concentrates (e.g., grains), can help maintain energy levels during recovery. Feeding fat supplements can also provide a concentrated energy source.

   Balanced protein diet: Protein is necessary for tissue repair and immune function. Feeding rumen-degradable protein (RDP) and rumen-undegradable protein (RUP) in balanced amounts ensures that the cow has sufficient building blocks for recovery. A crude protein content of 14-16% in the diet is generally recommended.

b) ***Anti-inflammatory nutrients:*** Inflammation in the peritoneal lining due to TRP can cause significant pain and slow recovery. Including anti-inflammatory nutrients in the diet can help reduce inflammation and promote healing.

Omega-3 fatty Acids: Omega-3 fatty acids, found in flaxseed or fish oil, have natural anti-inflammatory properties. Supplementing the diet with these sources can help reduce systemic inflammation.

Vitamin E and selenium: Vitamin E and selenium are powerful antioxidants that support the immune system and help reduce oxidative stress in the body. Adequate levels of these nutrients (1,000-2,000 IU/day of vitamin E and 0.3 ppm selenium) can aid in the recovery process.

c) ***Vitamins and minerals for immune support:*** Vitamins and minerals play a crucial role in supporting the immune system, which is vital for fighting infection and promoting recovery from TRP.

Vitamin A: Vitamin A is important for maintaining the health of epithelial tissues, including the lining of the reticulum and peritoneum. Supplementing with 50,000-100,000 IU/day of vitamin A helps promote tissue repair.

Zinc: Zinc is essential for tissue repair and immune function. Providing organic zinc (e.g., zinc methionine) in the diet can help speed up recovery from TRP.

d) ***Probiotics and prebiotics for gut health:*** Cattle/ buffalo recovering from TRP may suffer from reduced rumen function and impaired digestion due to inflammation and stress. Probiotics and prebiotics can help restore healthy gut flora, improve digestion, and support overall recovery.

Probiotics: Supplementing with probiotics such as *Lactobacillus* and *Bifidobacterium* species can help restore rumen microbial balance and improve feed efficiency.

Prebiotics: Prebiotics, such as oligosaccharides (mannon oligosaccharides; MOS and fructo oligosaccharides; FOS), can feed beneficial bacteria in the gut, promoting a healthy rumen environment and aiding digestion during recovery.

3. **Hydration and electrolyte management:** Cows/ buffaloes with TRP often suffer from dehydration due to reduced water intake and fluid loss from inflammation. Ensuring that affected animals remain hydrated is critical to their recovery.

a) ***Access to clean water:*** Providing constant access to clean, fresh water is essential to prevent dehydration. Dehydrated cows are more prone to complications, and adequate hydration helps maintain proper blood circulation and nutrient delivery to healing tissues.

b) ***Electrolyte supplementation:*** In some cases, electrolyte supplementation may be necessary to help maintain fluid balance and support cellular function during recovery from TRP. Products containing sodium, potassium, chloride, and bicarbonate can help restore electrolyte balance in affected cattle.

4. **Preventing secondary infections:** Cattle/ buffalo with TRP are at high risk of developing secondary bacterial infections due to the initial trauma and the introduction of pathogens into the abdominal cavity. Nutritional strategies that boost immune function can help prevent these complications.

   a) ***Immune-boosting supplements:*** Supplements like yeast extracts, beta-glucans, and plant-derived compounds with antimicrobial properties can help boost the cow's immune response and prevent secondary infections.

   b) ***Maintaining clean feeding areas:*** Ensuring that feeding areas and water troughs are clean can help reduce the risk of introducing additional pathogens that may cause infections in cows already weakened by TRP.

Traumatic reticulo-peritonitis (TRP) is a serious condition in ruminants that requires careful management to prevent and mitigate its effects. Nutritional strategies are a critical component of both preventing TRP and supporting recovery from the condition. Preventive measures such as ensuring proper feed management, administering rumen magnets, and maintaining clean feeding environments are essential for reducing the risk of TRP. For animals already affected by TRP, providing a diet rich in energy, protein, anti-inflammatory nutrients, and immune-supporting vitamins and minerals is key to promoting healing and recovery. Additionally, ensuring adequate hydration and preventing secondary infections further supports the cow's return to health. When combined with appropriate veterinary care, these nutritional strategies can help reduce the incidence and severity of TRP in dairy herds, improving overall productivity and animal welfare.

6

Hypothermia

Hypothermia is a life-threatening condition that occurs when an animal's body loses heat faster than it can produce it, causing its core temperature to drop below normal levels. Animals, like humans, rely on maintaining a stable internal temperature to function. This condition can affect pets, livestock, and wildlife, and understanding its causes, effects, and treatment is critical for their survival and well-being.

In animals, body temperature varies by species. For example, the average temperature for dogs and cats ranges from 101°F to 102.5°F, while for livestock like cows, it is around 101.5°F. Hypothermia occurs when their body temperature falls below these ranges, impairing physiological processes and potentially leading to organ failure.

Stages of hypothermia

Mild hypothermia: A slight drop in body temperature, with symptoms like shivering, lethargy, and a cold nose or extremities.

Moderate hypothermia: A significant temperature drop that can result in muscle stiffness, weak pulse, and slowed breathing.

Severe hypothermia: Critical condition where the animal may lose consciousness, and the heart rate and breathing slow dramatically, potentially leading to death.

Causes of hypothermia

Several factors can lead to hypothermia in animals:

1. ***Cold weather exposure***: Prolonged exposure to freezing or near-freezing temperatures is a common cause, especially for animals not acclimated to such conditions or without adequate shelter.
2. ***Immersion in cold water***: Falling into icy water can rapidly lower an animal's body temperature, as water conducts heat away from the body 25 times faster than air.
3. ***Age and health***: Young, elderly, or sick animals are more susceptible due to their limited ability to regulate body temperature effectively.

4. ***Breed and size***: Smaller animals and those with short or thin coats (e.g., Chihuahuas or Greyhounds) are more prone to hypothermia than larger, well-insulated breeds like Huskies.
5. ***Inadequate nutrition***: Malnourished animals lack the fat reserves necessary to generate and retain heat.
6. ***Shock or trauma***: Injured animals may develop hypothermia due to blood loss or impaired circulation.

Mechanism of action of *hypothermia*

Hypothermia occurs when heat loss surpasses heat production, resulting in a drop in core body temperature. This can be caused by cold weather, immersion in cold water, or inadequate insulation. The body reacts through a series of mechanisms:

1. ***Peripheral Vasoconstriction***: Blood vessels in the extremities constrict to redirect blood flow to vital organs, preserving core temperature. This reduces heat loss but can lead to frostbite in extremities.
2. **Shivering thermogenesis**: Muscles contract involuntarily to generate heat through metabolic activity. This is an energy-intensive process, depleting glycogen stores quickly.
3. **Reduced metabolic rate**: As hypothermia progresses, metabolic processes slow down to conserve energy, impairing organ function. Enzymatic activities, including those involved in digestion and energy production, are significantly reduced.
4. ***Neurological impairment***: Lowered brain temperature affects coordination, decision-making, and consciousness. Prolonged hypothermia can lead to coma or death due to failure of respiratory and cardiovascular centers in the brain.

Effects of hypothermia

Hypothermia can disrupt an animal's normal physiological functions, leading to:

Reduced circulation: Peripheral blood vessels constrict to preserve core temperature, potentially causing frostbite in extremities.

Metabolic slowdown: Enzymatic and cellular functions slow, impairing organ performance.

Neurological impairment: Confusion, loss of coordination, and unconsciousness may occur as the brain cools.

Cardiac arrhythmias: Severe hypothermia can affect the heart's electrical activity, increasing the risk of arrhythmias or cardiac arrest.

In wildlife, hypothermia can weaken animals, making them more susceptible to predators or further environmental stressors.

Signs of hypothermia

Recognizing hypothermia early can save an animal's life. Key signs include:

Shivering (initially, as a heat-generating response)

Cold extremities (ears, paws, or tail)

Pale or blue mucous membranes

Lethargy or weakness

Rapid breathing (early stages) or slow breathing (severe cases)

Stiff muscles

Loss of consciousness (advanced stages)

Pet owners, farmers, and wildlife rehabilitators should monitor animals closely during extreme weather for these signs.

Prevention of hypothermia

Prevention is key to avoiding hypothermia in animals:

1. ***Provide shelter***: Ensure animals have access to warm, dry, and insulated areas during cold weather. For livestock, barns or windbreaks can offer critical protection.
2. ***Appropriate clothing***: Use coats or sweaters for pets, especially those with thin fur or small body size.
3. ***Limit exposure***: Avoid prolonged outdoor activities in extremely cold weather, especially for vulnerable animals.
4. ***Proper nutrition***: Provide high-calorie diets during winter to help animals generate sufficient body heat.
5. ***Monitor water sources***: Ensure drinking water does not freeze, as dehydration can exacerbate hypothermia risk.
6. ***Awareness of vulnerabilities***: Be mindful of age, health, and breed-specific susceptibilities to cold.

Treatment of hypothermia

Prompt intervention is crucial for animals suffering from hypothermia. Treatment involves gradually restoring their body temperature and addressing any underlying health issues.

Immediate actions

1. ***Remove animal from cold***: Bring the animal indoors or into a warmer environment.
2. ***Dry the animal***: If wet, use towels to dry its fur thoroughly.
3. ***Provide warmth***: Use blankets, heat pads, or warm water bottles (wrapped in towels to prevent burns) to gently warm the animal. Focus on the chest and abdomen rather than extremities.
4. ***Monitor temperature***: Use a rectal thermometer to measure body temperature. Do not rely on external body warmth as an indicator.
5. ***Offer warm fluids***: If the animal is conscious and able to swallow, provide warm (not hot) water or electrolyte solutions.

Nutritional management of hypothermia

Nutrition plays a critical role in both preventing and recovering from hypothermia by providing the energy and nutrients needed for thermogenesis, tissue repair, and immune function.

Prevention

1. ***Increased caloric intake:*** Animals exposed to cold environments require more calories to maintain body heat. High-energy diets rich in fats (the most calorie-dense macronutrient) are ideal for cold-weather conditions.
2. ***Frequent feeding***: Smaller, more frequent meals help sustain metabolic heat production. For livestock, providing roughage like hay promotes fermentation in the rumen, a natural heat-generating process.
3. ***Hydration***: Animals have access to unfrozen water to prevent dehydration, which can impair thermoregulation.

Treatment and recovery

1. ***Warm fluids and energy-rich diets:*** Warm fluids with electrolytes can help rehydrate and support circulation. Offer energy-dense, easily digestible foods to replenish glycogen stores (e.g., cooked grains, meat broths).
2. ***Supplementation***: *Vitamins and minerals*: Vitamin B-complex supports energy metabolism, while antioxidants like Vitamin E combat oxidative stress caused by hypothermia.

   Omega-3 fatty acids: Promote tissue repair and reduce inflammation.
3. ***Rewarming***: Gradual rewarming ensures nutrients are efficiently absorbed and utilized as circulation improves.

Veterinary care: For moderate to severe hypothermia, immediate veterinary attention is essential. Treatment may include:

Intravenous fluids: Warmed fluids help stabilize temperature and restore hydration.

Oxygen therapy: Supports animals with impaired breathing.

External warming devices: Veterinary clinics may use specialized equipment like warm air blankets.

Monitoring: Continuous assessment of heart rate, breathing, and organ function.

Special considerations for wildlife

Wildlife often suffers from hypothermia during harsh winters. Rehabilitators must handle these cases with care:

Minimize handling to reduce stress, which can worsen the condition.

Use species-specific techniques for warming and care.

Avoid releasing animals prematurely into harsh environments.

Hypothermia in animals is a serious and preventable condition that requires vigilance, especially during colder months. By recognizing the risks, symptoms, and treatment options, animal caretakers can protect pets, livestock, and wildlife from its potentially fatal effects. With appropriate shelter, nutrition, and care, animals can thrive even in challenging weather conditions.

7

Hyperthermia

Hyperthermia is a condition in which an animal's body temperature rises above the normal range, often due to excessive environmental heat or a failure in the body's thermoregulation mechanisms. Unlike fever, where the body actively increases its temperature in response to infection, hyperthermia results from external or internal factors that overwhelm the body's ability to dissipate heat. This condition is life-threatening and requires prompt recognition and treatment to prevent severe organ damage or death.

The normal body temperature of animals varies across species. For example, dogs and cats typically maintain a body temperature of 101°F to 102.5°F, while livestock like horses and cattle have average temperatures of around 99°F to 101.5°F. Hyperthermia occurs when the body temperature exceeds these ranges, potentially leading to heat exhaustion or heatstroke.

Types of hyperthermia

1. ***Heat stress/ heat exhaustion***: A mild to moderate rise in temperature, accompanied by symptoms like panting, lethargy, or drooling.
2. ***Heatstroke***: A severe and potentially fatal rise in body temperature that causes multi-organ dysfunction or failure.

Causes of hyperthermia

Hyperthermia in animals can occur due to various factors:

1. ***Environmental heat***: Prolonged exposure of animals to high temperatures especially in humid conditions where evaporative cooling (panting or sweating) is less effective. Being left in hot cars or confined spaces without adequate ventilation.
2. ***Physical overexertion***: Excessive exercise or work, particularly in hot weather, can lead to hyperthermia. Breeds with thick coats or brachycephalic (flat-faced) animals like Bulldogs and Persian cats are especially prone due to reduced cooling efficiency.
3. ***Underlying medical conditions***: Certain conditions, such as hyperthyroidism in cats or heart diseases in dogs, can predispose

animals to overheating. Malignant hyperthermia, a genetic condition, can be triggered by stress, anesthesia, or certain drugs.

4. ***Dehydration***: Lack of adequate water reduces the body's ability to dissipate heat through sweating or panting, exacerbating hyperthermia.
5. ***Poor ventilation or housing conditions***: Animals housed in poorly ventilated barns, kennels, or transport vehicles are at risk, particularly during warm weather.

Effects of hyperthermia

Hyperthermia can severely affect an animal's body, leading to systemic complications:

Dehydration: Excessive heat causes water loss, reducing blood volume and impairing circulation.

Organ failure: Prolonged heat can damage the brain, kidneys, heart, and gastrointestinal tract.

Neurological impairment: High body temperature can cause seizures, disorientation, or coma.

Coagulation disorders: Heatstroke can trigger disseminated intravascular coagulation (DIC), leading to uncontrolled bleeding.

Cardiac stress: Increased heart rate and reduced oxygen delivery may result in cardiovascular collapse.

Signs of hyperthermia

Recognizing hyperthermia early can prevent severe complications. Symptoms vary depending on the severity of the condition:

1. ***Mild to moderate hyperthermia (heat stress)***:

   Panting or rapid breathing,

   Restlessness or lethargy,

   Excessive drooling (in dogs) or grooming (in cats),

   Warm ears or paw pads,

   Reduced activity or reluctance to move,

2. ***Severe hyperthermia (heatstroke)***:

   Vomiting or diarrhea, often with blood,

   Bright red or pale gums,

   Rapid or irregular heartbeat,

   Disorientation, wobbling or seizures,

   Collapse or unconsciousness,

Owners, caretakers, and veterinarians should be vigilant, especially during hot weather, to detect these signs early.

Mechanism of action of hyperthermia

Hyperthermia occurs when heat production exceeds heat dissipation. This condition is often caused by high environmental temperatures, strenuous physical activity, or underlying health issues.

Key mechanisms include

1. ***Increased heat production***: Physical exertion or fever leads to heightened metabolic activity, generating excessive heat.
2. ***Impaired Heat Dissipation:*** In humid conditions, evaporative cooling (e.g., sweating or panting) becomes less effective, trapping heat within the body.
3. ***Cellular damage***: Proteins and enzymes begin to denature at elevated temperatures, disrupting metabolic pathways. Oxidative stress increases, leading to damage to DNA, lipids, and cellular membranes.
4. ***Neurological effects***: Elevated temperatures impair brain function, causing confusion, seizures, or loss of consciousness. If untreated, hyperthermia can result in multi-organ failure or death.

Prevention of hyperthermia

Preventing hyperthermia involves proactive measures to reduce heat exposure and ensure proper hydration:

1. ***Provide shade and ventilation***: Ensure animals have access to shaded, well-ventilated areas during hot weather. For livestock, barns or trees can provide essential relief. Avoid leaving animals in enclosed spaces like cars, even for short periods.
2. ***Hydration***: Always provide clean, fresh water for drinking.

   For livestock, regularly check water supplies to ensure they are sufficient and accessible.
3. ***Limit physical activity***: Avoid strenuous exercise during peak heat hours. Exercise pets in the early morning or late evening when temperatures are cooler.
4. ***Cooling aids***: Provide cooling mats, fans, or sprinklers for pets and livestock.

   Use lightweight blankets or towels soaked in cool water for additional relief.

5. ***Breed-specific care***: Be especially cautious with brachycephalic breeds, overweight animals, or those with medical conditions that impair thermoregulation.

Treatment of hyperthermia

Immediate intervention is critical for animals suffering from hyperthermia to reduce body temperature and minimize organ damage.

First aid for hyperthermia

1. ***Move to cool area***: Relocate the animal to a shaded or air-conditioned environment.
2. ***Cool the animal gradually***: Use cool (not ice-cold) water to wet the animal's body, focusing on areas like the belly, groin, and paw pads. Place cool, wet towels on the animal but avoid covering them entirely, as this may trap heat.
3. ***Offer water***: If the animal is conscious, offer small amounts of cool water to drink. Avoid forcing them to drink.
4. ***Monitor temperature***: Use a rectal thermometer to track the animal's body temperature. The goal is to reduce it to around 103°F, at which point active cooling should stop to prevent hypothermia.

Nutritional management of hyperthermia

Managing hyperthermia requires reducing metabolic stress, supporting hydration, and replenishing lost electrolytes and nutrients.

Prevention

1. ***Hydration and electrolyte balance***: Provide access to clean, cool water at all times to support evaporative cooling and hydration.

   For livestock, adding electrolyte solutions to drinking water can replenish sodium, potassium, and chloride lost through sweating.
2. ***High-moisture diets***: Incorporate water-rich foods like fruits and vegetables (e.g., cucumbers, melons) for pets. For livestock, ensure forage is not overly dry and provide silage or soaked hay to increase water intake.
3. ***Adjusted feeding times***: Feed animals during cooler parts of the day (early morning or late evening) to prevent metabolic heat production during peak temperatures.
4. ***Fat and protein balance***: Reduce excessive protein intake, as protein metabolism generates more heat than fats or carbohydrates. Focus on

moderate-fat diets that provide sustained energy without excessive heat production.

Treatment and recovery

1. ***Rehydration therapy***: Offer oral rehydration solutions with electrolytes to address dehydration and restore electrolyte balance. Severe cases may require intravenous fluids administered by a veterinarian.
2. ***Easily digestible foods***: Offer small, frequent meals with easily digestible carbohydrates to provide quick energy without increasing metabolic heat. Avoid high-fat or high-protein meals initially, as they require more energy for digestion.
3. ***Nutritional supplements***: *Antioxidants*: Vitamins C and E help mitigate oxidative damage caused by hyperthermia.

   Essential minerals: Calcium and magnesium are critical for muscle and nerve function, especially in animals experiencing heat stress.
4. ***Cooling strategies with nutrition***: For pets, frozen treats or ice cubes made with low-sodium broth or diluted juice can help reduce body temperature.

   For livestock, providing cool water troughs or wet forage can aid in cooling from the inside out.

Veterinary care: Severe cases of hyperthermia require immediate veterinary attention. Treatments may include:

Intravenous fluids: Rehydrate and support circulation.

Oxygen therapy: Aid breathing and oxygen delivery to tissues.

Medications: Manage shock, seizures, or inflammation.

Blood tests: Monitor organ function and assess for complications like DIC. Veterinarians will also address any underlying conditions contributing to hyperthermia.

Special considerations for wildlife

Wildlife often faces challenges during heatwaves, particularly in urban areas where water sources may be scarce. To assist:

Provide shallow water dishes in shaded areas for birds and small mammals.

Avoid handling wildlife unless they appear in distress, as stress can worsen hyperthermia.

Wildlife rehabilitators must handle hyperthermic animals carefully, focusing on gradual cooling and hydration.

Hyperthermia is a serious and potentially fatal condition in animals, but it is also preventable. By understanding its causes, effects, and treatment options, caregivers can protect pets, livestock, and wildlife from heat-related illnesses. Proactive measures like providing shade, ensuring hydration, and limiting exposure during hot weather are essential. Prompt intervention during early signs of hyperthermia can save lives and prevent long-term complications.

8

Spontaneous Hypothermia

Spontaneous hypothermia in animals occurs when their body temperature drops below the normal range due to internal or environmental factors, without direct exposure to extreme cold. Unlike hypothermia caused by environmental factors, spontaneous hypothermia often results from metabolic, physiological, or pathological conditions that impair thermoregulation. It is more challenging to predict and manage as it can occur in a wide range of scenarios, including illness, trauma, or systemic dysfunction. Proper nutritional strategies play a critical role in both prevention and recovery.

Spontaneous hypothermia is a drop in core body temperature due to the body's inability to generate, conserve, or regulate heat, independent of external temperature changes.

Mechanism of action

The body maintains its temperature through a balance of heat production (via metabolism and muscular activity) and heat loss (via radiation, convection, conduction, and evaporation). Spontaneous hypothermia occurs when:

1. ***Heat production fails***: Conditions such as malnutrition or metabolic disorders limit the body's ability to generate heat.
2. ***Heat loss increases***: Trauma or vascular conditions can lead to excessive heat loss.
3. ***Thermoregulatory failure***: Diseases affecting the nervous system or endocrine glands disrupt the mechanisms that regulate body temperature.

Causes of spontaneous hypothermia

1. ***Metabolic disorders***

   Hypothyroidism: Low levels of thyroid hormones reduce metabolic rate and heat production, making animals prone to hypothermia.

   Hypoglycemia: A lack of glucose limits energy production, affecting thermogenesis.

2. ***Systemic illness***

   Sepsis: Severe infections can lead to a dysregulated immune response, causing peripheral vasodilation and heat loss.

   Chronic kidney disease: Impaired kidney function can alter electrolyte balance and thermoregulation.

3. ***Shock or trauma:*** Animals in shock lose heat rapidly due to poor circulation and altered blood flow patterns. Injuries may expose internal tissues to heat loss.
4. ***Neonates and geriatric animals:*** Young animals have immature thermoregulatory systems, while older animals may have reduced metabolic activity and fat reserves.
5. ***Neurological disorders:*** Diseases affecting the hypothalamus (the brain's thermoregulation center) can impair temperature control.
6. ***Malnutrition:*** Poor nutrition depletes energy reserves, making it difficult for the body to maintain heat production.

Clinical signs

Spontaneous hypothermia presents with similar signs to environmentally induced hypothermia, including:

Lethargy or weakness,

Shivering (in early stages) or absence of shivering (in advanced stages),

Cold extremities (ears, paws or tail),

Slow heart rate (bradycardia),

Pale or blue mucous membranes,

Impaired consciousness or coma,

Nutritional management of spontaneous hypothermia

Nutrition plays a vital role in preventing spontaneous hypothermia and supporting recovery by replenishing energy reserves, restoring metabolic balance, and improving overall health. The following strategies are tailored to the underlying causes and specific needs of the animal:

1. Prevention through adequate nutrition

1. **High-calorie diets:** Provide diets rich in fats and carbohydrates to supply the energy needed for heat production.

   For livestock, supplement with high-energy feeds like grains or molasses during colder months or periods of stress.

2. **Balanced macronutrient intake:** Ensure a balanced diet with adequate protein to support metabolic functions and tissue repair.

   Include high-quality fat sources (e.g., fish oil, flaxseed) to serve as a dense energy reserve.

3. **Thermal nutrition:** Offer warm meals to help maintain body temperature, especially in young, ill, or elderly animals. Avoid cold or frozen foods, which can exacerbate heat loss.

4. **Micronutrient support**

   Vitamin E and selenium: Support cellular metabolism and protect against oxidative stress.

   B-Complex: Essential for energy metabolism and maintaining body temperature.

5. **Adequate hydration:** Ensure access to clean, warm water to prevent dehydration, which can impair thermoregulation.

Nutritional management dring hypothermia

When an animal presents with spontaneous hypothermia, nutritional interventions should be combined with Vitamins gradual rewarming techniques to restore homeostasis.

1. **Energy-dense foods:** Offer easily digestible, high-calorie foods to replenish depleted energy reserves. For pets, provide small, frequent meals of energy-rich options like cooked chicken, rice, or broth.

   For livestock, consider feeding warm bran mash or energy supplements.

2. **Electrolyte solutions:** Administer oral rehydration solutions to restore fluid and electrolyte balance. These are especially important in cases of hypothermia caused by sepsis or kidney disease.

3. **Warm fluids:** Warm milk replacers for neonates or energy drinks for livestock can provide both hydration and calories.

4. **Supplements for Specific Conditions**

   Hypoglycemia: Administer glucose or honey dissolved in warm water to quickly raise blood sugar levels.

   Hypothyroidism: Provide iodine-rich foods (e.g., seaweed, eggs) or synthetic thyroid hormones as prescribed by a veterinarian.

Recovery Nutrition

Post-hypothermia recovery focuses on restoring normal metabolism, repairing damaged tissues, and preventing recurrence.

1. ***Gradual caloric increase:*** After recovery, increase calorie intake gradually to avoid overwhelming the digestive system.

   Focus on nutrient-dense meals with adequate fiber to support digestion and energy absorption.

2. ***Immune Support***: Include antioxidants like Vitamin C and zinc to boost immune system,

   Support gut health with probiotics, especially if antibiotics were used during treatment,

   Healthy gut flora ensures optimal nutrient absorption, vital for recovery.

3. **Monitor protein Intake:** Provide high-quality protein sources like lean meats, fish, or legumes to support tissue repair and metabolic balance. Avoid excessive protein, as it can strain the kidneys in animals with underlying renal conditions.

Special considerations for At-risk animals

1. ***Neonatal animals:*** Neonatal animals require colostrum within the first few hours of life for energy and immune protection. Use commercial milk replacers if maternal milk is unavailable.
2. ***Geriatric Animals:*** Older animals need diets tailored to their slower metabolism and reduced activity levels, emphasizing nutrient density over quantity.
3. ***Livestock and wildlife:*** Provide windbreaks and additional feed during cold spells to prevent spontaneous hypothermia. Monitor wildlife rehabilitants closely for adequate caloric intake during recovery.

Spontaneous hypothermia in animals arises from internal factors that disrupt the balance between heat production and loss, making it a challenging condition to manage. Nutritional strategies are essential not only for prevention but also for recovery, as they address the metabolic deficits caused by hypothermia. High-calorie diets, energy-dense foods, electrolyte supplementation, and tailored feeding plans help restore body temperature, support organ function, and prevent recurrence. A proactive approach combining nutritional management with veterinary care ensures the best outcomes for affected animals.

9

Grain Engorgement Syndrome

Grain engorgement syndrome, also known as grain overload, lactic acidosis, ruminal acidosis, or carbohydrate overload, is a metabolic condition primarily affecting ruminant animals such as cattle, buffalo, sheep, and goats. It occurs when animals consume excessive amounts of readily fermentable carbohydrates, commonly found in grains such as wheat, barley, and corn, or feed concentrates. The sudden overload leads to rapid fermentation in the rumen, causing a cascade of physiological disturbances that can be life-threatening if not promptly managed.

This condition is of significant importance in livestock farming, particularly where intensive feeding systems, grain supplementation, or unintentional access to stored grains occurs. Proper understanding of the pathophysiology, clinical signs, diagnosis, and nutritional management is crucial for preventing mortality and economic losses in affected herds.

Pathophysiology

Ruminants rely on a microbial ecosystem in their rumen to digest fibre and carbohydrates. The rumen contains a diverse population of bacteria, protozoa, and fungi that ferment ingested feed, producing volatile fatty acids (VFAs) such as acetate, propionate, and butyrate. These VFAs provide energy to the animal. When ruminants ingest a sudden excess of high-starch grains or other fermentable carbohydrates, the balance of the rumen ecosystem is disrupted:

Rapid fermentation: The high starch content is rapidly fermented by gram-positive bacteria such as *Streptococcus bovis*. This leads to a sudden and excessive production of lactic acid.

Rumen pH decline: Lactic acid accumulation lowers the rumen pH drastically, often below 5.0 (normal range: 6.0–7.0). This acidic environment suppresses the growth of beneficial rumen microbes, further exacerbating lactic acid accumulation.

Microbial shifts: Acid-tolerant bacteria such as *Lactobacillus* spp. proliferate in the acidic rumen, worsening the production of lactic acid. The sudden increase in acidity also kills off other rumen microorganisms, leading to rumen dysbiosis.

Osmotic effect: The accumulation of lactic acid in the rumen increases the osmotic pressure, drawing water from surrounding tissues into the rumen. This results in ruminal distension and dehydration of the animal.

Rumen wall damage: The acidic environment damages the rumen epithelium, allowing bacteria and toxins to enter the bloodstream, leading to systemic acidosis, endotoxemia, and possible sepsis.

Risk factors for grain engorgement syndrome

Several scenarios can predispose animals to grain overload:

Sudden access to large quantities of grain: For example, animals breaking into grain storage areas.

Inadequate adaptation to grain diets: Gradual dietary transition is critical when introducing high-concentrate diets.

Feeding errors: Overfeeding of grain concentrates in feedlot settings.

Inadequate fibre in the diet: Insufficient roughage decreases rumination and saliva production, which normally help buffer rumen acidity.

Clinical signs of grain overload

The severity of clinical signs depends on the quantity of grain consumed and the extent of rumen fermentation and acidosis. Animals are classified as mild, moderate, or severe cases.

Mild cases

Reduced appetite or anorexia

Mild ruminal distension

Reduced or absent rumen motility

Soft feces

Moderate cases

Anorexia and depression

Severe ruminal distension or bloat

Diarrhoea with a characteristic sour odour (from lactic acid)

Dehydration (sunken eyes, dry mucous membranes)

Weakness and in-coordination (ataxia) due to systemic acidosis

Severe cases

Recumbency (inability to stand)

Weak or absent ruminal contractions

Tachycardia (rapid heart rate) and tachypnea (rapid breathing)

Severe dehydration and shock

Neurological signs: Staggering, head pressing, or blindness due to metabolic derangements

Death may occur within 24–48 hours in untreated severe cases.

Diagnosis of grain engorgement syndrome

Diagnosis is based on clinical history, physical examination findings, and laboratory tests:

1. ***History***: Sudden access to grains or recent changes in diet.
2. ***Clinical examination***: Presence of bloat, rumen stasis, diarrhea, dehydration, and systemic signs of acidosis.
3. ***Rumen fluid analysis***

   Low rumen pH (below 5.0) indicates acidosis.

   Microscopic examination reveals a lack of normal protozoa and overgrowth of gram-positive bacteria.

   Foul odour due to lactic acid fermentation.
4. ***Bloodwork***

   Metabolic acidosis (low blood pH and bicarbonate levels)

   Dehydration indicators (increased hematocrit and total plasma proteins)

   Increased lactate levels in severe cases
5. ***Necropsy (post-mortem)***

   Rumen contents appear watery, with a strong sour odour.

   Ulceration or haemorrhages on the rumen wall may be visible.

Treatment of grain engorgement syndrome

The treatment strategy aims to address ruminal acidosis, dehydration, and systemic shock. Treatment must be prompt to improve survival rates:

1. ***Rumen decompression***: If severe bloat occurs, a stomach tube or trochar can be used to relieve gas accumulation.
2. ***Rumen lavage***: In cases of significant grain overload, rumen lavage (flushing out rumen contents) may be performed via a rumenotomy.
3. ***Alkalinizing agents***: Oral administration of magnesium hydroxide or sodium bicarbonate can help buffer the rumen acidity.
4. ***Rehydration therapy***: Intravenous fluids with bicarbonate to correct metabolic acidosis and dehydration, Oral rehydration may also be provided.

5. ***Rumen transfaunation***: Introducing rumen fluid from a healthy donor helps restore the microbial population in the affected rumen.
6. ***Antimicrobials***: Antibiotics (e.g., penicillin) may be used to prevent bacterial translocation and secondary infections due to damaged rumen walls.
7. ***Supportive care***: Anti-inflammatory drugs (NSAIDs) help manage pain, shock, and inflammation.

Prevention of grain engorgement syndrome

Prevention is the most effective way to manage grain overload:

1. ***Controlled grain feeding***: Gradual introduction of high-grain diets over 2–3 weeks allows rumen microbes to adapt.
2. ***Provide roughage***: Ensure adequate fiber intake to promote rumination and saliva production.
3. ***Limit access to grains***: Secure grain storage facilities to prevent accidental overeating.
4. ***Use feed additives***: Buffers like sodium bicarbonate can help mitigate acid build-up. Ionophores (e.g., monensin) can reduce the risk of lactic acidosis.
5. ***Monitor feed intake***: Feedlot animals should be observed for overeating or irregular feeding patterns.

Prognosis

The prognosis depends on the severity of the condition, the amount of grain ingested, and how quickly treatment is initiated. Mild cases have a favourable prognosis, whereas severe cases often result in fatalities despite treatment due to systemic complications such as endotoxemia, laminitis or liver abscesses.

Nutritional management of grain engorgement syndrome

Effective nutritional management plays a critical role in preventing, mitigating, and recovering from this condition. The focus lies on careful diet formulation, feeding strategies and post-recovery nutritional support to restore rumen function and overall animal health. A comprehensive approach for nutritional management is required that covers prevention, immediate intervention and long-term recovery strategies.

Prevention through dietary management

Preventing grain engorgement syndrome is far more effective and economical than treatment. Proper nutritional strategies include:

Gradual adaptation to grain diets

Rumen microbes require time to adapt to high-starch diets. A gradual increase in grain over 2–3 weeks allows the rumen environment to adjust to increased production of volatile fatty acids (VFAs). Introduce grains at low levels (0.5–1% of body weight) and increase by no more than 10–15% weekly. During this adaptation period, observe animals closely for signs of reduced appetite, bloat, or changes in fecal consistency.

Adequate fibre in the diet

High-fibre roughages (e.g., hay, straw, silage) are essential for maintaining rumen health.

Fibre stimulates rumination and saliva production, which contains bicarbonate—a natural buffer that neutralizes acid in the rumen. A minimum of 10–15% fibre in the diet as neutral detergent fibre (NDF) is recommended to prevent rumen acidosis. Long-stem roughage (>1 inch) should constitute at least 10–20% of the total dry matter intake.

Controlled grain access

Prevent animals from accessing large amounts of grains accidentally, such as spills or open grain storage. Secure all feeding areas and storage facilities. Limit concentrate feeding to a maximum of 2–2.5% of body weight per day. For example, a 500 kg cow should receive no more than 10–12.5 kg of grain daily.

Incorporating buffers in the diet

Adding dietary buffers helps stabilize rumen pH and reduce the risk of acid build-up.

Common buffers include:

Sodium bicarbonate (1–1.5% of the total diet)

Magnesium oxide (0.5–1% of the diet)

Calcium carbonate and bentonite clay may also provide buffering effects.

Buffers are particularly useful during feedlot transitions, periods of high-concentrate feeding, and stress.

Use of Ionophores

Ionophores such as monensin and lasalocid help improve rumen microbial efficiency and reduce the production of lactic acid. These feed additives inhibit the growth of lactic acid-producing bacteria (*Streptococcus bovis*) and promote the growth of beneficial microbes.

Monitor feed consistency

Ensure that grains are evenly mixed into the total mixed ration (TMR) to prevent animals from selectively consuming grain (sorting).

Process grains appropriately. Fine grinding increases the rate of fermentation and acid production, so rolling or coarsely grinding grains is preferred.

Immediate nutritional management in acute cases

When grain overload occurs, immediate nutritional interventions are critical to minimize damage and restore rumen function:

Fasting and feeding roughage

Withhold grain and concentrates for at least 24–48 hours to reduce further acid production in the rumen. Provide high-quality long-stem roughage (e.g., grass hay) during the recovery period. Roughage encourages rumination, increases saliva production, and helps buffer the rumen pH.

Rumen alkalinization

Administer oral alkalizing agents to neutralize excess lactic acid in the rumen:

Magnesium hydroxide: 400–800 g dissolved in water for large ruminants.

Sodium bicarbonate: 500–1,000 g orally for cattle.

These agents raise the rumen pH and alleviate acidic conditions.

Rumen lavage or rumenotomy

In severe cases, rumen lavage (flushing the rumen) or rumenotomy (surgical removal of rumen contents) may be performed to remove undigested grain. After removal, provide fresh hay and warm water to promote rumen recovery.

Reintroducing feed Gradually

Once animals stabilize, start feeding small amounts of hay (free-choice) and gradually reintroduce grains over 5–7 days.

Begin with very low amounts of concentrates (0.5–1% of body weight) and monitor for signs of relapse.

Probiotics and r umen Transfaunation

Probiotics: Provide live yeast cultures (*Saccharomyces cerevisiae*) or other microbial supplements to reestablish a healthy microbial population in the rumen.

Rumen transfaunation: Administer rumen fluid from a healthy donor to repopulate beneficial microbes in the affected animal's rumen.

Recovery nutritional management

The recovery phase focuses on restoring rumen function, maintaining body condition, and preventing long-term complications:

High-fibre diet

Continue feeding high-quality roughage (e.g., hay) for 1–2 weeks. This allows the rumen lining to heal and rumen motility to normalize. Gradually reintroduce concentrates, ensuring that grains are coarsely processed and evenly mixed into the diet.

Nutrient Support

Animals recovering from grain engorgement often experience weight loss and reduced energy reserves.

Provide balanced diets that include

Energy sources: High-quality forages, limited grains, and digestible fiber sources such as beet pulp.

Protein: Supplement with protein sources (e.g., soybean meal, canola meal) to support recovery and rumen microbial growth.

Vitamins and Minerals: Offer B vitamins (thiamine) to prevent polioencephalomalacia, a neurological condition caused by thiamine deficiency. Ensure adequate calcium, magnesium, and phosphorus supplementation.

Monitor Rumen Function

Observe animals for:

Normal rumination activity (chewing cud)

Consistency of faeces (return to normal solid form)

Appetite and energy levels

Long-term feeding management

Once animals recover fully, adopt preventive feeding strategies to avoid future episodes of grain overload. Introduce grains gradually, ensure adequate fibre intake, and monitor animals for abnormal feeding behaviour or symptoms.

Nutritional management in feedlots

In feedlot systems, where high-concentrate diets are common, nutritional strategies are crucial to minimize the risk of grain engorgement:

Use step-up programs to transition animals to high-energy finishing diets gradually.

Incorporate buffers (sodium bicarbonate) and ionophores (monensin) into the diet.

Feed total mixed rations (TMRs) to ensure uniform consumption and prevent grain sorting. Provide ample clean water to support hydration and rumen function.

Grain engorgement syndrome is a preventable yet potentially fatal condition in ruminants. It underscores the importance of proper feeding management and early intervention. Farmers, veterinarians, and livestock handlers must adopt precautionary practices to minimize the risks of grain overload while ensuring prompt treatment protocols are in place to mitigate its impact on animal health and productivity.

Nutritional management of grain engorgement syndrome involves a multi-faceted approach, encompassing prevention, acute treatment, and recovery strategies. Key principles include providing adequate fiber, gradual adaptation to high-grain diets, and buffering rumen acidity to maintain a stable microbial ecosystem. For animals recovering from grain overload, balanced diets rich in roughage, probiotics, and essential nutrients are crucial for restoring rumen function and overall health. Proper feeding practices, monitoring, and secure storage of grains are essential to prevent future cases, safeguard animal welfare, and optimize productivity in livestock operations.

10

Hypocalcaemia

Hypocalcaemia, commonly referred to as milk fever, is a metabolic disorder in dairy animals, particularly high-producing cows, that occurs due to low calcium levels in the blood (serum calcium less than 8 mg/dl in cows). It predominantly affects animals in the peri-parturient period, as the demand for calcium for colostrum and milk synthesis outpaces the body's ability to maintain adequate blood calcium levels. The disorder not only affects animal welfare but also results in significant economic losses due to reduced milk production, fertility issues, and increased veterinary costs. Nutritional management plays a pivotal role in preventing and controlling milk fever, making it a key focus for dairy farmers and veterinarians.

Hypocalcaemia in specific species

1. **Ruminants (Cattle, Buffalo, Sheep, and Goats)**

   Milk fever (Peri-parturient hypocalcaemia): Commonly affects dairy animals around calving.

2. **Small animals (Dogs and Cats)**

   Eclampsia (Puerperal tetany): Seen in lactating bitches and queens. It results from excessive calcium loss through milk.

   Clinical signs: Restlessness, panting, and muscle tremors, tetany, seizures, and hyperthermia if untreated.

3. **Horses:** Hypocalcaemia can occur due to heavy sweating, lactation, or gastrointestinal disorders (e.g., colic or enteritis). Clinical signs include muscle fasciculation's, stiffness, and synchronous diaphragmatic flutter commonly referred to as thumps.

4. **Birds and reptiles:** Egg-laying species are at risk due to calcium depletion for eggshell formation. Clinical signs include weakness, soft or malformed eggs, and reproductive failure.

Importance of calcium in animal physiology

Calcium is a critical mineral involved in numerous physiological processes, including:

1. ***Neuromuscular function***: Calcium is essential for muscle contraction and nerve impulse transmission.
2. ***Bone health***: It provides structural integrity to bones and teeth.
3. ***Enzymatic activity***: Many enzymes depend on calcium as a cofactor.
4. ***Milk production***: In lactating animals, calcium is a primary component of milk.

A disruption in calcium homeostasis can lead to severe health issues and economic losses in the livestock industry.

Causes of hypocalcaemia

1. ***Dietary Imbalance***: Inadequate dietary calcium or an improper calcium-to-phosphorus ratio,

   Excessive dietary potassium or magnesium interfering with calcium absorption,
2. ***Lactation-associated demand***: High-producing dairy animals are prone to hypocalcaemia due to increased calcium demand for milk synthesis, particularly during early lactation (e.g., milk fever in cows).
3. ***Endocrine disorders***: Dysfunction of the parathyroid gland leads to inadequate secretion of parathyroid hormone (PTH), which is critical for calcium mobilization.
4. ***Vitamin D deficiency***: Vitamin D facilitates calcium absorption from the gastrointestinal tract.
5. ***Other causes***: Renal failure, which impairs calcium metabolism,

   Chronic diseases, such as intestinal disorders that hinder calcium absorption,

Pathophysiology

Calcium is vital for muscle contraction, nerve function, and milk production. A dairy animal (cow/ buffalo) enters in lactation the demand for calcium increases sharply to support colostrum production. If this demand is not met through dietary intake or mobilization of calcium from bones, blood calcium levels drop, leading to hypocalcaemia.

Key factors in milk fever pathophysiology include

Insufficient calcium mobilization: Older cows are at greater risk because their bones become less responsive to parathyroid hormone (PTH), which mobilizes calcium.

Delayed PTH response: A sudden drop in calcium around calving does not always trigger an immediate and sufficient PTH response.

Dietary imbalances: High potassium diets, common in forages like alfalfa, interfere with calcium absorption and mobilization.

Failure of vitamin D activation: Vitamin D facilitates calcium absorption from the gut, but its activation may be impaired due to metabolic or dietary factors.

Clinical signs of milk fever

Milk fever typically progresses through three stages if untreated:

1. ***Stage-I (Mild hypocalcaemia)***

   Restlessness, reduced appetite, and muscle tremors,

   Slight drop in milk yield,

2. ***Stage-II (Moderate hypocalcaemia)***

   Recumbency (inability to stand), depression and cold extremities,

   Weak pulse and subnormal body temperature,

3. ***Stage-III (Severe hypocalcaemia):*** Complete muscle flaccidity, comatose state, and potential death within hours. Sub-clinical hypocalcaemia where calcium levels are low but clinical signs are absent is also a significant concern as it affects productivity and predisposes cows to secondary diseases like mastitis, ketosis and retained placenta.

Risk factors for milk fever

1. **Age**: Older cows have reduced bone calcium stores and slower calcium mobilization.
2. **Breed**: Jersey cows are more susceptible due to their higher milk production relative to body size.
3. **Parity**: Multiparous cows have a higher risk compared to first-lactation heifers.
4. **Dietary factors**: Excessive dietary calcium during the dry period reduces the cow's ability to mobilize calcium postpartum. High potassium and sodium in pre-partum diets increase the risk by inducing metabolic alkalosis, which impairs PTH function.
5. **High milk yield**: Cows producing larger volumes of milk or colostrum have higher calcium demands.

Nutritional management of milk fever

Proper nutritional strategies before and after calving are critical in preventing hypocalcaemia. These strategies focus on managing calcium intake, balancing minerals, and ensuring optimal metabolic responses. Maintain a balanced calcium-to-phosphorus ratio in the diet (e.g., 2:1 in ruminants). Avoid excessive potassium in the diet, particularly from high-potassium forages. Use anionic salts during the dry period in dairy cows to promote calcium mobilization.

1. Dietary calcium management

Low calcium diets during the dry period

Restrict dietary calcium to less than 50 grams per day during the dry period to prime the calcium homeostatic mechanisms. This stimulates PTH secretion and bone calcium mobilization. Feeding low-calcium forages (e.g., grass hay) is recommended.

Calcium supplementation post-calving

Provide calcium supplements (e.g., calcium boluses or gels) immediately after calving to meet the sudden increase in demand.

2. Dietary cation-anion difference (DCAD)

The DCAD approach focuses on balancing cations (potassium, sodium) and anions (chloride, sulphur) in the diet to influence blood pH and calcium metabolism.

Negative DCAD diets

Feeding anionic salts (e.g., calcium chloride, ammonium sulphate) during the last 2–3 weeks of the dry period induces a mild metabolic acidosis, enhancing the responsiveness of PTH and promoting calcium mobilization.

Target DCAD values: −50 to −150 mEq/kg dry matter (DM)

Monitor urine pH (ideal range: 6.0–6.5 in Holsteins and 5.5–6.0 in Jerseys) to evaluate the effectiveness of the DCAD diet.

3. Magnesium supplementation

Magnesium plays a crucial role in calcium metabolism by supporting PTH function.

Supplement diets with 0.4% magnesium (as magnesium oxide or sulphate) during the dry period.

4. Vitamin D supplementation

Provide vitamin D_3 injections or dietary supplements 3–7 days before calving to enhance intestinal calcium absorption.

5. Potassium and phosphorus management

Limit potassium in the pre-partum diet to less than 1.5% DM to reduce the risk of metabolic alkalosis.

Avoid phosphorus excess in the dry period as it can interfere with calcium metabolism. Maintain phosphorus at 0.3–0.4% DM.

6. Forage selection

Use forages with low potassium content (e.g., mature grasses) during the dry period.

Analyze forage mineral content and balance the diet accordingly.

Other management practices

1. ***Monitoring high-risk animals***: Older, multi-parous, and high-yielding cows should be monitored closely around calving. Regularly check serum calcium levels in susceptible animals.
2. ***Prompt treatment of sub-clinical hypocalcaemia***: Sub-clinical hypocalcaemia can be addressed through oral calcium supplements to prevent progression to clinical milk fever.
3. ***Minimizing stress***: Avoid unnecessary handling, transportation, or environmental stressors during the transition period. Avoid over-conditioning of animals during the dry period.
4. ***Maintaining body condition***: Ensure cows have an optimal body condition score (3.0–3.5) at calving to prevent metabolic disorders.

Treatment of milk fever

Treatment focuses on restoring blood calcium levels quickly and effectively:

1. ***Intravenous calcium therapy***: Administer calcium borogluconate slowly IV (500 mL of a 23% solution) under veterinary supervision. Monitor the heart rate during administration to avoid complications like arrhythmias.
2. ***Oral calcium supplements***: Provide calcium gels or boluses immediately after IV therapy to sustain calcium levels.
3. ***Supportive care***: Keep the animal warm and provide assistance in rising to prevent muscle damage.

Economic impact of milk fever

Milk fever results in direct and indirect losses:

Reduced milk yield: Affected dairy cows produce less milk during the entire lactation cycle.

Secondary disorders: Increased risk of displaced abomasum, mastitis, and retained placenta.

Reproductive issues: Prolonged calving intervals and reduced fertility.

Higher culling rates: Severe cases may lead to premature culling.

Prevention through nutritional management is cost-effective compared to treating advanced cases of milk fever.

Future perspectives

Emerging strategies to prevent hypocalcaemia include:

1. ***Precision feeding***: Apply advanced feed analysis tools to optimize mineral balance in the diet of dairy animals.
2. ***Biotechnological approaches***: Development of slow-release calcium supplements and advanced vitamin D analogs to combat the milk fever in affected dairy animals.
3. ***Genetic selection***: Breeding programs focusing on selecting animals with lower susceptibility to milk fever.

Hypocalcaemia remains a significant challenge in dairy farming, with nutritional management being the cornerstone of its prevention and control. By adopting practices such as low-calcium diets during the dry period, negative DCAD feeding, and proper mineral balance, farmers can minimize the risk of hypocalcaemia and improve the health, welfare, and productivity of their dairy herds. Early diagnosis, prompt treatment, and a focus on transition cow nutrition are essential to mitigating the economic and health impacts of this metabolic disorder.

11

Toxic Minerals in the Rumen/ Rumino-reticulum

Ruminants, including cattle, buffalo, sheep and goats are highly efficient at digesting fibrous plant materials, thanks to their specialized stomach anatomy comprising the rumen, reticulum, omasum, and abomasum. However, these animals are also susceptible to the accumulation of toxic minerals in the rumen or rumino-reticulum, which can significantly affect their health, productivity, and survival. This chapter delves into toxic minerals, their sources, mechanisms of toxicity, clinical manifestations, and potential mitigation strategies.

The rumen and reticulum, as integral parts of a ruminant's digestive system, are central to the efficient fermentation and digestion of fibrous feeds. However, these chambers are also vulnerable to toxic minerals from environmental, dietary, and water sources. Toxic minerals can disrupt the rumen ecosystem, impair digestion, and negatively impact animal health.

The rumen and reticulum serve as fermentation chambers, hosting billions of microbes that break down complex carbohydrates, synthesize volatile fatty acids (VFAs), and produce microbial proteins essential for ruminant health. However, toxic minerals can interfere with this delicate balance, leading to reduced microbial efficiency, impaired digestion, and systemic toxicity.

Toxic minerals are naturally occurring or anthropogenically introduced elements that become harmful when consumed in excessive quantities. Toxic minerals, also known as heavy metals or trace mineral contaminants, are naturally occurring elements that can become hazardous when their concentrations exceed the physiological tolerance of ruminants. The following are some of the most problematic toxic minerals for ruminants:

Lead (Pb)

Arsenic (As)

Cadmium (Cd)

Mercury (Hg)

Fluorine (F)

Selenium (Se) (in excessive amounts)

Copper (Cu) (in excessive amounts)

Sources of toxic minerals

Toxic minerals enter the ruminant system primarily through feed, water, and environmental exposure.

a) **Feed contamination**

Industrial waste: Crops grown near industrial areas may absorb heavy metals from contaminated soils.

Pesticides and fertilizers: Arsenic and cadmium are commonly found in agricultural inputs.

Feed supplements: Over-supplementation of minerals like copper or selenium can lead to toxicity.

b) **Water contamination**

Contaminated groundwater often contains lead, arsenic, or fluoride.

Effluents from mining activities can introduce mercury and cadmium.

c) **Soil and environmental pollution:** Grazing on pastures contaminated with heavy metals from industrial runoff or air deposition contributes to mineral toxicity.

d) **Accidental ingestion:** Consumption of lead-based paints or batteries is a known cause of lead toxicity in ruminants.

Mechanisms of toxicity in the rumen/rumino-reticulum

The rumen, a fermentation chamber housing a diverse microbial population, plays a critical role in the digestion of feed. Toxic minerals interfere with this process in various ways:

Microbial inhibition: Heavy metals like lead and cadmium disrupt the metabolic activities of ruminal microbes, impairing fermentation.

Bioaccumulation: Minerals like mercury and cadmium can accumulate in the rumen wall, causing localized toxicity.

Impaired absorption: Toxic minerals may compete with essential nutrients for absorption, leading to deficiencies.

Chemical reactions: For example, excessive fluoride forms insoluble compounds with calcium, affecting skeletal integrity.

4. Toxic minerals and their effects

a) Lead (Pb)

Sources: Paints, batteries, contaminated water, and industrial waste.

Mechanism: Lead binds to sulfhydryl groups in enzymes, disrupting metabolic functions. It also accumulates in soft tissues and bones.

Clinical signs

Neurological symptoms: blindness, convulsions, and ataxia.

Gastrointestinal issues: anorexia, constipation, or diarrhea. It also reduced growth and milk production.

b) Arsenic (As)

Sources: Pesticides, industrial effluents, and contaminated water.

Mechanism: Arsenic interferes with cellular respiration by inhibiting enzyme activity.

Clinical signs

Acute toxicity: vomiting, abdominal pain, and sudden death.

Chronic exposure: skin lesions, reduced weight gain, and poor coat condition.

c) Cadmium (Cd)

Sources: Phosphate fertilizers, industrial pollution, and contaminated feeds.

Mechanism: Cadmium disrupts renal and hepatic function by replacing essential metals like zinc.

Clinical signs

Kidney damage: Proteinuria and reduced urine output. It leads to anemia and weakened immunity.

d) Mercury (Hg)

Sources: Contaminated water, fungicides, and industrial waste.

Mechanism: Mercury binds to proteins and enzymes, causing cellular and neurological damage.

Clinical signs

Neurological symptoms: tremors and ataxia. It also reduced feed intake and growth rates.

e) Fluoride (F)

Sources: Industrial emissions, contaminated water, and rock phosphates.

Mechanism: Fluoride replaces hydroxyl groups in hydroxyapatite, weakening bones and teeth.

Clinical signs

Dental fluorosis: discolored and brittle teeth.

Skeletal fluorosis: lameness and stiffness.

f) Selenium (Se) (Toxic in excess)

Sources: Seleniferous soils and over-supplementation.

Mechanism: Excess selenium generates reactive oxygen species, causing oxidative damage.

Clinical signs

Acute toxicity: respiratory distress and sudden death.

Chronic toxicity (alkali disease): hair loss, hoof deformities, and reduced fertility.

g) Copper (Cu) (Toxic in excess)

Sources: Over-supplementation and consumption of copper-rich feeds.

Mechanism: Excess copper damages liver cells and releases free radicals.

Clinical Signs

It leads to hemolytic anemia, jaundice, and dark urine. It also causes liver damage and sudden death.

Diagnosis of mineral toxicity

Accurate diagnosis involves a combination of clinical observation, laboratory tests, and environmental assessment.

- ***a) Clinical examination:*** Symptoms like neurological deficits, lameness, or poor growth.
- ***b) Laboratory tests:***

  Blood and tissue analysis for mineral concentrations,

  Liver biopsy for copper and selenium,

  Urine analysis for cadmium and lead
- ***c) Feed and water testing:*** Testing for mineral contamination in feed, water, and pasture samples.

Management and mitigation strategies

- ***a) Preventive measures***

  Feed quality control: Regular testing of feed for contaminants.

  Water treatment: Ensuring access to clean, uncontaminated water.

Balanced supplementation: Avoiding over-supplementation of trace minerals.

b) Detoxification

Chelating agents: Substances like EDTA bind to toxic metals, facilitating their excretion.

Activated charcoal: Effective for binding arsenic and mercury in acute cases.

c) Supportive therapy

Nutritional support: Supplementing with essential minerals like zinc and calcium to counteract deficiencies.

Antioxidants: Administering vitamin E and selenium to reduce oxidative damage.

d) Environmental management: Reducing industrial pollution and improving waste management in grazing areas.

Nutritional management of toxic minerals

Effective nutritional management strategies can mitigate the adverse effects of toxic minerals on ruminants. These strategies focus on reducing exposure, enhancing detoxification, and supporting overall health.

Prevention of mineral contamination

Feed quality control: Regular testing of feedstuffs for heavy metals and other contaminants is critical.

Water management: Providing clean, uncontaminated water prevents exposure to waterborne toxic minerals.

Avoidance of contaminated areas: Restricting grazing on pastures near industrial zones or polluted sites reduces exposure.

Nutritional strategies

1. ***Chelation therapy:*** Chelating agents like ethylenediaminetetraacetic acid (EDTA) bind to toxic metals, forming complexes that are excreted in faeces or urine. Supplementing with chelating agents can effectively reduce lead and cadmium toxicity.
2. ***Mineral antagonism:*** Supplementation with competing minerals can reduce the absorption of toxic minerals. For example: Zinc competes with cadmium for absorption sites. Calcium reduces fluoride absorption. Sulphur reduces selenium toxicity.

3. ***Activated charcoal and bentonite:*** Activated charcoal and bentonite clay bind to toxic minerals in the rumen, reducing their bioavailability.
4. ***Antioxidant supplementation:*** Selenium and vitamin E can mitigate oxidative stress caused by heavy metals like mercury and cadmium. Other antioxidants like vitamin C and beta-carotene also support detoxification.
5. ***Fibre and protein:*** Adequate fibre intake helps sequester toxic minerals in the rumen, preventing their absorption. Protein supplementation supports tissue repair and immune function.
6. ***Feed additives***

   Yeast cultures: Yeast probiotics improve rumen health and enhance the detoxification of heavy metals.

   Microbial enzymes: Adding enzymes to feed helps maintain microbial balance in the rumen, reducing the impact of toxic minerals.
7. ***Detoxification support:*** *Liver protectants*: Silymarin and other hepatoprotective agents help the liver detoxify accumulated heavy metals.

   Electrolyte therapy: Maintaining electrolyte balance supports kidney function in excreting toxic minerals.

Monitoring and diagnosis

a) ***Clinical monitoring:*** Regular observation of animals reared on toxic minerals contaminated pasture, feed and water for signs of mineral toxicity, such as reduced appetite, lameness, or neurological symptoms.

b) ***Laboratory testing:*** Analysis of feed, water, and soil for toxic mineral concentrations,

   Routine analysis of blood and urine samples and liver function tests to assess toxic mineral status,

c) ***On-farm records:*** Maintaining records of feed sources, mineral supplementation, and animal health helps identify and address potential issues.

Case studies

a) ***Lead toxicity in cattle***

   A herd consuming water from a lead-contaminated source exhibited reduced milk production, anorexia, and neurological symptoms. Nutritional management included:

   Removal of the contaminated water source,

Supplementation with calcium and zinc to reduce lead absorption,

Administration of EDTA as chelation therapy,

Recovery was observed after three months.

b) ***Selenium toxicity in sheep***

A flock grazing on selenium-rich pastures showed hair loss, hoof deformities, and weight loss. Interventions included:

Supplementation with sulphur and vitamin E to reduce selenium toxicity,

Provision of alternative grazing areas with low selenium content,

Long-term management and prevention

Sustainable grazing practices: Rotating pastures and avoiding overgrazing help reduce environmental contamination.

Integrated farming systems: Collaboration between crop and livestock production can reduce industrial pollution exposure.

Policy and regulation: Enforcing limits on industrial emissions and agricultural chemical usage can minimize environmental contamination.

Toxic minerals in the rumen/ruminoreticulum pose a significant threat to ruminant health, productivity, and welfare. Understanding the sources, mechanisms, and clinical manifestations of these toxins is crucial for effective prevention and management. Nutritional management strategies, including feed quality control, mineral antagonism, chelation therapy, and antioxidant supplementation, are essential for mitigating these effects. A combination of preventive measures, timely interventions, and regular monitoring ensures the health and productivity of ruminants, contributing to sustainable livestock farming. By adopting these strategies, producers can safeguard their animals from the harmful effects of toxic minerals and enhance overall herd performance.

12

Parakeratosis in the Rumen

Parakeratosis is a pathological condition characterized by the abnormal keratinization of epithelial tissues. In ruminants, particularly cattle, buffaloes, sheep, and goats, parakeratosis primarily affects the rumen epithelium. This condition is often associated with dietary imbalances, particularly in animals fed high-concentrate, low-fibre diets. Left unchecked, it can lead to significant health issues, including impaired nutrient absorption, secondary infections, and reduced productivity.

The rumen is a vital component of the ruminant digestive system, serving as a fermentation chamber where microbes break down fibrous plant material into volatile fatty acids (VFAs), which are absorbed and used as energy. The rumen wall consists of four layers: the stratum basale, stratum spinosum, stratum granulosum, and stratum corneum. The stratum corneum is the outermost layer, consisting of keratinized epithelial cells that protect the underlying tissues. Under normal conditions, the keratinization of the rumen epithelium is tightly regulated. However, certain dietary and environmental factors can disrupt this process, leading to parakeratosis.

Nutritional management is a cornerstone of both the prevention and treatment of parakeratosis. Proper dietary practices ensure the maintenance of a healthy rumen environment, optimal microbial activity, and the prevention of epithelial damage. The link between diet and parakeratosis is well established.

Causes of rumen parakeratosis

The primary cause of rumen parakeratosis is a high-concentrate, low-fibre diet. Specific factors include:

1. **Dietary imbalance:** Ruminants fed large amounts of readily fermentable carbohydrates (e.g., grains) leads to excessive VFA production and a decrease in rumen pH, creating an acidic environment conducive to parakeratosis. A lack of sufficient dietary fibre reduces rumen motility and the mechanical stimulation needed to maintain normal epithelial turnover.

2. **Insufficient fibre:** Low-fibre diets reduce rumen motility, saliva production, and mechanical stimulation of the rumen lining, disrupting normal keratinization.
3. **Rapid dietary changes:** Sudden transitions from high-forage to high-concentrate diets can overwhelm the rumen's adaptive capacity, leading to acidosis and epithelial damage.
4. **Micronutrient imbalances:** Deficiencies in certain vitamins and minerals (e.g., zinc and vitamin A) impair epithelial health and repair mechanisms.
5. **Rumen acidosis:** Sub-acute or chronic rumen acidosis often precedes parakeratosis. The acidic environment damages the epithelial layers, altering the keratinization process.
6. **Disruption of rumen microflora:** High-starch diets favour the proliferation of acid-producing bacteria (e.g., *Streptococcus bovis* and *Lactobacillus spp.*), further lowering pH and impairing microbial diversity.
7. **Environmental and management factors:** Inadequate roughage or poor-quality forages,

   Rapid dietary changes without adequate adaptation time,

   Overcrowding or stress, which may reduce feed intake consistency.

Pathophysiology

Parakeratosis begins with dietary or environmental factors disrupting the normal keratinization process of the rumen epithelium. Key events include:

1. ***Acidic rumen environment:*** Excessive VFA production and accumulation lower rumen pH, damaging the epithelial layers. Chronic exposure to low pH results in hyperkeratosis (thickening of the stratum corneum) and parakeratosis (retention of nuclei in keratinized cells).
2. ***Altered epithelium:*** Thickened, irregular epithelium impairs nutrient absorption, particularly VFAs. Increased keratinized debris accumulates, creating a favourable environment for bacterial invasion and secondary infections.
3. ***Inflammatory response:*** Bacterial infiltration may trigger localized inflammation or abscess formation in severe cases.
4. ***Impact on rumen function:*** Reduced absorptive capacity leads to systemic energy deficits and lower production efficiency.

Clinical signs

The clinical manifestation of rumen parakeratosis varies in severity, depending on the extent of epithelial damage and associated complications. Common signs include:

1. ***Digestive disturbances***

   Reduced feed intake and appetite,

   Decreased rumination and cud chewing,

   Chronic bloat due to impaired rumen motility,

2. ***Production losses***

   Weight loss or poor weight gain,

   Reduced milk yield and quality in dairy animals,

3. ***Secondary bacterial infections:*** Secpndary bacterial infections increased the susceptibility to rumenitis, liver abscesses and bacterial translocation.

4. ***Behavioural changes:*** Lethargy or discomfort, particularly after feeding

Diagnosis

Diagnosing rumen parakeratosis involves a combination of clinical evaluation, dietary assessment, and laboratory analysis.

1. ***History and dietary assessment:*** Detailed review of feeding practices, particularly the ratio of concentrates to forages.

2. ***Clinical examination***

   Palpation of the rumen and observation of digestive function,

   Assessment of body condition and productivity,

3. ***Laboratory analysis***

   Measurement of rumen pH (normal range: 5.5–7.0; lower values indicate acidosis),

   Microscopic examination of rumen epithelial samples for keratinization abnormalities,

4. ***Post-mortem findings:*** In severe cases, necropsy may reveal thickened, yellowish-brown rumen epithelium with areas of erosion or ulceration.

Principles of nutritional management

Effective nutritional management for rumen parakeratosis involves three main objectives:

1. Stabilizing rumen pH.
2. Promoting proper keratinization of the rumen epithelium.
3. Supporting microbial balance and overall digestive function.

1. Restoration of dietary balance

a) Adequate fibre intake

Role of fibre

Fibre stimulates chewing and saliva production, which buffers rumen pH and supports microbial fermentation.

Recommended sources

Long-stemmed forages, such as alfalfa hay, grass hay, and silage, are ideal for promoting rumen motility and maintaining a healthy epithelial lining.

Fibre length

Physical fibre length is crucial for stimulating rumination. Forages should have a minimum particle size of 1.5 cm to encourage cud chewing.

b) Controlled concentrate feeding

Balanced diet

Limit the proportion of concentrates to no more than 40% of the total dry matter intake (DMI) in most production systems.

Grain processing

Avoid finely ground grains, as they ferment rapidly, exacerbating rumen acidosis. Coarsely rolled or whole grains are better options.

2. Gradual dietary transitions

Ruminants require time to adapt to changes in diet, especially when increasing concentrate levels.

Recommended transition period

Gradually increase concentrates over 2–3 weeks, allowing rumen microflora to adjust and minimize the risk of acidosis.

3. Rumen buffering strategies

a) Feed buffers

Incorporating buffers into the diet helps stabilize rumen pH by neutralizing excess acids. Commonly used buffers include:

Sodium bicarbonate: Administered at 1–1.5% of DMI, it effectively combats acidosis.

Magnesium oxide: Works synergistically with sodium bicarbonate to improve buffering capacity.

b) Alkalinizing agents

Calcium carbonate and other alkalis can also be added to feeds to reduce acidity.

c) Monitoring rumen health

Regular assessment of rumen pH and microbial activity,

Early intervention in cases of sub-acute acidosis

d) Good management practices

Reducing overcrowding and providing consistent access to feed and water and minimize environmental stressors.

4. Supplementation with feed additives

a) Probiotics

Probiotics containing *Lactobacillus* and *Saccharomyces cerevisiae* improve microbial balance, enhance fibre digestion, and reduce acid accumulation.

b) Yeast cultures

Yeast stimulates the growth of beneficial bacteria that metabolize lactate, preventing ruminal acidosis. Include 10–20 g/day of live yeast in the diet for optimal results.

c) Ionophores

Compounds like monensin help modulate fermentation, reducing the production of lactic acid and maintaining stable rumen pH.

5. Micronutrient management

a) Zinc

Role: Zinc plays a vital role in maintaining the structural integrity of epithelial tissues and promoting healing.

Sources: Supplement diets with 40–60 ppm of zinc from organic or inorganic sources.

b) Vitamin A

Function: Essential for epithelial cell repair and regeneration.

Dosage: Ensure adequate intake of 2,000–5,000 IU/kg of feed.

c) Selenium

Supplementation of Se supports immune function and tissue repair. Include selenium at 0.3 ppm in the diet.

6. Water quality and access

Importance: Adequate water intake promotes saliva production, a natural buffer for rumen pH.

Guidelines: Ensure clean, fresh water is available at all times, particularly after feeding concentrates.

Preventive nutritional strategies

a) ***High-forage diets:*** Prioritize diets with a high proportion of forages to prevent excessive acid production.

b) ***Feed mixing and consistency:*** Uniform mixing of rations ensures consistent nutrient intake and prevents animals from sorting for concentrates. Total Mixed Rations (TMRs) are ideal for maintaining diet consistency.

c) ***Avoid overcrowding:*** Provide sufficient feeding space to reduce competition and ensure even feed intake.

d) ***Regular monitoring:*** Periodically measure rumen pH using rumenocentesis or other diagnostic methods to identify sub-acute acidosis early. Observe animals for signs of digestive disturbances or changes in productivity.

Case study: Implementing nutritional management

A dairy farm reported cases of parakeratosis in a group of high-yielding cows fed a 60:40 concentrate-to-forage diet. Clinical signs included reduced feed intake, bloat and decreased milk yield.

Intervention

1. ***Diet Adjustment:*** Increased forage to concentrate ratio (70:30) and incorporated long-stemmed hay.
2. ***Buffer supplementation:*** Buffering agent such as sodium bicarbonate should be added @ 1% of DMI.
3. ***Probiotic inclusion:*** Yeast-based probiotic should be introduced at the rate of 20 g/day per cow.
4. ***Zinc supplementation:*** Zinc should be administered @ 50 ppm of zinc in the feed.

Economic impact

Rumen parakeratosis can have significant economic implications for livestock producers. Reduced feed efficiency, lower growth rates, and decreased milk yield contribute to financial losses. Furthermore, the costs associated with treatment and management of complications, such as liver abscesses, further burden producers.

Outcome

Within four weeks, rumen health improved significantly, evidenced by increased appetite, stable rumen pH, and higher milk yield.

Parakeratosis in the rumen is a preventable condition with proper dietary management and attention to animal welfare. By understanding its causes, pathophysiology, and clinical implications, livestock producers can implement effective prevention strategies, ensuring optimal rumen health and productivity. The nutritional management of rumen parakeratosis hinges on the careful balance of dietary components to maintain a stable rumen environment and prevent epithelial damage. Strategies such as providing adequate fibre, managing concentrate levels, using buffers and probiotics, and ensuring sufficient micronutrient intake are essential. By implementing these practices, producers can prevent the onset of parakeratosis, enhance animal health, and optimize productivity. Early diagnosis and treatment are crucial to mitigating the long-term impact of this condition. As the livestock industry continues to evolve, integrating scientific knowledge with practical management will remain essential in addressing challenges like rumen parakeratosis.

13

Ruminitis

Ruminitis is an inflammatory condition affecting the rumen wall, primarily caused by disruptions in the rumen's microbial environment and metabolic processes. It is often linked to excessive consumption of rapidly fermentable carbohydrates, leading to rumen acidosis and subsequent damage to the ruminal epithelium. If untreated, ruminitis can progress to severe systemic complications such as liver abscesses and endotoxemia, significantly impacting animal health, productivity, and profitability in livestock systems. Nutritional management plays a central role in both preventing and treating ruminitis.

Causes

1. ***High-starch diets:*** Diets rich in grains and low in fibre promote rapid fermentation, leading to an overproduction of volatile fatty acids (VFAs) and lactic acid, lowering rumen pH (rumen acidosis).
2. ***Low-fibre diets:*** Insufficient fibre intake reduces chewing and saliva production, which are essential for buffering rumen acidity.
3. ***Inconsistent feeding practices:*** Irregular feeding schedules or abrupt dietary changes disrupt microbial populations in the rumen.
4. ***Other factors:*** Overfeeding of easily fermentable feeds (e.g., silage or molasses), Stress or illness reducing feed intake and altering rumen function.

Pathophysiology

1. ***Rumen acidosis:*** Excessive carbohydrate fermentation leads to a rapid drop in rumen pH, favouring the proliferation of acid-producing bacteria like *Streptococcus bovis* and *Lactobacillus*.
2. ***Epithelial damage:*** Low pH damages the ruminal epithelium, causing inflammation, ulceration, and, eventually, necrosis.
3. ***Bacterial translocation:*** Damaged epithelium allows bacteria to enter the bloodstream, potentially causing systemic infections such as liver abscesses or sepsis.

4. ***Chronic inflammation:*** Persistent epithelial damage results in chronic inflammation, impairing nutrient absorption and rumen function.

Clinical signs

1. Appetite and feed intake decreased.
2. Reduced rumination and cud-chewing activity.
3. Weight loss or poor body condition.
4. Signs of acidosis, such as diarrhoea, bloating and lethargy.
5. In severe cases, signs of systemic infection like fever or jaundice.

Nutritional management of ruminitis

Managing ruminitis requires a multi-pronged approach focused on stabilizing rumen pH, supporting microbial balance, healing the ruminal epithelium, and improving overall nutrition.

1. Dietary fibre optimization

a) ***Importance of fibre:*** Fibre stimulates chewing and saliva production, a natural buffer for rumen acidity. Long-stemmed forages also slow feed passage, allowing microbes to process feed more effectively.

b) ***Fibre sources:*** High-quality forages, such as grass hay, alfalfa hay, and silage, should form the basis of the diet. Forage should account for at least 60–70% of the dry matter intake (DMI) in ruminants.

c) ***Particle size:*** Ensure adequate forage particle size to encourage cud-chewing and promote proper rumen motility. A minimum particle size of 1.5 cm is recommended.

2. Controlled carbohydrate feeding

a) ***Gradual introduction of concentrates:*** Introduce high-energy feeds, such as grains, gradually over 2–3 weeks to allow rumen microbes to adapt. Avoid abrupt dietary changes that disrupt microbial populations.

b) ***Limit starch intake:*** Restrict concentrate levels to no more than 40% of the total diet. Replace some starch-rich feeds with alternative energy sources, such as fats or high-quality by-products (e.g., beet pulp or soybean hulls).

c) ***Grain processing:*** Avoid finely ground grains, as they ferment too rapidly. Opt for coarsely rolled or cracked grains to slow fermentation.

3. Buffering agents

a) ***Sodium bicarbonate***: It is acts as an effective buffer to neutralize excess acid production in the rumen. Include at 1–1.5% of DMI.

b) ***Magnesium oxide***: It enhances buffering capacity when used alongside sodium bicarbonate.

c) ***Other buffers***: Calcium carbonate and sodium sesqui-carbonate can also help maintain rumen pH.

*4. **Feed additives to stabilize rumen microflora***

a) ***Probiotics***: Probiotic supplements containing beneficial bacteria, such as *Lactobacillus* and *Bifidobacterium*, can improve microbial balance and reduce lactic acid accumulation.

b) ***Yeast cultures***: Yeast (*Saccharomyces cerevisiae*) supports the growth of fibre-digesting bacteria and reduces the build-up of lactic acid. Include 10–20 g/day per animal.

c) ***Ionophores***: Ionophores like monensin reduce the proliferation of lactic acid-producing bacteria, stabilizing rumen pH and preventing further epithelial damage.

*5. **Nutritional support for healing***

a) ***Zinc***: Zinc is crucial for epithelial repair and immune function. Supplement diets with 40–60 ppm of zinc from organic or inorganic sources.

b) ***Vitamin A***: Vitamin A supports epithelial cell regeneration and healing. Ensure 2,000–5,000 IU/kg of feed.

c) ***Selenium***: Selenium boosts immune function and helps control oxidative stress. Include selenium at 0.3 ppm in the diet.

*6. **High-quality water availability***

a) ***Clean water***: Provide access to clean, fresh water to support saliva production and maintain hydration.

b) ***Adequate access***: Ensure water availability at multiple points to prevent competition among animals.

*7. **Ration balancing and consistency***

a) ***Total mixed rations (TMR)***: TMR ensures consistent intake of nutrients, preventing selective feeding of concentrates over forages.

b) ***Uniform mixing***: Properly mix rations to avoid clumps of grain or uneven distribution of feed ingredients.

Preventive strategies

1. ***Balanced feeding***: Maintain a forage-to-concentrate ratio of at least 60:40 to prevent excessive acid production.

2. ***Gradual dietary transitions***: Allow sufficient time for animals to adapt the dietary changes, particularly when introducing high-energy feeds.
3. ***Regular monitoring***: Periodically monitor rumen pH to detect and address subacute acidosis early.
4. ***Minimize stress***: Provide a stress-free environment to encourage consistent feed intake and rumen function.
5. ***Avoid overfeeding***: Prevent overeating by controlling access to high-energy feeds and avoiding sudden feed availability after fasting.

Case study: Nutritional management of ruminitis

Scenario

A dairy farm reported decreased milk production and signs of acidosis in a herd of high-yielding cows. Clinical examination and dietary assessment revealed a high-starch, low-fibre diet with inconsistent feeding practices.

Intervention

1. ***Dietary adjustments:*** Increased forage content to 70% of DMI using high-quality alfalfa hay. Reduced concentrate intake and replaced a portion of corn grain with soybean hulls.
2. ***Buffer supplementation:*** Sodium bicarbonate should be added @ 1% of the ration.
3. ***Feed additives:*** Live yeast culture as feed additives should be included in the ration @ 15 g/day/ cow.
4. ***Micronutrient supplementation:*** Zinc and vitamin A should be supplemented to promote epithelial healing.

Outcome

Within four weeks, cows showed improved appetite, increased milk production, and stable rumen pH. Long-term changes in feeding practices prevented recurrence.

Ruminitis is a multi-factorial condition that can be effectively managed and prevented through proper nutritional strategies. Key approaches include optimizing dietary fibre, controlling concentrate intake, stabilizing rumen pH with buffers and probiotics, and supporting epithelial healing with micronutrients. Consistent feeding practices and early intervention are crucial to maintaining rumen health, improving animal productivity, and reducing economic losses in ruminant production systems.

14

Milk Impaction in Neonatal Calves

Milk impaction in neonatal calves is a condition characterized by the improper digestion or accumulation of milk in the abomasum, leading to digestive disturbances, discomfort, and potential health complications for the calf. This condition is most commonly seen in newborn or very young calves; typically within the first few days to weeks of life, as their digestive systems are still developing and may be more susceptible to dysfunctions.

Causes of milk impaction

Several factors can contribute to milk impaction in neonatal calves. Understanding these causes is essential for prevention and treatment. The primary causes are:

1. ***Overfeeding or undigested milk:*** One of the most common causes of milk impaction is overfeeding, where the calf is fed more milk than it can digest properly. Overfeeding can overwhelm the capacity of the abomasum, leading to a backup of undigested milk. Undigested milk in the stomach can ferment, creating an acidic environment that may disrupt normal digestion and lead to impaction.
2. ***Inappropriate milk temperature:*** Milk fed to calves should be at the proper temperature, usually between 38-39°C (100-102°F). Milk that is too cold can cause a delay in the digestive process and lead to impaction, as it may not be digested as quickly as milk at the ideal temperature. Similarly, milk that is too hot can cause digestive stress or irritation.
3. ***Incorrect feeding techniques:*** Poor feeding practices, such as force-feeding or feeding calves in a stressful environment, can contribute to digestive issues. If a calf drinks too quickly or gulps down a large volume at once, it may not be able to properly process the milk, leading to impaction. Additionally, poorly designed feeding systems or unclean feeding equipment can introduce bacteria that disrupt normal digestion.
4. ***Poor quality or contaminated milk:*** Milk that is not fresh or contains pathogens can disrupt the calf's digestion. Contaminated milk can lead to digestive upset, including milk impaction. The calf's immune system

is still developing, making it vulnerable to infections and digestive disturbances.

5. ***Calf's physical condition:*** Weak or stressed calves are more likely to experience digestive issues, including milk impaction. Factors such as low birth weight, illness, or inadequate colostrum intake can affect the calf's ability to properly digest milk. Calves born under stressful conditions, such as difficult deliveries or extreme weather, are more susceptible to digestive disturbances.
6. ***Congenital abnormalities:*** In some cases, calves may have congenital defects that impair their digestive function. These may include anatomical issues such as pyloric stenosis (narrowing of the stomach outlet) or other abnormalities in the gastrointestinal tract that impede the normal flow of milk from the stomach.

Symptoms of milk impaction

Recognizing the symptoms of milk impaction is essential for prompt intervention and treatment. The signs of milk impaction may vary depending on the severity of the condition, but typical symptoms include:

1. ***Reduced appetite or anorexia***: One of the first signs of milk impaction is a loss of appetite. A calf with impaction may refuse to suckle or may not drink as much milk as usual. This can lead to weight loss and dehydration.
2. ***Bloating and distended abdomen***: A calf with milk impaction may show signs of abdominal distension or bloating. This is caused by the accumulation of milk or gas in the stomach, which cannot be digested or passed through the digestive tract.
3. ***Colic or abdominal pain***: Impaction can cause discomfort or pain in the calf's abdomen, leading to signs of colic. The calf may show signs of restlessness, kicking at its abdomen, or lying down and getting up frequently.
4. ***Diarrhea or abnormal faeces***: In some cases, milk impaction may result in diarrhea. The stool may be watery, foul-smelling, or abnormal in color. Alternatively, the calf may experience constipation, where no feces are passed for an extended period.
5. ***Dehydration***: A calf with milk impaction may become dehydrated due to reduced milk intake and digestive dysfunction. Signs of dehydration include sunken eyes, dry mucous membranes, and lethargy.

6. ***Weakness and lethargy***: Impaction can lead to a decrease in the calf's energy levels. Weakness, lethargy, and a lack of interest in the environment or food are common signs.

Diagnosis of milk impaction

Diagnosing milk impaction involves a combination of clinical signs, physical examination, and diagnostic tools. The veterinarian will typically perform a thorough examination of the calf, including:

1. ***History and clinical signs***: The veterinarian will gather information on the calf's feeding history, including the amount and frequency of milk feeding, any recent changes in diet or feeding practices, and whether the calf is experiencing any digestive distress.
2. ***Physical examination***: A careful palpation of the abdomen may reveal signs of bloating or discomfort. The veterinarian may also assess the calf's hydration status, body condition, and overall health.
3. ***Ultrasound or X-ray***: In some cases, imaging techniques like ultrasound or X-rays may be used to assess the abomasum and the presence of undigested milk or gas. These tools can help determine the severity of the impaction and rule out other potential causes of gastrointestinal distress.
4. ***Laboratory tests***: Blood tests may be used to assess the calf's hydration levels, electrolyte balance, and overall health. Abnormalities in these tests may provide additional clues to the presence of milk impaction.

Treatment and management of milk impaction

The treatment of milk impaction in neonatal calves focuses on relieving the obstruction and restoring normal digestive function. Some common treatment options include:

1. ***Rehydration and electrolyte replacement***: Dehydration is a common complication of milk impaction, so rehydration is the first step in treatment. Oral rehydration solutions containing electrolytes may be given to the calf to restore proper fluid balance. In severe cases, intravenous (IV) fluids may be necessary.
2. ***Milk adjustment***: Adjusting the amount and frequency of milk feeding can help manage impaction. Reducing the milk volume or dividing the feedings into smaller, more frequent meals can aid digestion. Additionally, the temperature of the milk should be carefully monitored to ensure it is at the optimal level for digestion.

3. ***Prokinetic drugs***: In some cases, veterinarians may prescribe prokinetic drugs to stimulate motility in the abomasum and promote the movement of milk through the digestive system. These drugs help enhance the calf's ability to process and pass the milk more effectively. For example,
   a) **Metaclopramide** @ 0.1-0.3 mg /kg S/c or I/V every 8-12 hours in cattle and 0.04-0.08 mg/kg I/V as a constant rate infusion.
   b) **Erythromycin** @ 8.6-10 mg.kg orally every 8-12 hours in cattle and 0.5 – 1 mg.kg I/V every 6-8 hours.
   c) **Neostigmine** @ 0.02-0.04 mg/kg S/c or I/M and 0.01-0.02 mg/kg I/V or I/M in horses.
   d) **Domperidone** @ 0.1-0.2 mg/kg orally every 12-24 hours in cattle and horses.
4. ***Manual decompression***: In severe cases of milk impaction, manual decompression of the abomasum may be necessary. This involves using a tube to relieve the pressure caused by milk buildup in the stomach, allowing the milk to pass through the digestive system.
5. ***Antibiotics***: If there is a bacterial infection or the impaction has led to secondary complications, antibiotics may be prescribed to prevent or treat infections in the gut or other areas of the body.

Prevention of milk impaction

Preventing milk impaction in neonatal calves involves careful management of feeding practices and overall calf care. Some key preventive measures include:

1. ***Proper feeding practices***: Ensuring that calves are fed the correct amount of milk at appropriate intervals is essential. Overfeeding should be avoided, and milk should be offered at the proper temperature.
2. ***Clean feeding equipment***: Maintaining clean and sanitized feeding equipment is crucial to prevent contamination and ensure proper digestion.
3. ***Monitoring calf health***: Regularly checking calves for signs of illness or discomfort and providing prompt care when needed can help prevent complications, including milk impaction.
4. ***Quality colostrum***: Ensuring that calves receive adequate amounts of high-quality colostrum within the first few hours of life is essential for boosting the calf's immune system and supporting healthy digestion.
5. ***Minimizing stress***: Reducing stress factors, such as temperature extremes or overcrowding, can help maintain healthy digestion and prevent milk impaction.

Milk impaction is a significant health concern for neonatal calves that can lead to discomfort, poor growth, and potential long-term issues. By understanding the causes, symptoms, and treatment options for this condition, farmers and veterinarians can take steps to minimize the risk of milk impaction and ensure the health and well-being of young calves. Early intervention, proper feeding practices, and good management can help prevent this issue and ensure that neonatal calves thrive.

15

Acute Tympany

Acute tympany, commonly referred to as bloat, is a serious digestive disorder in ruminants characterized by the rapid accumulation of gas in the rumen, leading to ruminal distension. If untreated, this condition can cause severe discomfort, respiratory distress, and even death due to circulatory failure or suffocation. Tympany is categorized into two types: frothy bloat and free-gas bloat, each requiring distinct management approaches. Nutritional management is central to preventing and treating acute tympany. This chaper provides an in-depth discussion on the causes, pathophysiology, clinical signs, and nutritional strategies to manage and prevent acute tympany in ruminants.

Types of tympany

1. ***Frothy bloat***: Occurs when gas produced during fermentation becomes trapped in a stable foam, preventing its release through eructation (belching). It is associated with diets high in lush legumes (e.g., alfalfa, clover) or finely ground grains.
2. ***Free-gas bloat***: It results from physical or functional obstruction of the esophagus or impaired eructation mechanisms. Common causes include esophageal obstruction, hypocalcemia, or damage to rumen motility.

Causes of acute tympany

1. ***High-risk diets:*** Diets high in legumes (e.g., clover, alfalfa) produce stable foam in the rumen. Finely ground grains and high-starch diets lead to rapid fermentation and excessive gas production.
2. ***Low-fibre intake:*** Insufficient fibre reduces cud-chewing and saliva production, impairing rumen buffering and motility.
3. ***Abrupt dietary changes:*** Sudden transitions to high-risk diets overwhelm the rumen's adaptive capacity, increasing the risk of tympany.
4. ***Other contributing factors:***

   Physical obstructions (e.g., foreign bodies in the esophagus),

   Poorly formulated total mixed rations (TMR),

   Stress or illness reducing feed intake or motility.

Pathophysiology

1. ***Gas production***: During normal fermentation, rumen microbes produce gases (e.g., methane and carbon dioxide), which are released through eructation.

   In frothy bloat, the gas becomes trapped in foam, preventing expulsion.

   In free-gas bloat, physical or functional blockages hinder gas release.
2. ***Rumen distension***: Gas accumulation causes rumen distension, which compresses surrounding organs.
3. ***Respiratory and circulatory compromise***: Severe rumen distension exerts pressure on the diaphragm, reducing lung capacity and impeding blood flow, leading to hypoxia and shock.

Clinical signs

1. Rapid rumen distension, especially on the left side.
2. Signs of discomfort, such as restlessness, kicking at the abdomen, or grunting.
3. Reduced feed intake and cessation of rumination.
4. Difficulty breathing, drooling, and extended neck posture in severe cases.
5. Collapse and death if untreated.

Nutritional management of acute tympany

Nutritional management focuses on preventing conditions that lead to excessive gas accumulation, managing high-risk diets, and supporting rumen health.

1. High-fiber diets to reduce risk

a) ***Role of fibre***: Fibre stimulates cud-chewing and saliva production, promoting the natural release of gas and maintaining rumen pH.

b) ***Recommended sources***: Provide long-stemmed forages such as grass hay or straw to balance lush or high-starch diets. Avoid over-reliance on legume forages like alfalfa or clover.

c) ***Fibre proportion in diet***: Ensure a forage-to-concentrate ratio of at least 60:40 to maintain rumen function.

2. Gradual dietary transitions

a) ***Importance***: Abrupt changes in diet disrupt the rumen microbial population, increasing the risk of tympany.

b) ***Transition period***: Gradually introduce high-risk feeds, such as legumes or grains, over 10–14 days.

3. Avoiding overconsumption of high-risk feeds

a) ***Lush pastures***: Limit access to lush legume pastures by:

Introducing animals to pastures after they have been partially filled with hay,

Using rotational grazing to reduce overconsumption,

b) ***Grain feeding***: Avoid finely ground grains and opt for coarsely rolled or cracked grains to slow fermentation. Restrict grain feeding to prevent excessive gas production.

4. Use of anti-foaming agents

a) ***Poloxalene***: It is a widely used surfactant that reduces foam stability in frothy bloat. Administer through feed blocks, loose minerals, or water systems, especially during high-risk periods.

b) ***Mineral oil or vegetable oil***: Adding oils to the diet can reduce foam formation in frothy bloat.

5. Feed additives to enhance rumen function

a) ***Probiotics***: Probiotics containing *Saccharomyces cerevisiae* and *Lactobacillus* help stabilize rumen microflora and reduce gas production.

b) ***Yeast cultures***: Yeast promotes the growth of beneficial microbes, enhancing fiber digestion and reducing the risk of acidosis and tympany.

c) ***Ionophores***: Ionophores like monensin regulate fermentation patterns and reduce gas production.

6. Water management

a) ***Clean and consistent access***: Provide clean, fresh water at all times to support rumen function and saliva production.

b) ***Avoid rapid water intake***: Prevent animals from consuming excessive water after feeding high-risk diets, as it can exacerbate frothy bloat.

7. Mineral and vitamin supplementation

a) ***Magnesium***: Low magnesium levels in lush pastures can impair rumen motility. Supplement diets with magnesium oxide.

b) ***Zinc and vitamin E***: Zinc and vitamin E support epithelial integrity and immune function, reducing the risk of secondary infections.

Preventive strategies

1. ***Pasture management***: Avoid grazing on immature, lush pastures. Use mixed-species pastures to dilute legume intake.

2. ***Ration formulation***: Ensure balanced rations with adequate fibre and controlled levels of fermentable carbohydrates.
3. ***Regular monitoring***: Observe animals for early signs of tympany, particularly when grazing on lush pastures or consuming high-starch diets.
4. ***Minimize stress***: Stress can disrupt feeding patterns and exacerbate digestive disorders. Provide a calm environment and consistent routines.

Case study: Nutritional management of tympany

Scenario

A dairy herd grazing on lush clover pastures experienced multiple cases of frothy bloat, with some animals showing severe rumen distension.

Intervention

1. ***Pasture management***: Rotated animals to grass-dominant pastures and limited grazing time on clover to 2–3 hours per day. Animals should be provided long-stemmed hay before they turning out to graze.
2. ***Use of anti-foaming agents***: Anti-foaming agent such as poloxalene should be administered through feed blocks during high-risk grazing periods.
3. ***Dietary adjustments***: Fibre intake should be increased by supplementing the ration with grass hay.
4. ***Probiotic supplementation***: Added live yeast cultures to the diet to stabilize rumen fermentation.

Outcome

Incidence of bloat decreased significantly, with no further cases of severe tympany observed.

Acute tympany is a life-threatening condition that requires careful dietary management and preventive strategies. Key approaches include optimizing fibre intake, controlling access to high-risk feeds, using anti-foaming agents, and supporting rumen health with feed additives. By implementing these practices and monitoring animals closely, producers can prevent tympany, improve animal health, and reduce economic losses. Early intervention and consistent management are critical to successfully mitigating the risks associated with acute tympany.

16

Radiation Injury in Livestock

Radiation injury occurs when living tissues are exposed to harmful ionizing radiation. This radiation can damage cellular structures and genetic material, leading to acute and chronic health issues. Sources of ionizing radiation include nuclear accidents, radiation therapy, industrial exposure, and environmental contaminants. Radiation injury can have devastating effects on livestock and wildlife health, necessitating effective veterinary treatment and nutritional management to mitigate its consequences.

Mechanism of radiation injury

Radiation injury occurs when high-energy particles or electromagnetic waves interact with biological tissues. The mechanisms involved include:

1. ***Ionization of molecules***:Ionizing radiation removes electrons from atoms, forming free radicals such as hydroxyl radicals (OH) that damage cellular components.
2. ***DNA damage***: Radiation causes single- or double-strand breaks in DNA, leading to mutations, impaired replication, or cell death.
3. ***Lipid peroxidation***: Free radicals attack lipid membranes, disrupting cellular integrity and causing oxidative stress.
4. ***Protein denaturation***: Radiation alters the structure and function of proteins, affecting enzymatic activities and cell signalling.
5. ***Inflammatory response***: Damaged cells release inflammatory cytokines, leading to tissue inflammation and fibrosis.

Effects of radiation injury

The effects of radiation injury depend on the dose, duration, and type of radiation exposure. These effects are categorized as:

1. ***Acute radiation syndrome (ARS):*** High doses of radiation over a short period cause ARS, which affects rapidly dividing tissues such as:

   Bone marrow: Suppression of hematopoiesis, leading to anaemia, leukopenia, and immune-suppression.

Gastrointestinal tract: Damage to intestinal lining, causing nausea, vomiting, diarrhea, and mal-absorption.

Skin: Burns, ulcers, and delayed wound healing.

2. ***Chronic effects:*** Long-term exposure to lower doses may lead to:

   Cancer: Increased risk of leukemia, thyroid cancer, and other malignancies.

   Cardiovascular diseases: Radiation-induced damage to blood vessels.

   Neurological effects: Cognitive decline or neurological deficits.

3. **Radiation in livestock:** In animals, radiation exposure may result in:

   Reduced fertility and reproductive performance,

   Impaired growth and milk production,

   Genetic mutations passed to offspring.

Symptoms of Radiation Injury

Common symptoms of radiation injury include:

1. ***Mild exposure***: Fatigue, headache, and mild skin redness.
2. ***Moderate exposure***: Nausea, vomiting, hair loss, and skin burns.
3. ***Severe exposure***:

   Bleeding, infections, organ failure, and death,

   Economic and Environmental Impacts,

Radiation injury can have severe economic & ecological consequences

1. **Animal health costs**: Animal healthcare expenses should be increased for treating cancer, genetic disorders and radiation-related diseases.
2. **Agricultural losses**: Contamination of crops, soil, and water supplies,

   Reduced productivity and culling of exposed livestock.

3. **Environmental damage**: It causes long-term ecological imbalance due to radiation contamination affecting flora and fauna.

Nutritional management of radiation injury

Nutrition plays a critical role in mitigating the effects of radiation injury. A well-balanced diet can support the body's ability to repair damage, counteract oxidative stress, and enhance immune function.

1. Antioxidants

Antioxidants neutralize free radicals, reducing oxidative damage caused by radiation.

Key nutrients

Vitamin C: Found in citrus fruits, strawberries and bell peppers.

Vitamin E: Present in nuts, seeds and vegetable oils.

Selenium: Found in Brazil nuts, fish and eggs.

Carotenoids: Found in carrots, spinach and sweet potatoes.

2. Protein

Adequate protein intake is essential for tissue repair and immune function.

Sources

Animal-based: Lean meats, eggs and dairy products.

Plant-based: Lentils, beans and tofu.

3. Essential fatty acids

Omega-3 fatty acids reduce inflammation and support cellular repair.

Sources

Fish oil, flaxseed and walnuts.

4. Minerals

Zinc: Supports DNA repair and immune function. Found in oysters, red meat and legumes.

Iron: Essential for hemoglobin production, particularly in radiation-induced anemia. Found in spinach, red meat and fortified cereals.

5. Probiotics and prebiotics

Maintain gut health and support recovery from gastrointestinal damage.

Sources

Probiotics: Yogurt, kefir and fermented foods.

Prebiotics: Garlic, onions and whole grains.

6. Phytochemicals

Compounds in plants, such as flavonoids and polyphenols, have radioprotective properties.

Sources

Tree leaves, leaf meal (LM) and their leaf meal mixture (LMM), etc.

7. Hydration

Adequate water intake flushes toxins and supports cellular recovery.

Dietary strategies for radiation-exposed livestock

1. ***High-quality forage:*** Provide uncontaminated, nutrient-rich forage to support recovery.
2. ***Supplementation:*** Add antioxidant-rich supplements such as vitamin E and selenium to livestock diets.
3. ***Detoxifying agents:*** Incorporate clay or activated charcoal in feed to bind and eliminate radioactive particles.
4. ***Immune boosters:*** Include immune-enhancing supplements such as zinc and omega-3 fatty acids.
5. ***Monitor feed sources:*** Avoid feed grown in contaminated areas to minimize radiation exposure.

Preventive nutritional strategies

1. ***Radioprotective Nutrients***: Diets rich in antioxidants, vitamins, and minerals help build resilience against radiation.
2. ***Balanced diet***: Ensure a balanced diet with sufficient macronutrients and micronutrients to maintain overall health.
3. ***Fortified foods***: Use fortified foods or supplements in high-risk areas.

Research and advances

Emerging research focuses on:

1. ***Radioprotective compounds***: Natural compounds like curcumin, resveratrol, and green tea catechins show promise in mitigating radiation effects.
2. ***Genetic engineering***: Development of crops with enhanced radioprotective properties for human and animal consumption.
3. ***Therapeutic diets***: Specialized diets must be formulated for patients undergoing radiation therapy.

Radiation injury has severe implications for human health, livestock, and the environment. While medical interventions are crucial, nutritional management offers a complementary approach to mitigate the effects of radiation. A diet rich in antioxidants, proteins, essential fatty acids, and other key nutrients can help repair tissue damage, enhance immune function, and counteract oxidative stress. For livestock, careful feed management and supplementation can minimize the economic and health impacts of radiation exposure. By integrating these strategies, the long-term consequences of radiation injury can be significantly reduced.

17

Effect of Antibiotic Abuse and Antibiotic Supplements/Additives

Antibiotics are widely used in animal husbandry to treat diseases, prevent infections, and promote growth. While they have revolutionized livestock production, their misuse and overuse have raised serious concerns about antibiotic resistance, environmental contamination, and public health. Antibiotic abuse in animal feed, often as growth promoters or preventive measures, has profound effects on animals, humans, and ecosystems. This topic examines the effects of antibiotic abuse, the role of antibiotic supplements/additives in animal feeds, and strategies for responsible use to safeguard public and animal health.

Mechanism of action of antibiotics

Antibiotics are chemical compounds that kill or inhibit the growth of bacteria. They work through various mechanisms:

1. ***Inhibition of cell wall synthesis***: Penicillins and cephalosporins disrupt bacterial cell wall formation, leading to lysis.
2. ***Disruption of protein synthesis***: Antibiotics like tetracyclines and macrolides target bacterial ribosomes, halting protein production.
3. ***Interference with nucleic acid synthesis***: Fluoroquinolones and rifampin block DNA replication or RNA transcription in bacteria.
4. ***Alteration of cell membrane integrity***: Polymyxins disrupt bacterial membranes, causing cell death.

Effects of antibiotic abuse in animal foods

1. **Antimicrobial resistance (AMR)**

   Mechanism: Overuse of antibiotics in livestock selects for resistant bacterial strains. These bacteria can transfer resistance genes to other bacteria through horizontal gene transfer.

   Impact: AMR compromises the effectiveness of antibiotics in treating human and animal diseases, posing a global health crisis.

Examples: Methicillin-resistant *Staphylococcus aureus* (MRSA), Multidrug-resistant *Escherichia coli* and *Salmonella*, etc.

2. **Residues in animal products**

   Mechanism: Antibiotics administered to animals can accumulate in tissues, milk, eggs, or honey.

   Impact: Consumption of these products may expose humans to low doses of antibiotics, promoting resistance and allergic reactions.

3. **Impact on gut microbiota**

   Mechanism: Prolonged antibiotic use disrupts the natural balance of gut microbiota in animals.

   Impact: Reduced microbial diversity, impairing digestion and immunity and overall increased growth of harmful bacteria leading to diseases like colitis.

4. **Environmental contamination**

   Mechanism: Antibiotics excreted by animals enter soil and water, affecting microbial ecosystems.

   Impact: Disruption of soil fertility and aquatic ecosystems. Spread of resistance genes through environmental pathways.

5. **Economic consequences**

   Increased costs for treating antibiotic-resistant infections,

   Loss of market access for products with antibiotic residues, impacting international trade,

Role of antibiotic supplements/additives in animal feeds

Antibiotic supplements in animal feeds are used primarily for three purposes:

1. ***Disease treatment:*** Antibiotics are included in feed to treat bacterial infections in large herds or flocks where individual treatment is impractical. Example: Tylosin to treat liver abscesses in cattle.
2. ***Disease prevention:*** Prophylactic use of antibiotics helps prevent infections in high-stress conditions like weaning or transportation. Example: Coccidiostats in poultry to prevent coccidiosis.
3. ***Growth promotion:*** Sub-therapeutic doses of antibiotics promote growth by altering gut microbiota to enhance nutrient absorption and reduce subclinical infections. Example: Bacitracin or virginiamycin in pig and poultry feeds.

Risks associated with antibiotic supplements/ additives

1. ***Selection for resistant pathogens***: Prolonged exposure to antibiotics creates selective pressure for resistant bacteria.
2. ***Impact on human health***: Resistant bacteria in animal products can infect humans or transfer resistance genes to human pathogens.
3. ***Disruption of natural microbial ecosystems:*** Antibiotics in animal feeds can affect microbial communities in the environment and the animal gut.
4. ***Residue concerns***: Inadequate withdrawal periods can result in antibiotic residues in animal products.

Strategies for responsible use of antibiotics in animal feeds

1. ***Regulations and policy implementation***

   Ban the use of antibiotics as growth promoters where alternatives exist.

   Enforce strict withdrawal periods to prevent residues in animal products.

   Monitor and regulate the sale and use of veterinary antibiotics.

2. ***Antibiotic stewardship programs:*** Educate farmers and veterinarians on the judicious use of antibiotics. Promote veterinary oversight in prescribing antibiotics.
3. ***Alternatives to antibiotics***

   Probiotics: Beneficial microbes that enhance gut health and immunity. Example: *Lactobacillus* and *Bifidobacterium* strains.

   Prebiotics: Non-digestible fibers that stimulate the growth of beneficial gut bacteria. Example: Fructo oligosaccharides (FOS) and inulin.

   Phytogenics: Plant-derived compounds with antimicrobial and antioxidant properties. Example: Essential oils like oregano, thyme, or garlic.

   Vaccination: Prevent bacterial infections through immunization, reducing the need for antibiotics.

4. ***Improved farm management practices:*** Enhance biosecurity to reduce disease outbreaks. Provide clean water, balanced nutrition, and stress-free environments to promote animal health.
5. ***Research and innovation:*** Develop new antibiotics and alternative growth promoters. Study the role of precision nutrition in reducing the reliance on antibiotics.

Economic and environmental benefits of reducing antibiotic abuse

1. ***Improved product quality***: Antibiotic-free labels attract premium prices and consumer trust.
2. ***Sustainable livestock production***: Reducing antibiotic use aligns with sustainable agriculture goals, minimizing environmental contamination.
3. ***Cost savings***: Treatment expenses should be lowered for treating resistant infections in both animals and humans by using locally available natural alternatives such as plant secondary metabolites (PSMs).
4. ***Global market access***: Meeting international standards for antibiotic residue levels enhances export opportunities.

Case Studies and global perspectives

1. ***European Union*:** The EU banned antibiotics as growth promoters in 2006 and promotes alternatives like probiotics and improved hygiene.
2. ***United States*:** The FDA implemented the Veterinary Feed Directive (VFD) to regulate antibiotic use in livestock.
3. ***Developing countries*:** Challenges persist due to lack of regulation, inadequate veterinary infrastructure, and cost constraints.

The abuse of antibiotics in animal feeds poses significant risks to human and animal health, the environment, and global food security. While antibiotics remain essential for treating diseases, their indiscriminate use as supplements or additives must be curbed to prevent antimicrobial resistance and residue contamination. By adopting responsible use policies, improving farm management practices, and exploring alternative growth promoters, the livestock industry can reduce its reliance on antibiotics while ensuring animal welfare and public health. A balanced approach involving farmers, veterinarians, policymakers, and consumers is crucial to achieving sustainable and ethical animal husbandry.

18

Deficiencies Causing Ruminal Disorders

Nutritional deficiencies are a major cause of ruminal disorders, impacting animal health, productivity, and farm profitability. Balanced diets tailored to the specific needs of ruminants can prevent these deficiencies and ensure optimal rumen function. By employing preventive strategies, using high-quality feedstuffs, and incorporating appropriate supplements, farmers can safeguard their livestock from the adverse effects of nutrient deficiencies. Combining traditional knowledge with modern technology and research is essential for sustainable livestock production.

The rumen is a complex and vital organ in ruminants, responsible for fermenting feed into nutrients that support growth, reproduction, and production. Maintaining optimal rumen function depends heavily on a balanced diet that meets the animal's nutritional requirements. Deficiencies in key nutrients can lead to ruminal disorders, disrupting the fermentation process, impairing nutrient absorption, and compromising overall animal health and productivity.

a) Cobalt deficiency causes ruminal disorders

Cobalt deficiency is a significant concern in ruminants, as it plays a vital role in their health and productivity. This trace element is an essential component of vitamin B_{12} (cobalamin), which is synthesized by rumen microorganisms and is critical for various metabolic processes. Cobalt deficiency leads to a cascade of health issues, prominently manifesting as ruminal disorders. Here explores the mechanisms, symptoms, and management of cobalt deficiency in ruminants, focusing on its impact on ruminal health.

Role of cobalt in ruminants

Cobalt is not directly utilized by the animal but is required by the rumen microflora for the synthesis of vitamin B_{12}. This vitamin is essential for:

1. ***Energy metabolism***: Vitamin B_{12} is a cofactor for enzymes involved in the conversion of propionate to glucose through gluconeogenesis. Propionate, a volatile fatty acid (VFA) produced during fermentation, is a primary energy source in ruminants.

2. ***Amino acid metabolism***: It aids in the synthesis and breakdown of amino acids, contributing to protein metabolism.
3. ***Nucleic acid synthesis***: It supports DNA and RNA production, essential for cellular growth and repair.
4. ***Methionine synthesis***: Vitamin B_{12} is involved in the methylation cycle, which is crucial for many biochemical processes.

   Without sufficient cobalt in ruminants' diet, ruminal microbes cannot synthesize enough vitamin B_{12}, leading to metabolic dysfunctions.

Effects of cobalt deficiency on the rumen

Cobalt deficiency disrupts the rumen's microbial ecosystem and metabolic activities, causing a range of disorders:

1. ***Reduced microbial fermentation:*** The rumen's microbial population relies on cobalt to produce vitamin B_{12}, which supports metabolic pathways. A deficiency results in:

   Impaired synthesis of VFAs, especially propionate, reducing energy availability,

   Accumulation of unfermented feed, as microbial activity diminishes,

   Lowered down the efficiency of fibre digestion, leading to poor feed utilization,
2. ***Altered rumen pH:*** Cobalt deficiency can lead to sub-acute ruminal acidosis (SARA) due to imbalanced fermentation. The lack of propionate utilization may cause an increase in lactic acid, lowering ruminal pH and affecting microbial diversity. This imbalance may favour the growth of acid-tolerant bacteria over fibre-degrading microbes.
3. ***Anorexia and reduced feed intake:*** Ruminants with cobalt deficiency exhibit a marked reduction in feed intake due to poor rumen function and energy metabolism. This anorexia exacerbates the deficiency, as the animal's ability to ingest and digest feed diminishes.
4. ***Secondary complications***

   Impaired gluconeogenesis: Reduced propionate conversion leads to low glucose levels, affecting overall energy metabolism.

   Weight loss: The animal fails to gain or maintain weight due to poor nutrient absorption and energy deficiency.

   Weak immunity: Cobalt deficiency weakens the immune system, making the animal prone to infections.

   Metabolic disorders: Conditions like ketosis and fatty liver disease may arise due to energy deficits.

Clinical signs of cobalt deficiency

The symptoms of cobalt deficiency are subtle initially but become pronounced over time. Key clinical signs include:

Loss of appetite and weight loss

Poor growth rates in young animals

Pale mucous membranes, indicative of anemia

Rough coat and poor body condition

Reduced milk production in lactating animals

Reproductive failures, such as poor conception rates

Behavioural changes, including lethargy

In severe cases, ruminants may display signs of pica, where they chew on unusual materials like soil or wood, in an attempt to satisfy their mineral deficiency.

Diagnosis of cobalt deficiency

Diagnosing cobalt deficiency involves:

1. ***Clinical evaluation***: Observing symptoms such as poor growth, anorexia, and anaemia.
2. ***Blood tests***: Measuring serum vitamin B_{12} levels or methylmalonic acid (MMA), an indicator of vitamin B_{12} deficiency.
3. ***Rumen fluid analysis***: Assessing microbial activity and fermentation patterns.
4. ***Liver biopsies***: Evaluating cobalt and vitamin B_{12} stores in the liver.
5. ***Pasture and soil analysis***: Determining cobalt levels in the animal's environment, as deficiency is often linked to cobalt-poor soils.

Management and prevention

Effective management of cobalt deficiency focuses on supplementation and improving the animal's environment.

1. **Cobalt supplementation:** Cobalt can be supplemented through:

   Mineral licks or blocks: Providing cobalt-enriched salt licks.

   Feed additives: Incorporating cobalt salts, such as cobalt sulfate or cobalt carbonate, into the feed.

   Injectable vitamin B_{12}: Direct supplementation of vitamin B_{12} bypasses the need for ruminal synthesis, offering immediate relief.

   Slow-release boluses: These provide a sustained supply of cobalt or vitamin B_{12} over several months.

2. **Pasture Management:** Improving cobalt levels in pastures involves:

   Applying cobalt-containing fertilizers to deficient soils,

   Rotational grazing to ensure adequate forage intake and prevent overgrazing,

3. **Monitoring and adjustment:** Regularly monitor the health of the herd and cobalt levels in the diet. Adjust supplementation based on the specific needs of the animals and environmental factors.

Cobalt deficiency in specific regions

Certain regions, such as sandy or acidic soils, are naturally deficient in cobalt. These areas require targeted interventions, as animals grazing on these pastures are at a higher risk of deficiency.

Cobalt deficiency has a profound impact on ruminal health and overall productivity in ruminants. By impairing vitamin B_{12} synthesis, it disrupts critical metabolic processes, leading to ruminal disorders and systemic health issues. Early detection and intervention are essential to prevent long-term consequences. Regular supplementation, combined with strategic pasture management, can mitigate the risks associated with cobalt deficiency, ensuring optimal health and performance in ruminants.

b) Sulphur deficiency causes ruminal disorder

Sulphur is an essential mineral for ruminants, playing a crucial role in ruminal microbial activity, protein synthesis, and overall metabolism. When sulphur is deficient in the diet, it can cause significant disruptions in ruminal function, leading to poor health, reduced productivity, and various metabolic disorders. This topic explores how sulphur deficiency impacts ruminants, focusing on its effects on the rumen, the clinical signs of deficiency, diagnostic methods, and strategies for prevention and management.

Importance of sulphur in ruminants

Sulphur is a vital component of several biological molecules and processes in ruminants, including:

1. ***Amino acids***: Sulphur is a key element in the amino acids methionine, cysteine, and cystine, which are essential for protein synthesis.
2. ***Vitamins***: It is a component of vitamins like thiamine (B_1) and biotin (vitamin H), which are involved in energy metabolism.
3. ***Microbial activity***: Sulphur is required by rumen microorganisms to synthesize sulphur-containing amino acids and other metabolites.

4. ***Structural components***: Sulphur contributes to the structural integrity of proteins, keratin (in wool and hair), and cartilage.

Ruminants obtain sulphur primarily from dietary sources, including sulphur-containing proteins in forage, grains, and supplements. In the rumen, sulphur is converted into hydrogen sulphide, which rumen microbes use for their metabolic processes.

Effects of sulphur deficiency on the rumen

Sulphur deficiency disrupts ruminal microbial activity, leading to a cascade of disorders affecting digestion, nutrient absorption, and overall health.

1. ***Impaired microbial protein synthesis:*** Rumen microbes depend on sulphur to synthesize sulphur-containing amino acids, which are crucial for microbial protein production. A deficiency results in:

   Reduced microbial growth and activity,

   Decreased production of microbial protein, an important protein source for the host animal,

   Poor digestion of fibrous feeds due to decline in cellulolytic (fibre-digesting) microbial populations as well as their fibre degrading enzymes.

2. ***Decreased volatile fatty acid (VFA) production:*** Sulphur deficiency negatively impacts the fermentation process in the rumen, leading to reduced production of VFAs, particularly acetate, propionate, and butyrate. These VFAs are the primary energy sources for ruminants. A decrease in VFA production results in:

   Reduced energy availability,

   Poor growth, weight gain, and lowered milk production.

3. ***Impaired synthesis of sulphur-containing compounds:*** Sulphur is necessary for the production of thiamine and biotin by rumen microbes. These vitamins are critical for energy metabolism and carbohydrate digestion. Deficiency leads to suboptimal energy metabolism, resulting in reduced productivity. A predisposition to metabolic disorders like polioencephalomalacia (PEM), especially in animals fed on high-concentrate diets.

4. ***Altered rumen environment:*** A lack of sulphur can lead to an imbalance in the rumen microbial ecosystem. The population of sulphur-dependent microbes declines, causing:

   Poor feed conversion efficiency and accumulation of undigested feed in the rumen.

Clinical signs of sulphur deficiency

The symptoms of sulphur deficiency in ruminants can range from subtle to severe, depending on the extent of the deficiency. Common signs include:

1. ***Reduced feed intake and weight loss***

   Animals exhibit poor appetite due to impaired digestion and reduced energy availability.

   Weight loss and poor body condition are common.

2. ***Decreased milk production:*** Lactating animals show a drop in milk yield and quality due to inadequate energy and protein supply.
3. ***Poor growth rates:*** Young animals fail to grow at expected rates due to reduced protein synthesis and energy availability.
4. ***Rough hair coat and weak hooves:*** Sulphur is a key component of keratin. Deficiency leads to brittle hooves, weak hair, and a rough coat.
5. ***Metabolic disorders:*** Deficiency in sulphur-derived vitamins (thiamine and biotin) can cause metabolic issues like PEM, characterized by neurological symptoms such as blindness, staggering, and seizures.
6. ***Decreased reproductive performance:*** Poor fertility and extended intervals between pregnancies may occur due to compromised energy and protein metabolism.

Causes of sulphur deficiency

Sulphur deficiency can arise from several factors:

1. ***Low-sulphur diets***

   Diets based on low-protein forage or grain often lack sufficient sulphur.

   Corn silage and certain grasses have naturally low sulphur content.

2. ***Imbalanced diets:*** Diets high in non-protein nitrogen (NPN), such as urea, without adequate sulphur supplementation can exacerbate deficiency.
3. ***Soil deficiency:*** Sulphur-poor soils produce forage with low sulphur content, especially in regions with heavy rainfall where sulphur leaches from the soil.
4. ***Competition with antagonists:*** High dietary molybdenum or selenium can interfere with sulphur metabolism, reducing its bioavailability.

Diagnosis of sulphur deficiency

Diagnosing sulphur deficiency requires a combination of clinical observation, dietary analysis, and laboratory testing:

1. ***Clinical signs:*** Observation of symptoms like poor growth, weight loss, and rough coat provides initial clues.
2. ***Dietary analysis:*** Measuring sulphur levels in feed and forage can determine if the diet is deficient.
3. ***Rumen fluid analysis:*** Evaluating ruminal hydrogen sulphide concentration can indicate sulphur status.
4. ***Blood and tissue testing:*** Sulphur levels in blood or liver tissue can confirm deficiency.

Management and prevention

Addressing sulphur deficiency involves dietary adjustments and careful management practices:

1. ***Sulphur supplementation:*** Sulphur can be supplemented in several forms:

   Elemental sulphur: Direct addition to feed or water.

   Sulphate salts: Sodium sulphate or magnesium sulphate are commonly used.

   Protein sources: High-quality proteins like soybean meal provide organic sulphur.

   Care must be taken not to over-supplement, as excessive sulphur can lead to toxicity and conditions like polioencephalomalacia.
2. ***Balanced diet formulation:*** Ensure diets have an appropriate sulphur-to-nitrogen (S:N) ratio, typically 1:10 - 1:15, to support microbial protein synthesis. Avoid over-reliance on NPN sources without sulphur supplementation.
3. ***Pasture management:*** Apply sulphur-containing fertilizers to sulphur-deficient soils to improve forage quality. Monitor sulphur levels in pastures and adjust supplementation accordingly.
4. ***Regular monitoring:*** Periodically test feed, forage, and water for sulphur content to ensure adequate intake.

Monitor animal health and productivity for early signs of deficiency.

Sulphur deficiency poses a significant risk to ruminal health and overall productivity in ruminants. It disrupts microbial activity, impairs protein synthesis, and reduces energy availability, leading to a cascade of health and metabolic issues. Early diagnosis, proper diet formulation, and targeted supplementation are essential to prevent and manage sulphur deficiency effectively. By addressing sulphur needs, producers can ensure optimal health, growth, and productivity in their herds.

c) Any other nutrient decficiency causes ruminal disorders

Other nutrients deficiency both directly or indirectly also causes ruminal disorders and their nutritional and preventive measures are given below:

Energy deficiency

Cause: Insufficient carbohydrate intake, particularly fermentable energy sources like starch or sugar.

Effect: Reduced microbial activity in the rumen, leading to inefficient fermentation and reduced volatile fatty acid (VFA) production.

Disorders: Poor growth, low milk yield, and impaired rumen motility (hypomotility).

Protein deficiency

Cause: Low dietary protein or poor-quality protein sources that fail to meet ruminal microbial needs.

Effect: Insufficient ammonia for microbial protein synthesis, reducing microbial populations and feed digestion efficiency.

Disorders: Suboptimal growth, poor feed conversion efficiency, and reduced fiber digestion.

Fibre deficiency

Cause: Feeding low-roughage diets or excessive concentrates.

Effect: Decreased rumination and saliva production, leading to reduced buffering capacity and acidic rumen conditions.

Disorders: Subacute ruminal acidosis (SARA), rumenitis, and parakeratosis.

Vitamin deficiencies

Vitamin A: Essential for epithelial health. Deficiency causes impaired ruminal mucosa, increasing the risk of infections.

Vitamin D: Regulates calcium and phosphorus metabolism. Deficiency may lead to poor feed digestion due to impaired microbial function.

Vitamin E: Acts as an antioxidant. Deficiency results in oxidative stress and compromised immune function, predisposing the rumen to infections.

Other mineral deficiencies

Calcium and phosphorus: Imbalance leads to reduced microbial activity and weak skeletal structures. Disorders occur such as milk fever and secondary ruminal hypomotility.

Magnesium: Deficiency causes grass tetany and disrupted microbial metabolism.

Selenium: Deficiency causes oxidative stress, impairing rumen health.

Potassium: Low levels impair water balance and rumen motility, causing reduced feed intake.

Water deficiency

Cause: Inadequate water intake due to poor access or low water quality.

Effect: Reduced saliva production and impaired digestion.

Disorders: Reduced rumen motility and increased risk of impaction.

Nutritional management to prevent and address ruminal disorders

Energy management

Provide adequate fermentable carbohydrates (e.g., corn, barley) to meet energy demands.

Avoid excessive concentrates to prevent acidosis.

Incorporate fibrous by-products like beet pulp to balance energy and fiber.

Protein optimization

Use high-quality protein sources like soybean meal and canola meal.

Include non-protein nitrogen (NPN) sources, such as urea, in balanced quantities to support microbial protein synthesis.

Monitor protein-to-energy ratio to avoid imbalances.

Adequate Fibre

Ensure at least 18-20% of the diet is effective fibre (e.g., long-stem hay or silage).

Feed fibrous crop residues, such as wheat or rice straw, to maintain rumen function.

Adjust fibre levels based on forage quality and total dietary composition.

Vitamin supplementation

Add fat-soluble vitamins (A, D, E) to diets, especially in high-producing animals.

Include vitamin supplements in drought or winter when fresh forage is limited.

Provide injectable vitamins in severe cases of deficiency.

Mineral balancing

Use balanced mineral premixes that include macro-minerals (calcium, phosphorus, magnesium) and trace minerals (cobalt, selenium, zinc).

Provide free-choice mineral blocks or licks.

Incorporate chelated minerals for better bioavailability.

Water management

Ensure access to clean and adequate water at all times.

Regularly clean water troughs to avoid contamination.

Increase water intake during hot weather or high-feed-intake periods.

Feed additives

Buffers: Sodium bicarbonate can neutralize excess acidity and stabilize rumen pH.

Probiotics: Enhance microbial populations, improving fibre digestion and nutrient absorption.

Yeast cultures: Stabilize rumen fermentation and reduce the risk of acidosis.

Preventive measures

Regularly monitor nutrient levels in feed through laboratory analysis.

Adjust diets based on life stage, production level, and environmental conditions.

Train farmers and staff on proper feeding practices and the importance of balanced nutrition.

19

Torson/Displacement of Rumen, Omasum and Abomasum

The torsion or displacement of the rumen, reticulum, omasum, and abomasum are significant conditions affecting ruminants, particularly dairy cattle. These gastrointestinal disorders result in severe digestive disruptions, metabolic imbalances, and, if untreated, life-threatening complications. This chapter delves into the causes, clinical manifestations, diagnostic approaches, and management strategies for these conditions, focusing on each stomach compartment of ruminants.

Anatomy and function of rumen, reticulum, omasum, and abomasum

Ruminants have a complex stomach with four compartments:

1. ***Rumen***: The largest compartment responsible for microbial fermentation and digestion of fibrous feed.
2. ***Reticulum***: Works closely with the rumen in fermentation and acts as a sorting chamber.
3. ***Omasum***: Absorbs water, electrolytes, and VFAs while filtering digesta.
4. ***Abomasum***: Functions as the true stomach, secreting acid and enzymes for protein digestion.

Torsion or displacement in any of these compartments disrupts their normal functioning, leading to severe health issues.

Displacement or torsion of the rumen

Rumen displacement or torsion is less common than issues in the abomasum but can still occur under specific circumstances.

Causes

Overfeeding grains or highly fermentable feeds: Rapid fermentation produces excess gas, causing bloating and possible displacement.

Physical trauma or rolling: Sudden movement or trauma can result in rumen torsion.

Secondary to other gastrointestinal disorders: Conditions like abomasal displacement can alter ruminal positioning.

Clinical signs

Severe abdominal distension (bloat),

Painful abdomen, reluctance to move, and vocalization,

Decreased or absent rumen motility,

Rapid dehydration and systemic shock if blood supply is compromised,

Diagnosis

Physical examination reveals a distended left flank.

Percussion and auscultation detect a resonant "ping" on the left side.

Ultrasound or X-rays may confirm torsion or abnormal positioning.

Management

Immediate decompression with a stomach tube or trocar to relieve gas,

Surgical correction may be required to untwist or reposition the rumen,

Antibiotics and anti-inflammatory drugs to manage infection and pain,

Dietary modifications to prevent recurrence,

Displacement or torsion of the omasum

Omasal torsion or displacement is rare but can occur, leading to obstruction and compromised function.

Causes

Sudden dietary changes that increase gas production,

Physical trauma or overcrowding in pens,

Adhesions or scar tissue from previous abdominal infections,

Clinical signs

Reduced feed intake and failure to thrive,

Severe dehydration and electrolyte imbalances,

Dry feces due to decreased water absorption in the omasum,

Pain and distension in the right side of the abdomen,

Diagnosis

Rectal examination may detect a distended omasum.

Imaging techniques like ultrasound can help identify abnormal positioning.

Blood tests reveal dehydration and electrolyte imbalances.

Management

Surgical intervention is often necessary to correct torsion or displacement,

Fluid therapy to address dehydration and restore electrolytes,

Gradual reintroduction of a balanced diet after recovery,

Displacement or Ttorsion of the abomasum

Displacement of the abomasum is one of the most common gastrointestinal disorders in dairy cattle, classified into left-displaced abomasum (LDA), right-displaced abomasum (RDA), and abomasal torsion (volvulus).

Causes

Dietary factors: High-concentrate, low-fiber diets reduce ruminal fill, allowing the abomasum to move.

Postpartum physiology: Rapid changes in abdominal space after calving increase the risk of displacement.

Hypocalcemia: Reduced muscle tone predisposes the abomasum to displacement.

Concurrent illnesses: Conditions like ketosis or retained placenta can increase susceptibility.

Clinical signs

Left-displaced abomasum (LDA)

Decreased milk production and appetite, especially for grain,

Reduced rumen motility and "ping" on the left flank,

Mild dehydration and ketosis (sweet-smelling breath),

Right-displaced abomasum (RDA) and Torsion

More severe signs compared to LDA,

Acute abdominal pain, rapid dehydration, and shock,

A higher-pitched "ping" on the right flank,

Complete obstruction of digesta flow in torsion cases,

Diagnosis

Percussion and auscultation detect characteristic "ping" sounds on the affected side.

Rectal examination may reveal displacement in advanced cases.

Blood tests show metabolic alkalosis, hypochloremia, and hypokalemia.

Management

Non-surgical management

Rolling the cow to reposition the abomasum (for LDA),

Administering oral fluids and electrolytes to stimulate motility,

Adjusting the diet to increase fiber intake and ruminal fill,

Surgical management

Laparotomy: The abomasum is manually repositioned and anchored to the abdominal wall (omentopexy or abomasopexy).

Minimally invasive techniques: Such as toggle pin fixation for LDA.

Supportive therapy

Intravenous fluids to correct dehydration and electrolyte imbalances,

Treating underlying metabolic conditions like ketosis or hypocalcemia,

Pain management and antibiotics to prevent infection,

Complications and prognosis

Complications

Ischemia: Torsion can cut off blood supply, causing necrosis.

Septic peritonitis: Leakage of digesta can lead to fatal infections.

Recurrence: Poor dietary management increases the risk of recurrence.

Prognosis

LDA typically has a good prognosis with prompt treatment.

RDA and abomasal torsion carry a guarded prognosis due to higher risks of complications.

Prevention strategies

1. ***Dietary management***

   Provide a balanced diet with adequate fiber to maintain rumen fill.

   Avoid abrupt dietary changes, particularly after calving.

2. ***Postpartum care***

   Monitor cows closely for signs of metabolic disorders or decreased feed intake.

   Administer calcium supplements to prevent hypocalcemia.

3. ***General management***

   Ensure adequate space and minimize stress during feeding and calving.

   Regularly monitor herd health and address underlying conditions promptly.

Torsion and displacement of the rumen, omasum, and abomasum are significant gastrointestinal disorders in ruminants, with potentially life-threatening consequences if untreated. Understanding the causes and symptoms is critical for timely diagnosis and intervention. While surgical correction and supportive therapies are effective treatments, prevention through proper dietary and herd management remains the cornerstone of reducing the incidence of these conditions. By maintaining a focus on nutrition, postpartum care, and overall animal welfare, producers can ensure better health and productivity in their herds.

20

Hyperactivity of the Rumino-reticulum

The ruminoreticulum, part of the ruminant digestive system, plays a pivotal role in food processing, fermentation and nutrient absorption in animals such as cattle, buffaloes, sheep, goats, and deer. This compartmentalized stomach, comprising the rumen and reticulum, is critical for fermentative digestion, enabling ruminants to utilize fibrous plant materials. While its normal function is essential for health and productivity, hyperactivity of the ruminoreticulum can significantly impact an animal's well-being, physiology, and economic value in agricultural contexts.

Structure and function of the ruminoreticulum

The ruminoreticulum is a fermentation chamber harbouring billions of microorganisms, including bacteria, protozoa, and fungi. It is designed to break down complex carbohydrates, such as cellulose and hemicellulose, into volatile fatty acids (VFAs), which provide a significant portion of a ruminant's energy. The coordinated motility of the rumen and reticulum ensures proper mixing of ingested feed with saliva and microbes, enhancing fermentation efficiency.

Rhythmic contractions of the ruminoreticulum known as primary and secondary cycles, aid in mixing, stratification of digesta, and eructation of fermentation gases. When these contractions increase in frequency or intensity beyond physiological norms, it constitutes hyperactivity of the ruminoreticulum.

Causes of hyperactivity in the ruminoreticulum

Several factors can lead to hyperactivity in the ruminoreticulum, including:

1. **Dietary factors**

   High concentrate diets: Feeding excessive grains can stimulate hypermotility as the ruminoreticulum strives to manage rapid fermentation and acid production.

   Sudden dietary changes: Abrupt transitions from forage to concentrates or vice versa can disrupt the microbial population and provoke increased motility.

Ingestion of irritants: Feed contaminated with molds, toxins, or chemicals can trigger hyperactivity as the system reacts to expel irritants.

2. **Metabolic imbalances**

   Acidosis: Subacute or acute ruminal acidosis, resulting from rapid fermentation of carbohydrates, can induce hyperactivity as the ruminoreticulum attempts to buffer and expel acidic contents.

   Electrolyte disturbances: Hypocalcemia or imbalances in potassium and sodium levels can alter motility patterns.

3. **Physical irritation**

   Foreign bodies: Ingestion of sharp objects can irritate the ruminal or reticular lining, prompting excessive contractions.

   Bloat (Tympany): Accumulation of gas in the rumen may result in compensatory hyperactivity to facilitate eructation.

4. **Parasites and infections**

   Parasitic burdens: Certain gastrointestinal parasites can cause inflammation and irritation, resulting in hypermotility.

   Bacterial and viral infections: Infections affecting the gastrointestinal tract or systemic health may lead to altered motility patterns.

5. **Neurological stimulation**

   Stress and excitement: Stress-induced sympathetic stimulation can influence motility patterns, increasing activity temporarily.

   Neurological disorders: Lesions or dysfunction in the vagus nerve, responsible for regulating ruminoreticular motility, may contribute to abnormal contractions.

Pathophysiology of hyperactivity

The hyperactivity of the ruminoreticulum is characterized by exaggerated, frequent, or forceful contractions. This increased motility disrupts the delicate balance of fermentation and digestion, leading to a cascade of effects:

1. ***Increased fermentation***: Hyperactivity can accelerate fermentation rates, resulting in excess production of VFAs and gases such as methane and carbon dioxide. This may exacerbate conditions like acidosis and bloat, causing further complications.
2. ***Mechanical stress***: Prolonged hypermotility may lead to wear and tear on the ruminoreticular walls, increasing susceptibility to injuries or infections. Excessive contractions can also disturb the placement of digesta, reducing stratification and hindering effective fermentation.

3. ***Disruption of microbial balance***: Frequent contractions can affect the stability of microbial populations, particularly cellulolytic bacteria essential for fibre digestion.
4. ***Altered feed passage rates***: Hyperactivity may lead to faster passage of feed through the digestive system, reducing nutrient absorption and leading to weight loss or malnutrition.

Clinical signs and diagnosis

Animals with hyperactivity of the ruminoreticulum may exhibit a range of signs, including:

1. ***Behavioural changes***: Behavioural changes includes restlessness, frequent lying down and getting up and signs of discomfort such as kicking at the abdomen.
2. ***Gastrointestinal signs***

   Increased audible ruminal sounds,

   Rapid, exaggerated ruminoreticular contractions on palpation or auscultation,

   Symptoms of indigestion, such as diarrhoea or abnormal faeces,
3. ***Secondary issues***

   Signs of bloat, such as distended abdomen and respiratory distress,

   Evidence of weight loss or poor growth,

   Diagnosis is based on clinical examination, history and observation of motility patterns. Techniques include:

Auscultation: Using a stethoscope to evaluate contraction frequency and intensity.

Rumen fluid analysis: Assessing pH, microbial populations, and fermentation activity.

Imaging: Ultrasound or radiography to detect foreign bodies or structural abnormalities.

Management and treatment

Addressing hyperactivity of the ruminoreticulum requires a multifaceted approach:

1. ***Dietary management***

   Gradual transition to balanced diets with appropriate fiber-to-concentrate ratios,

Incorporation of buffers (e.g., sodium bicarbonate) to stabilize ruminal pH,

Removal of contaminated or irritant feedstuffs,

2. ***Medications***

   Antispasmodics: Medications like buscopan can help reduce excessive contractions.

   Probiotics and prebiotics: To restore microbial balance and enhance fermentation stability.

   Antibiotics or antiparasitics: In cases where infections or parasitic infestations are underlying causes.

3. ***Correction of metabolic imbalances***:

   Intravenous calcium, magnesium, or other electrolytes to address deficiencies,

   Fluid therapy to maintain hydration and support systemic health,

4. ***Surgical Interventions***: Surgical intervention involves removal of foreign bodies by performing rumenotomy, if mechanical irritation is the main cause.

5. ***Supportive care***: Stress reduction measures, such as minimizing handling and providing a comfortable environment. Regularly monitor affected animals to ensure resolution of hyperactivity.

Prevention strategies

Preventing hyperactivity of the ruminoreticulum is the key to maintaining animal health and productivity:

1. ***Proper feeding practices***

   Gradual dietary changes and avoidance of excessive concentrates,

   Ensuring consistent feed quality and avoiding contamination.

2. ***Regular monitoring***

   Routine health checks to identify and address issues early.

   Rumination collars or boluses are used to monitor motility in high-value livestock.

3. ***Vaccination and parasite control***: Implementing vaccination programs and follows regular deworming schedules.

4. ***Stress management***: Try to reduce stress during transportation, handling, and environmental changes.

Hyperactivity of the rumino-reticulum is a multi-factorial condition that can significantly impact the health and productivity of ruminants. Early identification and management are essential to prevent complications and ensure optimal digestive function. By understanding the causes and implementing appropriate preventative and therapeutic measures, farmers and veterinarians can effectively safeguard the well-being of their livestock.

21

Torsion of One or More Segments of Alimentary Canal

Torsion, or twisting, of one or more segments of the alimentary canal is a serious and often life-threatening condition in animals. It involves the abnormal rotation of the gastrointestinal tract along its axis, resulting in obstruction, vascular compromise, and subsequent tissue death. Torsion can affect different parts of the alimentary canal, including the stomach, intestines, and cecum, and is more commonly observed in certain species and breeds due to anatomical and environmental factors.

This condition requires prompt diagnosis and treatment to minimize morbidity and mortality. This discussion provides a comprehensive understanding of torsion, its pathophysiology, causes, clinical signs, diagnostic methods, and management strategies.

Pathophysiology of torsion

The alimentary canal is suspended in the abdominal cavity by mesenteric attachments that contain blood vessels, nerves, and lymphatics. When a segment of the canal twists along its mesenteric axis:

1. ***Mechanical obstruction***: The lumen becomes occluded, preventing the passage of ingesta or gas.
2. ***Vascular compromise***: The twist impairs venous drainage, causing congestion and edema, and eventually compromises arterial blood flow.
3. ***Ischemia and necrosis***: Prolonged loss of blood supply leads to tissue death, inflammation, and bacterial translocation, resulting in septicemia or endotoxemia.
4. ***Systemic shock***: If untreated, this can progress to systemic shock, organ failure, and death.

Common sites of torsion

1. ***Stomach***

   Gastric dilatation-volvulus (GDV): This type of condition/ torsion is most common in large, deep-chested dogs. The stomach twists along its axis, trapping gas and ingesta. This is seen in breeds like Great Danes, German Shepherds, and Setters.

2. ***Intestines***

   Intestinal volvulus: Occurs when a segment of the intestine rotates, causing obstruction. Intestinal volvulus is more common in horses and cattle.

   Mesenteric torsion: Rare but severe, involving a large portion of the small intestine.

3. ***Cecum***: Cecal torsion is observed in ruminants, especially dairy cattle, and can lead to colic and obstruction.

4. ***Colon***

   Large colon volvulus: Large colon volvulus is common in horses, leading to severe colic and abdominal distension.

Causes and risk factors

1. ***Anatomical predisposition***: Large, deep-chested breeds of dogs are prone to gastric dilatation-volvulus (GDV).

   Horses with a long mesentery are more susceptible to intestinal volvulus.

2. ***Dietary factors***: Large meals, especially those high in fermentable carbohydrates, can increase gas production and predispose to torsion.

   Feeding immediately before vigorous exercise in dogs is a risk factor.

3. ***Age and breed***: Gastric dilatation-volvulus (GDV) is more common in older dogs, while intestinal volvulus is often seen in young animals like foals or calves.

4. ***Stress and physical activity***: Stress or intense exercise can increase the risk of GDV in dogs or colonic torsion in horses.

5. ***Concurrent conditions***: Inflammatory bowel disease, intestinal motility disorders, or recent abdominal surgery may predispose to torsion.

Clinical signs

Clinical manifestations of torsion vary based on the segment affected and the duration of the condition:

General signs

Severe abdominal pain (colic) and restlessness,

Abdominal distension,

Rapid heart rate and respiratory distress,

Depression and reluctance to move,

GDV in dogs

Non-productive retching or vomiting,

Drooling and anxiety,

A distended, drum-like abdomen,

Weak pulse and pale mucous membranes (signs of shock).

Intestinal volvulus

Sudden onset of colic,

Reduced or absent fecal output,

Shock and rapid deterioration if untreated,

Cecal or colonic torsion (horses and cattle)

Severe colic in horses, often with rolling or pawing,

Decreased rumen motility and appetite in cattle,

Visible abdominal distension in severe cases,

Diagnosis

Accurate and timely diagnosis is critical for effective treatment:

1. ***Clinical examination***

   Evaluate vital signs, abdominal distension, and pain response,

   Rectal examination in horses and cattle to assess intestinal displacement,

2. ***Imaging techniques***

   Radiography: Shows gas distension and organ displacement in GDV or intestinal volvulus.

   Ultrasound: It is useful for detecting abnormal positioning, fluid accumulation or compromised blood flow.

3. ***Laboratory tests***

   Blood tests may reveal signs of dehydration, electrolyte imbalances, and metabolic acidosis.

   Elevated lactate levels indicate tissue ischemia.

4. ***Abdominocentesis***: Analysis of peritoneal fluid should be performed for evidence of ischemia or infection.

Treatment and management

1. ***Emergency stabilization***

   Fluid therapy: Administer intravenous (IV) fluids to address dehydration, electrolyte imbalances and shock.

   Decompression: Relieve gas build-up in GDV using a stomach tube or trocarization.

2. ***Surgical intervention***

   GDV: Surgery involves repositioning the stomach, removing necrotic tissue and performing gastropexy to prevent recurrence.

   Intestinal volvulus: Resection of necrotic segments and anastomosis may be required.

   Cecal or colonic torsion: Correction of the twist and the removal of compromised sections should be made to recover from caecal and colonic torsion..

3. ***Postoperative care***

   Monitor for complications such as peritonitis, sepsis or recurrence.

   Postoperative care should be done in such a manner that the gradual reintroduction of feeding with special focus on easily digestible diets.

   Pain management and antibiotics as needed.

Prevention strategies

1. ***Dietary management***

   Feed small, frequent meals to dogs prone to GDV.

   Avoid high-starch or rapidly fermentable diets in horses and cattle.

2. ***Minimize stress and activity post-feeding***:

   Prevent vigorous exercise immediately after meals in dogs.

   Reduce stress during transportation or handling of livestock.

3. ***Preventive surgeries***: Gastropexy in at-risk dog breeds to prevent GDV.
4. ***Regular health monitoring***: Monitor high-risk animals for early signs of gastrointestinal issues.

Prognosis

The prognosis for torsion depends on the severity, duration, and timely intervention:

GDV in dogs

Good if treated early; mortality increases with delayed treatment or necrosis.

Intestinal volvulus

Prognosis is guarded due to rapid progression and potential complications like septicemia.

Colonic torsion in horses

Early surgical intervention improves outcomes but carries risks of recurrence.

Economic and welfare implications

Torsion not only impacts animal welfare but also incurs significant economic losses in livestock due to treatment costs, decreased productivity, or fatalities. Early recognition, preventive strategies, and effective management are crucial to mitigate these impacts.

Torsion of one or more segments of the alimentary canal is a critical condition requiring immediate attention. By understanding the risk factors, clinical signs, and treatment options, veterinarians and animal owners can work together to reduce the incidence and severity of this life-threatening issue. Early intervention and preventive measures are essential for improving outcomes and safeguarding animal health.

22

Impaction of One or More Segments of Alimentary Tract: Focus on Rumen and Reticulum

Impaction of the alimentary tract is a condition where ingesta or foreign material accumulates excessively in one or more segments of the gastrointestinal system, causing a blockage. In ruminants, the rumen and reticulum are common sites of impaction due to their roles as primary fermentation chambers. This condition significantly impacts the animal's health, productivity, and welfare, and if left untreated, it can lead to severe complications or death.

Pathophysiology of impaction

Impaction occurs when the normal passage of ingesta is hindered due to excessive accumulation of indigestible, poorly digestible, or foreign materials. This leads to:

1. ***Decreased motility***: Overloading the rumen or reticulum impairs their motility, disrupting the normal mixing and fermentation processes.
2. ***Reduced absorption***: The accumulation hinders the absorption of volatile fatty acids (VFAs) and other nutrients.
3. ***Fermentation imbalance***: The retention of ingesta alters microbial populations, causing reduced fermentation efficiency.
4. ***Secondary issues***: Prolonged impaction may lead to dehydration, metabolic acidosis, rumenitis, and systemic effects like toxemia.

Common causes of rumen and reticulum impaction

1. ***Dietary factors***:

   Poor-quality forage: Fibrous, dry, or coarse feed materials can accumulate in the rumen.

   Overfeeding concentrates: Leads to excessive fermentation and compaction.

   Foreign materials: Ingestion of indigestible objects like plastic, wire, or cloth (ruminal foreign body syndrome).

2. ***Water deprivation***: Insufficient water intake reduces feed breakdown and movement through the digestive system.
3. ***Grazing behaviour***: Grazing on sparse pastures or drought-affected regions increases the ingestion of sand, soil, or indigestible plant material.
4. ***Concurrent diseases***: Conditions like hypocalcemia or ketosis reduce ruminal motility, predisposing to impaction.
5. ***Other factors***:

   Dental problems preventing proper mastication,

   Stress, overcrowding, or poor feeding management,

Clinical signs of impaction

The clinical presentation of rumen and reticulum impaction varies depending on the severity and duration of the condition:

1. ***Rumen impaction***

   Distended abdomen, especially in the left paralumbar fossa,

   Reduced appetite and progressive weight loss,

   Decreased or absent rumen motility and eructation (belching),

   Dry feces or complete absence of defecation,

   Mild dehydration and signs of systemic weakness,

2. ***Reticulum impaction***

   Signs are similar to rumen impaction but often associated with pain if foreign bodies are involved.

   Animal showed posture like arched back or reluctance to move in cases of traumatic reticulitis.

Diagnosis of impaction

1. ***Clinical examination***

   Assess rumen motility, abdominal distension, and fecal output,

   Rectal palpation to check for firm, dry ingesta in the lower digestive tract,

2. ***Laboratory tests***

   Rumen fluid analysis: Changes in pH, protozoal activity or bacterial population indicate fermentation imbalances.

   Blood tests: Indicate dehydration, metabolic acidosis or electrolyte disturbances.

3. ***Imaging***

   Radiography/X-ray: Useful for detecting foreign bodies or gas patterns,

   Ultrasound: Can help to evaluate the rumen wall and reticulum and/or for any fluid accumulation.

4. ***Exploratory rumenotomy***: In severe or unclear cases, surgical exploration of the rumen can confirm the diagnosis.

Treatment and management of impaction

1. ***Dietary and fluid management***

   High-quality forage: Replace coarse or fibrous feed with easily digestible forage to reduce rumen load.

   Water therapy: Provide free access to clean water or administer oral fluids to rehydrate the contents of the rumen.

   Laxatives: Administer mineral oil, magnesium sulfate (Epsom salts), or sodium sulfate to soften and move the impacted material.

2. ***Rumenotorics and digestive stimulants***

   Drugs like neostigmine or prokinetic agents can stimulate rumen motility.

   Transfaunation (transfer of healthy rumen fluid) helps restore microbial balance.

3. ***Surgical intervention***:

   Rumenotomy: A surgical procedure to manually remove impacted or foreign material from the rumen,

   Indicated in severe cases where medical treatment fails.

4. ***Treatment of secondary conditions***

   Address dehydration, electrolyte imbalances, or acidosis with intravenous fluids and supportive care.

   Administer antibiotics if bacterial overgrowth or secondary infections are suspected.

Complications of rumen and reticulum impaction

1. ***Rumenitis***: Chronic impaction can lead to inflammation of the rumen wall, causing reduced absorption and chronic digestive issues.
2. ***Tympany (Bloat)***: Gas trapped in the rumen can lead to life-threatening bloat.
3. ***Traumatic reticuloperitonitis***: In cases where the impaction involves sharp foreign bodies, peritonitis or diaphragmatic puncture may occur.

4. ***Systemic toxemia***: Severe cases may progress to systemic inflammation and toxemia, with fatal outcomes.

Prevention of impaction

1. ***Proper feeding practices***:

   Ensure balanced diets with adequate fibre, avoiding over-reliance on poor-quality forage.

   Introduce dietary changes gradually to allow adaptation.
2. ***Clean feeding environments***: Prevent access to foreign materials or contaminated feed sources.
3. ***Adequate water supply***: Ensure free access to clean water to support digestion and fermentation.
4. ***Grazing management***: Avoid overgrazing or overstocking pastures to reduce the ingestion of soil or sand.
5. ***Routine health monitoring***: Regularly check for dental health, rumen motility, and body condition in animals.

Case studies in specific animals

Cattle

Scenario: Dairy cattle fed low-quality hay during a drought are prone to rumen impaction.

Management: Rehydrate with fluids, introduce better forage, and administer laxatives.

Sheep and goats

Scenario: Overgrazing leads to ingestion of indigestible plant material.

Management: Remove from affected pasture, provide palatable feed, and monitor for signs of bloat.

Camelids

Scenario: Llamas and alpacas fed overly fibrous diets may experience chronic impaction.

Management: Surgical intervention may be necessary in severe cases.

Economic and welfare implications

Impaction reduces productivity, especially in dairy animals, by decreasing feed intake and milk production. Treatment costs, potential fatalities, and decreased reproductive performance significantly impact farm profitability. Furthermore, impaction causes considerable pain and suffering, making prevention and timely intervention critical for animal welfare.

Impaction of the rumen and reticulum is a preventable yet potentially fatal condition in ruminants and other grazing animals. Proper feeding practices, access to clean water, and early detection are essential to reduce its incidence. When impaction occurs, prompt medical or surgical intervention can restore normal function and minimize complications. A proactive approach involving education, regular monitoring and veterinary collaboration ensures healthier, more productive livestock.

23

Bovine Pulmonary Emphysema

Bovine pulmonary emphysema (BPE), also known as atypical interstitial pneumonia (AIP) or fog fever is a respiratory condition in cattle characterized by the over-distension of alveoli in the lungs due to the destruction of alveolar walls. This results in impaired gas exchange and respiratory distress. The condition is more common in adult beef cattle, particularly those exposed to sudden dietary changes or grazing lush pastures rich in tryptophan. BPE can cause significant morbidity and mortality, leading to economic losses in affected herds.

Etiology

Bovine pulmonary emphysema (BPE) occurs when cattle are exposed to factors that damage the pulmonary system, particularly:

1. ***Lush pastures***: Sudden introduction to green, high-protein pastures such as alfalfa, clover, or brassicas leads to excess production of toxic metabolites.
2. ***Dietary tryptophan***: Lush pastures are high in L-tryptophan, which is converted in the rumen into 3-methylindole (3-MI), a toxic compound that damages lung tissue.
3. ***Dust and environmental irritants***: Prolonged exposure to airborne irritants like dust, molds, or smoke can cause chronic lung injury.
4. ***Viral infections***: Respiratory infections caused by viruses like bovine respiratory syncytial virus (BRSV) or parainfluenza virus can predispose cattle to BPE.
5. ***Other causes***: Allergens, toxic plants or exposure to chemical irritants like anhydrous ammonia.

Pathophysiology

The pathophysiology of BPE involves:

1. ***Metabolism of tryptophan***: Lush pastures contain high levels of L-tryptophan. Rumen microbes convert L-tryptophan into 3-MI. The 3-MI is absorbed into the bloodstream and transported to the lungs.

2. ***Lung damage***

   In the lungs, 3-MI is metabolized by club cells (Clara cells) in the bronchioles into reactive intermediates.

   These intermediates damage alveolar and bronchiolar epithelial cells, causing inflammation.

   Edema (fluid accumulation),

   Alveolar wall rupture, leading to emphysema,

3. ***Impaired gas exchange***: Damage to alveoli reduces oxygen exchange, leading to hypoxemia and respiratory distress.

Risk factors

1. ***Sudden dietary changes***: Transitioning cattle from dry feed or poor-quality pasture to lush, high-protein pasture increases the risk.
2. ***Environmental conditions***:

   Wet weather promotes the rapid growth of lush pastures.

   Dry, dusty environments exacerbate lung irritation.

3. ***Age and breed Susceptibility***:

   Adult cattle, particularly beef breeds, are more susceptible than dairy cattle.

   Bos taurus breeds appear more prone than *Bos indicus* breeds.

4. ***Management practices***: Overstocking and inadequate grazing rotation can increase exposure to lush pastures.

Clinical signs

The clinical presentation of BPE is often sudden, with affected cattle displaying signs of respiratory distress:

1. ***Acute cases***

   Severe dyspnea (difficulty breathing) with open-mouth breathing,

   Extended head and neck posture to facilitate breathing,

   Rapid, shallow respiration,

   Audible expiratory grunt,

2. ***Subacute or chronic cases***

   Mild respiratory distress,

   Decreased feed intake and lethargy,

   Reduced weight gain or productivity,

3. ***Additional signs***

   Frothy nasal discharge,

   Cyanosis (bluish discoloration of mucous membranes) due to hypoxia,

   In severe cases, sudden death without prior signs,

Diagnosis

Diagnosis of BPE involves a combination of clinical signs, history, and diagnostic tests:

1. ***Clinical history***

   Recent grazing on lush pastures,

   Sudden onset of respiratory distress in multiple animals,

2. ***Physical Examination***

   Observation of labored breathing, cyanosis and frothy discharge,

   Auscultation may reveal reduced lung sounds or crackles,

3. ***Post-mortem Examination***

   Affected lungs are large, heavy, and pale,

   Evidence of emphysema (air bubbles) and edema (fluid accumulation),

   Ruptured alveolar walls and interstitial thickening,

4. ***Laboratory tests***

   Rumen fluid analysis may indicate excess tryptophan metabolites,

   Blood gas analysis shows hypoxemia and hypercapnia (elevated CO_2),

5. ***Imaging***: Radiography or ultrasonography may reveal lung abnormalities in live animals.

Differential diagnosis

Conditions with similar respiratory signs must be ruled out:

Pasteurellosis: Bacterial pneumonia caused by *Pasteurella multocida.*

Bovine respiratory syncytial virus (BRSV): Viral pneumonia in cattle.

Tuberculosis: Chronic respiratory disease caused by *Mycobacterium bovis.*

Fog fever: A specific form of BPE caused by lush pasture grazing.

Treatment

Treatment focuses on managing respiratory distress and preventing further lung damage:

1. ***Remove from lush pastures***: Immediately move affected animals to dry, low-quality forage.
2. ***Supportive care***:

   Oxygen therapy: Administer oxygen if available to alleviate hypoxia.

   1. *Anti-inflammatory drugs*:

      Corticosteroids (e.g., dexamethasone) reduce inflammation and edema.

      Non-steroidal anti-inflammatory drugs (NSAIDs) like flunixin meglumine.

   2. *Bronchodilators*: Improve airflow by relaxing airway muscles.
3. ***Antibiotics***: Antibiotic may be necessary to prevent secondary bacterial infections.
4. ***Fluids and electrolytes***: Intravenous fluids support hydration and maintain electrolyte balance.
5. ***Minimize stress***: Handle affected animals gently to prevent worsening of respiratory distress.

Prevention

Prevention of BPE involves careful pasture and herd management:

1. ***Gradual transition to lush pastures***

   Slowly introduce cattle to green pastures over 7–10 days.

   Limit grazing time initially and supplement with dry forage.
2. ***Dietary modifications***

   Provide supplemental feed such as hay or silage to dilute the intake of lush forage.

   Add ionophores (e.g., monensin) to the diet to reduce 3-MI production in the rumen.
3. ***Pasture management***: Avoid overstocking and rotate grazing areas to prevent overconsumption of lush forage.
4. ***Prophylactic measures***

   Administer anti-foaming agents or surfactants in drinking water during high-risk periods.

   Monitor cattle closely during pasture transitions for early signs of respiratory distress.
5. ***Avoid risky pastures***: Keep cattle off pastures with high legume content or those rapidly growing after rainfall.

Economic implications

BPE results in significant economic losses due to:

Mortality of affected cattle

Reduced weight gain and productivity in survivors

Costs of veterinary care and treatment

Loss of market value for animals with chronic lung damage

Prognosis

The prognosis for BPE depends on the severity of lung damage and the timeliness of intervention:

Mild cases

Animals usually recover with supportive care and removal from lush pastures.

Severe cases

Prognosis is poor, especially if significant emphysema and fibrosis have developed.

Nutritional management

Bovine Pulmonary Emphysema (BPE), commonly associated with grazing lush pastures, requires targeted nutritional management to prevent its onset and support affected cattle. Nutritional interventions focus on minimizing risk factors, modifying feeding practices, and ensuring proper supplementation to maintain lung health and reduce toxic metabolite formation.

Nutritional strategies

1. ***Gradual transition to lush pastures:*** Sudden exposure to lush, high-protein pastures increases the production of 3-methylindole (3-MI), a toxic compound formed from L-tryptophan in the rumen. Introduce cattle to lush pastures gradually over 7–10 days to allow their rumen microbes to adapt. Limit grazing time initially and monitor cattle closely for signs of respiratory distress.
2. ***Dietary dilution:*** *Supplement with dry forage*: Providing dry forage like hay or straw dilutes the intake of high-protein forage and reduces the production of 3-methylindole (3-MI). Feed low-protein forage before turning cattle onto lush pastures to reduce their overall protein consumption.
3. ***Feed additives to modulate rumen function***

   Ionophores (e.g., Monensin): Adding ionophores to the diet modifies rumen microbial populations and reduces the conversion of L-tryptophan to 3-methylindole (3-MI).

Dose: Commercial form of monensin is rumensin.

Dairy cows: 11-33 mg/kg of DMI

Beef cattle: 5-40 mg per head per day

Growing and finishing cattle: 200-480 mg per head per day (depending on body weight and diet consumption).

Sheep and goats: 15-20 mg per head per day or 1.5 to 2.5 mg/kg body weight.

Buffalo: Similar to cattle; 11-33 mg/kg DMI.

Antioxidants: Vitamin E and selenium supplementation can help mitigate oxidative stress caused by lung tissue damage.

4. ***Controlled protein intake***

   Lush pastures are typically high in protein (18–25%), which contributes to excessive tryptophan levels.

   Balance the overall diet by including low-protein feedstuffs to reduce the rumen's substrate for 3-MI formation.

5. ***Mineral and vitamin supplementation***

   Magnesium: Adequate magnesium levels help improve overall metabolic efficiency and prevent grass tetany, which may co-occur with BPE.

   Vitamin A and zinc: Supplementation of vitamin A and zinc are essential for maintaining healthy epithelial tissues, including those in the respiratory system.

Management of affected animals

For cattle already exhibiting signs of BPE, nutritional management supports recovery:

1. ***Immediate dietary changes***:

   Remove affected cattle from lush pastures and provide a diet of dry, low-protein forage.

   Avoid feeding silage or high-moisture feeds, which can exacerbate respiratory issues.

2. ***Hydration***

   Ensure access to clean, fresh water to support hydration and recovery.

   Electrolyte supplements can help maintain fluid balance.

3. ***Energy-rich feeds***:

   Provide energy-dense feed such as grains to compensate for reduced feed intake and support immune function.

   Avoid excessive carbohydrates to prevent acidosis.

Preventive nutritional strategies

1. ***Pasture management***: Avoid grazing cattle on pastures with high legume content or those in rapid growth stages.

   Rotate grazing areas to prevent overconsumption of risky forage.

2. ***Prophylactic feed additives***: Administer surfactants or anti-foaming agents to prevent excessive gas production and ruminal fermentation issues.

Bovine pulmonary emphysema is a preventable respiratory condition that poses a significant threat to cattle health and productivity. Understanding its causes, pathophysiology, and clinical signs is essential for early diagnosis and effective management. Preventive strategies such as gradual dietary transitions, proper pasture management and dietary supplementation are crucial in reducing the incidence of bovine pulmonary emphysema (BPE). Early intervention and supportive care can improve outcomes and minimize economic losses for producers.

Effective nutritional management of BPE involves preventing exposure to dietary triggers, balancing protein intake, and supporting lung health with targeted supplements. By combining proper grazing practices, feed additives, and a balanced diet, producers can significantly reduce the risk of BPE and improve the outcomes for affected cattle. Early intervention and careful dietary adjustments are essential for minimizing economic losses and ensuring herd well-being.

24

Ketosis

Ketosis is a metabolic disorder that occurs when animals rely heavily on fat as an energy source due to inadequate glucose availability. This condition is common in high-producing dairy cows, small ruminants like sheep and goats, and occasionally other species. Ketosis is characterized by elevated levels of ketone bodies (acetone, acetoacetate and beta-hydroxybutyrate) in the blood, urine and milk. While ketosis can occur in various animals, it is most frequently observed during periods of negative energy balance, such as early lactation in dairy cattle. Proper understanding and management are critical to maintaining animal health and productivity.

The economic repercussions of ketosis in dairy herds are substantial. Affected dairy animals often exhibit reduced milk yield, diminished reproductive performance and an increased susceptibility to other diseases. Financial losses are attributed to decreased productivity, treatment expenses, and potential culling of severely affected animals. Recent studies estimate that the economic losses per case of ketosis can reach up to Rs. 10000– Rs. 20000 per cow per lactation due to ketosis, making prevention critical and a figure that may double when associated diseases are considered.

Early and accurate detection of ketosis is crucial for effective management. Traditional diagnostic methods include measuring ketone bodies in blood, urine or milk samples. Recent advancements have introduced point-of-care testing devices that allow for rapid on-farm detection of hyperketonemia, facilitating timely intervention. Additionally, continuous monitoring technologies and metabolic profiling are being explored to identify at-risk animals before the onset of clinical signs. Preventing ketosis involves comprehensive management practices aimed at minimizing NEB during the transition period.

Providing a well-balanced diet that meets the energy requirements of high-producing dairy cows is essential. Incorporating glucogenic precursors and ensuring adequate fibre intake can help maintain optimal rumen function and energy balance. Maintaining appropriate body condition scores (BCS) during the dry period and at calving is critical. Over-conditioned cows (BCS $\geq$ 4.0) are at higher risk for ketosis; thus, managing feed intake to prevent excessive fat accumulation is important.

While ketosis is extensively studied in dairy cattle, it can also occur in wild and exotic animals, particularly during periods of food scarcity or increased energy demands, such as migration or lactation. The pathophysiology mirrors that of domestic ruminants with NEB leading to fat mobilization and ketone body production. However, diagnosing and managing ketosis in these species pose unique challenges due to limited baseline metabolic data and the impracticality of conventional treatment methods. Preventive strategies focus on ensuring adequate nutrition during critical periods and minimizing stressors that exacerbate energy deficits.

Ketone bodies are derived from acetyl-CoA during periods of increased fat metabolism, such as fasting, prolonged exercise, or carbohydrate deficiency. The three main ketone bodies are:

1. ***Acetone***: A volatile compound contributing to the characteristic fruity smell in ketosis.
2. ***Acetoacetic Acid (AcAc)***: A precursor to other ketone bodies, including BHB.
3. ***Beta-Hydroxybutyric Acid (BHB)***: The most abundant and stable ketone body, serving as a reliable marker of ketosis.

Normal levels of ketone bodies in animals

In healthy animals, ketone body levels are maintained within physiological limits: These values vary slightly among species but remain low under normal metabolic conditions.

Pathophysiology

The pathophysiology of ketosis revolves around an imbalance between energy intake and expenditure. When the energy demands of an animal exceed its dietary intake, body fat reserves are mobilized. This leads to increased lipolysis and the release of non-esterified fatty acids (NEFAs) into the bloodstream. NEFAs are transported to the liver, where they undergo beta-oxidation to produce acetyl-CoA. In the absence of sufficient oxaloacetate (due to low glucose levels), acetyl-CoA is diverted to ketogenesis, resulting in the production of ketone bodies: acetone, acetoacetate, and beta-hydroxybutyrate (BHB). Thus, the clinicopathologic characterization of ketosis includes high serum concentrations of NEFAs and ketone bodies and low concentrations of glucose. Fat is broken down into free fatty acids (FFAs), which are transported to the liver.

In liver

1. FFAs are converted into acetyl-CoA.
2. Acetyl-CoA enters the citric acid cycle for energy production.
3. When glucose is limited, the citric acid cycle slows, leading to acetyl-CoA accumulation.
4. Excess acetyl-CoA is converted into ketone bodies.

While ketone bodies provide an alternative energy source, excessive accumulation leads to hyperketonemia, reduced feed intake and metabolic imbalances.

Types of ketosis

1. ***Primary ketosis***: Primary ketosis occurs when energy demands exceed dietary intake, often seen in high-yielding dairy cows during early lactation.
2. ***Secondary ketosis***: Secondary ketosis arises due to a primary condition (e.g., displaced abomasum, mastitis or lameness) that reduces feed intake.
3. ***Pregnancy toxaemia***: A form of ketosis occurs in pregnant ewes and does carrying multiple foetuses. The energy demands of rapidly growing foetuses lead to glucose depletion.

On the other hand the condition manifests in four forms: subclinical form, nervous form, alimentary or wasting form and milk fever type form (clinical ketosis, except subclinical), with observable symptoms and subclinical ketosis (SCK), which lacks overt signs but can significantly impact animal health and farm economics.

Causes and risk factors

1. ***Negative energy balance***: High-producing dairy cows require immense energy during early lactation, leading to energy deficits. Pregnant ewes and does experience similar deficits during late gestation.
2. ***Inadequate diets***: Low energy density in the diet or poor-quality forage can predispose animals to ketosis.
3. ***Obesity***: Over-conditioned animals are more prone to fat mobilization and ketosis.
4. ***Stress and concurrent diseases***: Illnesses like mastitis, metritis, or digestive disturbances reduce feed intake, triggering ketosis.
5. ***Genetic factors***: Some breeds and individuals may have a higher predisposition.

Clinical signs of ketosis

1. ***In-appetence***: Reduced feed intake, especially for concentrate feeds
2. ***Weight loss***: Rapid loss of body condition due to fat mobilization
3. ***Reduced milk production***: A hallmark sign in dairy cows
4. ***Neurological signs (in severe cases)***: Abnormal behaviours of ketosis affected animals such as head pressing, ataxia or chewing movements known as nervous form of ketosis.
5. ***Acetone Odour***: Sweet or fruity smell in breath, milk and/ or in urine of ketosis affected animals due to acetone excretion.

Ketone body levels in ketosis

In ketosis, the levels of ketone bodies in blood, milk and urine increase significantly:

Species-specific observations

1. ***Dairy cattle:*** Dairy cows, especially high-yielding ones, are highly prone to ketosis during the early lactation period due to negative energy balance.

Normal state	Ketosis-affected state
Blood BHB levels: <0.6 mmol/L	>1.4 mmol/L (clinical) or 0.6–1.2 mmol/L (sub-clinical)
Milk acetone: <0.05 mmol/L	>0.2 mmol/L
Urine ketone levels: Trace amounts	Moderate to high

2. ***Small ruminants (Sheep and goats):*** Small ruminants, particularly pregnant ewes and does, experience ketosis in the form of pregnancy toxaemia during the last trimester.

Normal state	Ketosis-affected state
Blood BHB levels: <0.6 mmol/L	>1.0 mmol/L
Milk acetone: Rarely detected	-
Urine ketone levels: Trace amounts	High, often used for field diagnosis, Severe cases: Acetone smell in breath due to elevated acetone levels

3. ***Equines:*** Horses are less prone to ketosis because their liver lacks key enzymes for efficient ketogenesis. However, ketone bodies can increase mildly during prolonged fasting or illness.

Normal state	Ketosis-affected state
Blood ketone levels: Negligible (<0.1 mmol/L)	>1.2 mmol/L (clinical) or 0.6–1.2 mmol/L (sub-clinical)

4. ***Swine:*** Pigs rarely experience ketosis under normal feeding practices. However, fasting or diseases like porcine stress syndrome can elevate ketone levels.

Normal state	Ketosis-affected state
Blood BHB levels: <0.3 mmol/L.	0.4–0.6 mmol/L
Milk and urine: Trace or absent	-

Diagnosis of ketosis

Clinical signs

Dairy animals: Reduced milk yield, anorexia, weight loss, and a distinct acetone-like odour in breath or milk.

Companion animals: Lethargy, vomiting, diarrhoea, and neurological signs (e.g., seizures in advanced cases

Serum NEFA levels: High levels indicate excessive fat mobilization.

Liver biopsy: Histological evidence of hepatic lipidosis in severe cases.

Biochemical examination reveals hypoglycemia, ketonemia and ketonuria, while, in sub clinical ketosis ketonemia with absence of clinical signs.

2. ***Laboratory tests***: *Blood ketone levels*: Elevated beta-hydroxybutyrate (BHB) levels (>1.4 mmol/L in cattle).

   Urine ketone test strips: Detect ketone bodies in urine.

   Milk ketone tests: Measure ketone levels in milk samples.

3. ***Differential diagnosis:*** It is crucial to distinguish ketosis from other metabolic disorders such as hypocalcaemia, fatty liver syndrome, and pregnancy toxaemia, etc. Wasting form of ketosis showed signs similar to traumatic reticulitis but no relation to calving and normal serum glucose. Vagus indigestion marked stasis of alimentary tract with bloat. Abomasal displacement (marked abomasal sounds and absence of ruminal sounds). Nervous form showed symptoms similar to rabies (mania and ascending paralysis). Hypomagnesaemia, hypocalcaemia, bovine spongiform encephalopathy and lead poisoning closely resemble nervous form in addition to blindness, convulsion history.

Nutritional management strategies for prevention and treatment

Effective management of ketosis relies on preventing negative energy balance, optimizing nutrition, and addressing underlying health issues.

1. Prevention strategies

1. ***Optimize pre-partum diets***:

   In dairy cows, feed a transition diet with adequate energy and fibre 3–4 weeks before calving.

   Avoid over-conditioning during the dry period to reduce fat mobilization risks postpartum.

2. ***Provide high-energy feeds post-partum***: Include energy-dense feeds like grains or by-products (e.g., molasses) in lactating animal diets. Maintain a balance to prevent acidosis or other digestive disturbances.

3. ***Balancing macronutrients***: Adequate proportions of nutrients (such as carbohydrates, fats and proteins) to prevent NEB in ketosis affected animals.

4. ***Feed additives***: Direct supplementation of glucose can quickly alleviate hypoglycemia and reduce ketone body production. Feed additives like propylene glycol, niacin, and monensin to support energy metabolism, thereby reducing the reliance on body fat mobilization. Oral drenching with propylene glycol or glycerol provides immediate energy sources and supports hepatic gluconeogenesis. Oral hyperglycaemic agent (propylene glycol or glycerine) in adult cattle @ 100 mg bid for 2 or 3 days Small cattle @ 50 mg bid for 2 or 3 days. Monensin sodium enhances the propionate production in the rumen at the rate of 25mg/day in grain feed mix.

5. ***Enhance forage quality***: Use high-quality forages with optimal digestibility to ensure sufficient energy intake.

6. ***Minimize stress***: Reduce overcrowding, ensure adequate space at feeding areas, and maintain proper herd management.

7. ***Rumen-protected choline:*** Helps in reducing fat mobilization and supports liver function.

Therapeutic interventions

When prevention is insufficient, prompt treatment of ketosis is necessary to restore energy balance and prevent further complications.

Treatment

1. **Glucose infusion***:* Restores blood glucose levels and reduces ketogenesis. Dextrose 50% solution @ 500-800 ml should be administered intravenously repeat every 12-24 hours as needed.

2. ***Corticosteroids and mineral:*** The administration of corticosteroids can stimulate gluconeogenesis in severe cases, aiding in the correction of energy deficits. Dexamethasone and/or Predinisolone @ 10 ml I/M. Sodium propionate (glucogenic effect) @ 110 -225 mg/ day for 3 days to cattle but it show very slow response. Lactates: Ca or sodium lactate 1 kg initially, later ½ kg / day for 7 days and sodium acetate @ 110-500 g/day.
3. ***Insulin @*:** 200 IU can be administered. It has antiketogenic effect.
4. ***Vitamin support*:** Address deficiencies exacerbating metabolic stress. Administration of vitamin B_{12} (@10-20 ml I/M daily for 3-5 days) and cobalt for the activation of coenzyme-A and cysteamine (precursor of co-enzyme A) @ 750 mg i/v for 3 doses at 1-3 days interval, whereas, sodium fumerate (precursor of co-enzyme A) 3 doses at 1-3 days interval. Niacin improves energy metabolism by reducing fat mobilization from adipose tissue. It can be provided in feed as a powdered supplement especially rumen-protected niacin is preferred as it by-pass the rumen degradation. Niacin supplemented @ 6-12 grams per day per animal as preventive dose, whereas, @ 12-16 grams per day per animal as therapeutic dose in ketosis affected animals.
5. **Chloral hydrate initially,** 30 g orally as capsule and later 7 g bid for several days as drench in molasses or water. It breaks the starch in the rumen and stimulates production and absorption of glucose.

Supportive care

Adequate hydration and electrolyte balance: Maintaining hydration and restoring electrolyte balance is crucial to counteract dehydration. Ringer's lactate or isotonic saline should be administered @ 5-10 litres I/V over 12-24 hours. It is used to address dehydration and electrolyte imbalances, particularly sodium and chloride deficits. Oral electrolytes for mild cases provide a balanced oral electrolyte solution. Ensure it contains sodium, potassium, and bicarbonate precursors Administered @ 60-100 grams of an electrolyte powder in 10-20 litres of water.

Addressing secondary infections or complications: Secondary infections such as mastitis, metritis, or pneumonia are common in ketotic animals and require prompt treatment with antibiotics: Choose based on infection site and severity:

Ceftiofur @ 2.2 mg/kg body weight IM or SC daily for 3-5 days or

Oxytetracycline (long-acting) @ 20 mg/kg body weight IM or SC every 72 hours or

Penicillin administered @ 10,000 IU/kg body weight IM daily for 3-5 days and Calcium borogluconate should be administered @ 300-500 ml I/V or S/C if hypocalcemia is suspected.

Anti-inflammatory therapy (if indicated): NSAIDs (e.g., Flunixin Meglumine) @ 2.2 mg/kg I/V once daily for 1-3 days. It reduces inflammation and improves overall recovery.

Ketosis in specific animals

Dairy cattle

Peak risk period: First 3–6 weeks of lactation.

Prevention: Effective transition cow management, regular body condition scoring, and monitoring postpartum energy intake.

Sheep and goats (Pregnancy toxemia)

Risk factors: Multiple foetuses, poor nutrition during late gestation, or cold weather.

Management

Supplement with high-energy feeds during late pregnancy.

Avoid stress or transportation in late gestation.

Horses

Ketosis is rare but can occur in severe cases of starvation or metabolic dysfunction.

Complications of ketosis

1. ***Fatty liver syndrome***: Excessive fat mobilization leads to fat accumulation in the liver, impairing its function.
2. ***Impaired reproductive performance***: Delayed oestrus, lower conception rates, and retained placentas in cattle.
3. ***Increased susceptibility to diseases***: Animals with ketosis are more prone to infections like mastitis and metritis.

Monitoring and early detection

1. ***Regular testing***: Test high-risk animals (e.g., postpartum cows) for ketone levels in blood, milk, or urine.
2. ***Body condition scoring (BCS)***: Maintain optimal BCS (3.0–3.5) to prevent obesity or excessive thinness.
3. ***Milk production trends***: Monitor for unexpected drops in milk yield as an early sign.

Economic impact of ketosis

Ketosis leads to significant economic losses in the livestock industry due to:

Reduced milk production

Increased veterinary costs

Poor reproductive performance

Increased culling rates

Preventing and managing ketosis effectively can improve profitability and animal welfare.

Economic and welfare impacts

Economic losses

Ketosis in dairy animals causes significant economic losses to farmers due to its direct and indirect effects on productivity and health. Here are the primary economic losses, estimated in rupees (Rs.):

Reduced milk production

Ketosis reduces milk yield by 3–7 litres per day, depending on severity. With an average milk price of Rs. 40 per litre, this can result in daily losses of Rs. 120– Rs. 280 per cow, accumulating to Rs.3600– Rs. 8,400 monthly.

Poor milk quality

Ketosis affects milk fat and protein percentages, leading to quality penalties from dairies. Farmers lose Rs. 200– Rs. 500 per animal per lactation due to reduced milk quality.

Reproductive losses

Subclinical ketosis causes delayed conception, lower fertility rates, and more insemination attempts. Each unsuccessful insemination costs Rs. 500– Rs.1500, while delayed calving intervals can cost Rs. 3,000– Rs. 6,000 per animal.

Treatment and veterinary costs

Treating ketosis involves veterinary fees, medications, and nutritional supplements, costing Rs. 1500– Rs. 5000 per case.

Increased culling

Severe cases lead to involuntary culling, resulting in losses of Rs. 25000– Rs. 40,000 per animal.

Indirect costs

Reduced immunity increases susceptibility to other diseases, adding Rs. 1000– Rs. 3,000 per animal to healthcare expenses.

Total estimated loss

On average, a farmer may lose Rs. 10000– Rs. 20000 per cow per lactation due to ketosis, making prevention critical.

Animal welfare concerns

Chronic discomfort and pain due to associated complications,

Increased risk of culling in severe cases,

Long-term health repercussions affecting overall well-being,

Control measures of ketosis

Animal should not be starved or over fat at calving. Adequate calorie intake should be ensured at early lactation. Feeding in preparation for the next lactation should not begin until about 4 weeks to calving. Four weeks prior to calving silage or hay or pasture should be fed along with 1 kg concentrate/ day for maintenance and then gradually increase to 5 kg concentrate/day at calving. After calving, increase concentrate as production increases i.e. 3 kg hay/ 100 kg body weight or 9 kg silage/ 100kg body weight as maintenance and 1 kg concentrate /3 kg milk produced. Protein in the ration should not exceed 16-18 %. Exercise is must in intensive rearing. Ration should contain adequate level of cobalt, phosphorus and iodine. Avoid wet silage or mouldy hay or dusty hay feeding as they have increased levels of butyrate. Prophylactic feeding of sodium propionate @ 110g daily for 6 weeks should be followed. Sodium propionate 110 g/ day for 6 weeks and propylene glycol @ 350 ml / day for 10 days after serving or 6% of concentrate ration for 2 months should be practiced. Routine blood glucose and milk ketone estimation should be performed in dairy animals.

Future perspectives

Emerging technologies, such as precision feeding and metabolic monitoring systems, hold promise for reducing the incidence of ketosis. Research into genetic predispositions and tailored nutrition can further improve outcomes across diverse species.

Ketosis is a common metabolic disorder in high-producing animals, particularly during periods of increased energy demand. It requires a comprehensive approach, including optimized nutrition, effective herd management, and early detection. By preventing negative energy balance and addressing risk factors, livestock producers can minimize the occurrence and impact of ketosis, ensuring better health and productivity for their animals.

25

Spontaneous Hyperglyceridemia

Spontaneous hyperglyceridemia is a metabolic condition characterized by abnormally high levels of glycerol and triglycerides in the bloodstream. This condition can occur in various animal species, including dogs, cats, horses, and ruminants, and is often linked to underlying metabolic or endocrine disorders. Left unmanaged, hyperglyceridemia can lead to severe health complications, including pancreatitis, hepatic lipidosis, and cardiovascular issues. Nutritional management is a cornerstone of treatment and prevention for this condition.

Hyperglyceridemia results from an imbalance in the metabolism of triglycerides and glycerol. Normally, dietary fats are broken down into glycerol and free fatty acids, which are absorbed, processed in the liver, and utilized for energy or stored. Hyperglyceridemia arises when triglyceride production exceeds utilization or clearance from the bloodstream.

Causes and risk factors

1. ***Genetic predisposition***: Breeds like Miniature Schnauzers, Shetland Sheepdogs, and Beagles are predisposed to hyperglyceridemia in dogs.
2. ***Obesity***: Excess fat tissue increases triglyceride synthesis and decreases clearance.
3. ***Endocrine disorders***: Conditions like diabetes mellitus, hypothyroidism, and Cushing's syndrome can impair lipid metabolism.
4. ***Dietary factors***: High-fat diets or excessive caloric intake may contribute to elevated triglycerides.
5. ***Stress and illness:*** Stress, infection, or concurrent illnesses can exacerbate the condition.

Species-specific manifestations

1. ***Dogs:*** Hyperglyceridemia is common in Miniature Schnauzers and can lead to secondary ancreatitis and liver dysfunction.
2. ***Cats:*** Often secondary to diabetes mellitus or obesity, with potential complications like hepatic lipidosis.

3. ***Horses***: Rare but can occur in conjunction with insulin resistance and equine metabolic syndrome.
4. ***Ruminants***: Occasional in cattle, often linked to high-energy diets or stress during lactation.

Pathophysiology of hyperglyceridemia

The liver plays a central role in lipid metabolism. When dietary fats are absorbed, they are transported to the liver for processing. Triglycerides are packaged into lipoproteins and released into the bloodstream. Hyperglyceridemia occurs when:

1. ***Excessive triglyceride production***: Excessive triglyceride production might stimulated by obesity, high-fat diets or hormonal imbalances.
2. ***Impaired lipoprotein clearance***: Due to enzymatic (lipoprotein lipase) deficiency or receptor dysfunctions impaired lipoprotein clearance.
3. ***Hormonal dysregulation***: Endocrine disorders can impair the normal regulation of lipid metabolism.

Clinical signs

The signs of hyperglyceridemia vary depending on its severity and duration:

1. ***Mild cases***

   Subclinical or asymptomatic

   Occasional lethargy or weight gain

2. ***Moderate to severe cases***

   Vomiting, diarrhea, abdominal pain and anorexia

   Neurological signs like seizures (in rare cases)

   Complications like pancreatitis, hepatic lipidosis or cardiovascular issues.

Diagnosis

1. ***Blood tests***: Elevated serum triglycerides and glycerol levels are hallmark indicators.

   Lipemia (cloudy plasma) may be visible in severe cases.

2. ***Underlying conditions***: Test for diabetes, hypothyroidism, or other endocrine disorders.
3. ***Liver function tests***: Assess liver health, as hyperglyceridemia can lead to hepatic complications.

Nutritional management of hyperglyceridemia

Nutritional intervention is a key strategy in both the management and prevention of spontaneous hyperglyceridemia. The primary goals are to reduce triglyceride levels, optimize energy balance, and address underlying metabolic issues.

Dogs

1. ***Low-fat diets***

   Reduce dietary fat to minimize triglyceride synthesis.

   Use diets with less than 10% fat on a dry matter basis.

2. ***High-fiber diets***

   Soluble and insoluble fibers help modulate lipid absorption and improve gut health.

   Include sources like psyllium, beet pulp, or specially formulated high-fiber dog foods.

3. ***Omega-3 fatty acids***

   Supplement with omega-3s (fish oil) to reduce triglycerides and inflammation.

   Provide appropriate dosages to avoid calorie overload.

4. ***Portion control***: Avoid overfeeding; monitor caloric intake based on the dog's energy needs.

5. ***Weight management***: Encourage weight loss in obese dogs through a controlled calorie deficit and regular exercise.

Cats

1. ***Balanced diets***

   High-protein, low-carbohydrate diets are beneficial for overweight or diabetic cats.

   Avoid sudden dietary changes to prevent stress and anorexia.

2. ***Essential nutrients***: Taurine and carnitine supplementation support lipid metabolism and liver function.

3. ***Continuous feeding***: Offer frequent small meals to stabilize blood glucose and lipid levels.

Horses

1. ***Low-sugar, low-starch diets***:

   Low sugar and/or low starch diets must be essential for horses with insulin resistance or metabolic syndrome.

   Use forage-based diets supplemented with low-glycemic feeds.

2. ***Controlled weight loss***: Reduce body weight gradually through restricted feeding and increased exercise.
3. ***Supplementation***: Add omega-3 fatty acids to reduce inflammation and improve lipid profiles.

Ruminants

1. ***Balanced rations***

   Avoid overfeeding energy-dense feeds, especially during lactation.

   Provide adequate fiber to support rumen health and reduce fat synthesis.

2. ***Energy supplements***: Use protected fats (e.g., calcium salts of fatty acids) to meet energy needs without overwhelming the liver.
3. ***Nutrient additives***: Supplement choline, niacin, or methionine to support liver function and fat metabolism.

General management practices

1. ***Monitor body condition***: Maintain optimal body condition scores (BCS) to prevent obesity.
2. ***Regular health checks***: Routine blood tests should be required to monitor triglyceride levels and overall health.
3. ***Minimize stress***: Reduce stressors through environmental enrichment and consistent routines.
4. ***Exercise***: Encourage regular physical activity to improve lipid metabolism and overall health.

Treatment of severe cases

In severe cases of hyperglyceridemia, medical intervention may be required:

1. ***Lipid-lowering medications***: Drugs like fibrates or statins may be prescribed in extreme cases (rarely used in animals).
2. ***Intravenous fluids***: Correct dehydration and support metabolic clearance.
3. ***Treatment of underlying Disorders:*** Manage endocrine or metabolic conditions like diabetes or hypothyroidism.

Prevention strategies

1. ***Tailored nutrition***: Design species-appropriate diets based on individual needs and health risks.
2. ***Weight control***: Prevent obesity through balanced diets and regular exercise.
3. ***Early detection***: Regular health screenings of all cases should be done for high-risk animals to catch hyperglyceridemia in early stage.
4. ***Owner education***: Teach pet owners about proper feeding practices and the risks of overfeeding.

Spontaneous hyperglyceridemia is a complex metabolic condition that can significantly impact animal health. Effective nutritional management, combined with regular monitoring and treatment of underlying conditions, can help control triglyceride levels and prevent complications. By addressing dietary factors, promoting healthy body weight, and ensuring proper care, veterinarians and animal caretakers can successfully manage this condition and improve the quality of life for affected animals.

26

Diabetes Mellitus

Diabetes mellitus is a chronic metabolic disorder characterized by persistent hyperglycemia due to impaired insulin production, insulin resistance or both. It affects various animal species, including dogs, cats, horses and occasionally other animals like birds and ferrets. This condition requires lifelong management and, if left untreated, can lead to severe complications such as ketoacidosis, neuropathy and organ damage. Understanding diabetes mellitus in animals, its clinical manifestations and appropriate management strategies is essential for improving the quality of life for affected animals.

Types of diabetes mellitus in Animals

There are two primary types of diabetes in animals, analogous to human diabetes:

1. ***Type 1 diabetes (Insulin-dependent)***:

   Most common in dogs,

   Characterized by the destruction of pancreatic beta cells, leading to insufficient insulin production,

   Typically immune-mediated or idiopathic,

2. ***Type 2 diabetes (Non-insulin-dependent)***:

   Common in cats and occasionally in horses,

   Involves insulin resistance, often associated with obesity, physical inactivity, or concurrent diseases,

   The pancreas initially produces insulin, but over time, beta-cell dysfunction occurs,

   Other forms of diabetes include gestational diabetes (during pregnancy) and secondary diabetes due to conditions like pancreatitis, Cushing's syndrome, or prolonged corticosteroid use.

Causes and risk factors

The development of diabetes mellitus is multifactorial and influenced by genetic, environmental and lifestyle factors:

1. ***Genetic predisposition***: Certain dog breeds (e.g., Samoyeds, Miniature Schnauzers, and Poodles) and cats with Burmese lineage are more prone.
2. ***Obesity***: Obasity increases the risk of insulin resistance, particularly in cats and horses.
3. ***Age and gender***: Middle-aged to older animals are more commonly affected. Female dogs and male cats have higher risks.
4. ***Concurrent diseases***: Pancreatitis, hyperadrenocorticism, and hypothyroidism can contribute to diabetes development.
5. ***Diet and lifestyle***: High-carbohydrate diets and sedentary lifestyles are significant risk factors, especially in cats.

Pathophysiology of diabetes mellitus

In healthy animals, insulin, a hormone produced by pancreatic beta cells, facilitates glucose uptake into cells for energy production or storage. In diabetes:

1. ***Type 1 diabetes***

   Insufficient insulin production leads to impaired glucose uptake and persistent hyperglycemia.

   Cells starve despite high blood glucose levels, triggering fat and protein catabolism.

2. ***Type 2 diabetes***

   Insulin resistance prevents effective glucose uptake,

   Over time, beta cells become exhausted, reducing insulin production,

   Persistent hyperglycemia damages blood vessels, nerves and organs, leading to complications.

Clinical signs

The clinical manifestations of diabetes mellitus are often gradual:

1. ***Classic signs***

   Polyuria (Excessive urination): Due to glucose spilling into the urine, causing osmotic diuresis.

   Polydipsia (Increased thirst): To compensate for water loss.

   Polyphagia (Increased appetite): Despite weight loss due to energy depletion.

   Weight loss: Fat and muscle breakdown for energy.

2. ***Other signs***

   Lethargy, dull coat, and recurrent infections (e.g., urinary tract infections),

   Cataracts (common in dogs),

   Peripheral neuropathy (weakness, particularly in cats),

3. ***Complications***

   Diabetic ketoacidosis (DKA): A life-threatening emergency with symptoms like vomiting, dehydration, and altered mental state.

Diagnosis

Diagnosing diabetes mellitus involves a combination of clinical history, physical examination, and laboratory tests:

1. ***Blood tests***: Persistent hyperglycemia and elevated fructosamine levels (indicating prolonged hyperglycemia).
2. ***Urinalysis***: Urine analysis must be performed for Glucosuria (glucose in urine) and possible ketonuria (ketones in urine).
3. ***Differential diagnosis***: Rule out other causes of hyperglycemia (e.g., stress hyperglycemia in cats).

Management of diabetes mellitus

Diabetes management aims to stabilize blood glucose levels, prevent complications, and improve the animal's quality of life. It requires a combination of insulin therapy, dietary management, weight control, and monitoring.

1. ***Insulin therapy:*** Insulin administration is the cornerstone of treatment for diabetic animals:
   1. **Dogs**: Most dogs require lifelong insulin injections (intermediate-acting or long-acting formulations).

      Common options: NPH, Vetsulin, or insulin glargine.
   2. **Cats**

      Long-acting insulin like glargine or PZI are preferred.

      Some cats may achieve remission with early and aggressive management.

      Dosing: Insulin doses are individualized based on blood glucose levels and monitored through glucose curves.
2. ***Dietary management:*** Diet plays a critical role in stabilizing blood glucose levels:
   1. **Dogs**:

High-fiber, moderate-carbohydrate diets slow glucose absorption.

Consistent feeding times synchronized with insulin injections.

2. **Cats**:

   High-protein, low-carbohydrate diets mimic natural feline metabolism.

   Prescription diabetic diets are often beneficial.

3. **Horses**:

   Low-sugar, high-fiber diets are essential for insulin-resistant horses.

   Grazing muzzles or restricted pasture access to control calorie intake.

3. ***Weight management:*** Obesity exacerbates insulin resistance. Weight loss in overweight animals improves insulin sensitivity and overall health.

4. ***Monitoring and adjustment:*** Regular monitoring is essential to ensure effective management:

   1. **Home monitoring**:

      Owners can use glucometers to measure blood glucose levels.

      Consistent tracking helps adjust insulin doses.

   2. **Veterinary monitoring**: Periodic glucose curves, fructosamine levels and urinalysis.

   3. **Signs of hypoglycemia:** Excess insulin can cause low blood sugar, leading to weakness, seizures or collapse. Owners should be trained to recognize and manage hypoglycemia.

Complications and their management

1. **Diabetic ketoacidosis (DKA)**: Emergency treatment includes IV fluids, insulin therapy and electrolyte correction.

2. **Cataracts (dogs)**: Surgery may restore vision in affected dogs.

   Cataracts should be managed by stabilizing blood glucose and supplementing with methylcobalamin (vitamin B_{12}).

Prevention strategies

1. ***Healthy diets***: Feed species-appropriate, balanced diets to prevent obesity and insulin resistance.

2. ***Neuropathy (Cats)***

3. ***Regular exercise:*** Encourage physical activity to maintain healthy body weight and improve insulin sensitivity.

4. ***Routine health checks:*** For early detection of metabolic disorders like obesity, pancreatitis or endocrine conditions the routine health checkups should be performed regularly.
5. ***Owner education:*** Teach owners about diabetes signs, management, and the importance of consistency in treatment.

Prognosis

With proper management, many diabetic animals can live normal, healthy lives. Dogs typically require lifelong insulin therapy, while some cats may achieve remission. Early diagnosis, consistent treatment, and regular veterinary care are critical for long-term success.

Nutritional management

Nutritional management plays a vital role in controlling diabetes mellitus in animals. Proper dietary adjustments help stabilize blood glucose levels, reduce insulin resistance, and minimize complications. The nutritional approach must be tailored to the species, underlying causes, and individual needs of the animal.

1. ***Stabilize blood glucose levels***: Ensure consistent glucose absorption from the gastrointestinal tract to avoid fluctuations.
2. ***Promote weight control***: Address obesity to enhance insulin sensitivity and overall health.
3. ***Support overall health***: Provide adequate nutrients to maintain a balanced diet while addressing diabetes-related issues.

Nutritional management in dogs

1. **High-fibre, complex carbohydrate diets**:

   Fibre slows glucose absorption, reducing postprandial blood glucose spikes.

   Soluble fibres like psyllium and insoluble fibres like beet pulp are beneficial.

   Avoid simple sugars or high-glycemic index carbohydrates.

2. **Moderate protein and fat**

   Ensure adequate protein for muscle maintenance.

   Fat intake should be moderate to prevent obesity or lipid metabolism disorders.

3. **Portion control and feeding schedule**

   Consistent feeding times aligned with insulin administration optimize glucose regulation.

4. **Commercial prescription diets**

   Veterinary-recommended diabetic dog foods are often ideal for managing the condition.

Nutritional management in cats

1. ***High-protein, low-carbohydrate diets***:

   Cats are obligate carnivores, and high-protein, low-carb diets support their natural metabolism.

   Low carbohydrate intake helps stabilize glucose and reduce insulin demand.

2. ***Canned food vs. dry food***:

   Wet or canned foods are preferable as they typically contain fewer carbohydrates than dry diets.

   Moist diets may also promote hydration, reducing the risk of urinary issues.

3. ***Weight control***: Obesity is a common risk factor for diabetes in cats. Controlled feeding and gradual weight loss improve insulin sensitivity.

4. ***Frequent, small meals***: Feeding smaller and more frequent meals throughout the day should be offered to aid in glucose stabilization.

Nutritional management in horses

1. ***Low-sugar, high-fibre diets***:

   Avoid feeds high in non-structural carbohydrates (e.g., grains).

   Use hay or forage-based diets with low starch content.

2. ***Grazing control***: Restrict access to lush pastures, especially during peak sugar content periods, or use a grazing muzzle.

3. ***Supplementation***: Add omega-3 fatty acids to reduce inflammation and support metabolic health.

Effective nutritional management is a cornerstone of diabetes mellitus treatment in animals. Tailoring diets to the species and individual needs combined with regular monitoring, supports glycemic control and improves overall health. Veterinary guidance is essential to ensure optimal outcomes.

Diabetes mellitus is a manageable condition in animals with the right combination of insulin therapy, dietary adjustments, and regular monitoring. Although it requires significant commitment from pet owners, proper management improves the quality of life and prevents serious complications. Early detection, proactive care, and collaboration between veterinarians and pet owners are essential for successful outcomes.

27

Fatty Liver

Fatty liver, also known as hepatic lipidosis is a metabolic disorder commonly seen in various animal species. It occurs due to excessive accumulation of fat (lipids) in liver cells (hepatocytes), leading to impaired liver function. This condition is particularly prevalent in dairy cattle, poultry, cats, and occasionally in dogs and other species. Fatty liver can significantly impact health, productivity, and overall well-being. Nutritional management plays a crucial role in both preventing and treating this condition. Fatty liver occurs when the liver's ability to metabolize and export fat is overwhelmed, leading to fat accumulation. This can happen due to various factors, including:

Imbalance in energy metabolism

Excess energy intake or inadequate utilization of energy during critical periods (e.g., early lactation in dairy cows) results in fat mobilization to the liver.

Obesity

Overweight animals are more prone to developing fatty liver due to an increased baseline fat deposit.

Nutritional deficiencies

Lack of essential nutrients like choline, methionine, and B vitamins can impair lipid metabolism.

Stress and disease

Stress or concurrent illnesses (e.g., ketosis or diabetes) can exacerbate lipid accumulation.

Species-specific manifestations

Dairy cattle

Fatty liver is commonly observed in high-producing dairy cows during the peri-parturient period (transition from late pregnancy to early lactation). Negative energy balance due to high milk production demands results in excessive fat mobilization, overwhelming the liver's metabolic capacity.

Poultry

Overfed broilers and layers may develop fatty liver due to excessive dietary energy or low protein intake, often exacerbated by heat stress or genetic predisposition.

Cats

Obese cats are prone to hepatic lipidosis, particularly during periods of anorexia or fasting, leading to rapid fat mobilization.

Other animals

In dogs, fatty liver is less common but may occur due to obesity, malnutrition, or metabolic disorders.

Pathophysiology of fatty liver

Fat mobilization from adipose tissue leads to an influx of non-esterified fatty acids (NEFAs) into the liver. Normally, these are oxidized for energy or converted into triglycerides and exported as very low-density lipoproteins (VLDLs). However, when the liver's capacity to oxidize or export fats is overwhelmed, triglycerides accumulate in hepatocytes, causing fatty liver. This can impair liver function, affecting protein synthesis, bile production, and detoxification. In severe cases, liver failure and systemic complications may occur.

Clinical signs

The symptoms of fatty liver depend on its severity and species but often include:

General symptoms: Lethargy, reduced appetite, weight loss, jaundice, and abdominal pain.

Species-specific signs

Dairy cows: Decreased milk production, poor appetite, and secondary infections.

Poultry: Reduced egg production, obesity, and sudden death.

Cats: Vomiting, diarrhoea, and severe lethargy.

Diagnosis

Diagnosis involves a combination of clinical history, physical examination, and diagnostic tests, including:

1. ***Blood tests***: Elevated liver enzymes (ALT, AST), bilirubin, and triglycerides.

2. ***Liver biopsy***: Confirms fat accumulation in hepatocytes.
3. ***Ultrasound or imaging***: Visualizes changes in liver size and texture.

Nutritional management

Nutritional management is the cornerstone of preventing and treating fatty liver. It aims to reduce fat mobilization, enhance fat metabolism, and support liver health.

Dairy cattle

1. ***Transition diets***

   Provide a balanced transition diet to minimize negative energy balance during the peri-parturient period.

   Increase energy density using highly digestible carbohydrates and fats (e.g., protected fats).

2. ***Protein management***

   Supply adequate protein to support milk production without promoting excessive fat mobilization.

   Use rumen-protected amino acids like methionine to support liver function.

3. ***Feed additives***

   Choline: Supplement rumen-protected choline to enhance VLDL synthesis and fat export from the liver.

   Niacin: Reduces fat mobilization and supports liver metabolism.

   Probiotics: Improve rumen health and nutrient utilization.

Poultry

1. ***Energy-protein balance***: Avoid overfeeding energy-rich diets, especially in broilers. Maintain an appropriate energy-to-protein ratio.
2. ***Essential nutrients***: Supplement diets with methionine, choline, and biotin to support lipid metabolism.
3. ***Fat quality***: Use high-quality fats in feed to ensure proper digestion and utilization. Avoid rancid fats.
4. ***Environmental management***: Mitigate heat stress by providing cool water, adequate ventilation, and electrolytes.

Cats

1. ***Weight management***

   Prevent obesity through portion control and balanced diets.

   Gradual weight loss plans for overweight cats to avoid sudden fat mobilization.

2. ***Feeding strategies***

   High-protein, moderate-fat diets are recommended.

   Ensure continuous food availability to prevent prolonged fasting.

3. ***Essential nutrients***: Supplement taurine, L-carnitine, and B vitamins to support liver health.

Other Animals

1. **Dogs:** Low-fat, high-quality protein diets with added omega-3 fatty acids to reduce inflammation and support liver function.
2. **Small ruminants**: Provide balanced energy and protein intake, especially during late pregnancy and lactation.

General management practices

1. ***Monitor body condition***: Maintain optimal body condition scores (BCS) to prevent obesity or excessive weight loss.
2. ***Regular health checks***: Monitor liver health through blood tests, particularly in high-risk animals.
3. ***Stress reduction***: Minimize environmental and physiological stress to reduce metabolic strain.
4. ***Adequate forage and fibre***: Ensure ruminants receive adequate fiber to maintain rumen health and prevent metabolic disorders.

Treatment of fatty liver

In severe cases, treatment may require medical intervention in addition to nutritional management:

1. ***Intravenous fluids***: To correct dehydration and electrolyte imbalances.
2. ***Medications***: Administer glucose precursors (e.g., propylene glycol), insulin, or corticosteroids in severe cases.
3. ***Nutritional Support***: Force-feeding or tube feeding in anorexic cats to prevent further fat mobilization.

Prevention strategies

1. ***Early intervention***: Identify at-risk animals early through regular monitoring.
2. ***Customized diets***: Tailor diets based on species, life stage, and metabolic demands.
3. ***Education***: Train animal caretakers to recognize early signs of fatty liver and implement preventive measures.

Fatty liver is a multifactorial condition influenced by genetics, diet, and management practices. Nutritional strategies play a vital role in preventing and managing the disorder. A comprehensive approach that includes balanced diets, appropriate supplementation, and stress reduction can significantly improve outcomes. Early recognition and intervention are key to minimizing the impact of fatty liver on animal health and productivity.

28

Diseases of Calcium Disorders

Calcium is a vital mineral playing an essential role in numerous physiological processes, including bone formation, muscle contraction, nerve impulse transmission, and blood clotting. Proper calcium metabolism in animals is regulated by a complex interplay of dietary intake, hormonal control (mainly by parathyroid hormone; PTH, calcitonin, and vitamin D), and organ systems such as the kidneys, intestines, and bones. Imbalances in calcium homeostasis, whether due to dietary deficiencies, genetic predispositions, hormonal dysfunctions, or systemic diseases, can lead to a wide range of disorders affecting the skeletal system and overall health.

Calcium-related disorders in animals can broadly be classified into those resulting from calcium deficiency, excess, or abnormalities in calcium regulation. Among these, skeletal manifestations are the most pronounced, as the bones act as a primary reservoir for calcium in the body. Various diseases related to calcium disorders and their nutritional management are discussed below:

a) Rickets

Rickets is a skeletal disorder that affects young animals and birds, characterized by improper bone mineralization. This condition occurs primarily due to deficiencies in vitamin D, calcium, and phosphorus, which are essential for proper bone development. Rickets can lead to deformities, pain, and reduced growth rates, and understanding its causes, symptoms, mechanism, and nutritional management is crucial for prevention and treatment.

Causes of rickets

The development of rickets in young animals and birds is mainly linked to deficiencies or imbalances in calcium, phosphorus, and vitamin D. These nutrients are critical for bone growth and mineralization.

Vitamin D deficiency

Vitamin D is essential for calcium and phosphorus absorption in the intestines. Young animals and birds that don't receive adequate sunlight (or UVB light) or dietary vitamin D are at risk.

Indoor animals and birds or those raised in areas with limited sunlight, are particularly susceptible to vitamin D deficiency.

Calcium deficiency

Calcium is a primary mineral in bones, and a lack of calcium in the diet leads to poor bone density.

Growing animals and birds with high nutritional demands for skeletal development are especially at risk if their diets lack adequate calcium.

Phosphorus deficiency or imbalance

Phosphorus works closely with calcium to form strong bones. Both calcium and phosphorus need to be present in the correct ratio; an imbalance can disrupt mineral absorption and utilization. A diet too rich in calcium can interfere with phosphorus absorption, and vice versa, leading to rickets.

Mal-absorption or metabolic disorders

Conditions that affect nutrient absorption or metabolism, such as gastrointestinal disorders, can lead to deficiencies in vitamin D, calcium or phosphorus.

Genetic predisposition

Certain breeds or species may have a higher genetic susceptibility to rickets, especially if they have rapid growth rates or specific metabolic requirements.

Symptoms of rickets

Early recognition of rickets is essential, as symptoms are more manageable in the initial stages. The severity and presentation of symptoms vary depending on the degree of deficiency and the animal's species, but common signs include:

Skeletal deformities: One of the most characteristic symptoms, these may include bowed legs, enlarged joints, and a softened skull in young animals.

Bone pain and weakness: Affected animals and birds may show signs of discomfort or pain, particularly in their legs, and they may be reluctant to move or bear weight.

Lameness and difficulty moving: As bones weaken, animals may become lame or show abnormal gaits.

Poor growth or stunted development: Rickets often leads to slow growth, as the body diverts energy toward maintaining existing bone structure.

Muscle weakness: Muscle weakness accompanies rickets, as bones fail to provide adequate support.

Fractures: Weakened bones can fracture easily, even with minor trauma.

Swollen or thickened joints: This is common in the legs, and affected birds may have enlarged wrists and ankles.

Beak and claw softening (birds): In birds, rickets can cause softening of the beak and claws due to insufficient mineralization.

These symptoms should prompt immediate veterinary consultation to prevent further complications.

Mechanism of action in rickets

The underlying mechanism of rickets revolves around disrupted bone mineralization, resulting from imbalances in calcium, phosphorus, and vitamin D.

Vitamin D and calcium absorption

Vitamin D obtained through sunlight exposure or dietary intake, is crucial for absorbing calcium and phosphorus in the intestines. In the absence of adequate vitamin D, calcium absorption decreases, leading to hypocalcaemia (low blood calcium). When blood calcium levels drop, the body compensates by drawing calcium from bones, leading to weakened, demineralized bones.

Calcium-phosphorus imbalance

A proper balance of calcium and phosphorus is necessary for bone mineralization. The ideal calcium-to-phosphorus ratio varies slightly among species, but it typically ranges from 1.2:1 to 2:1. If calcium is excessively high relative to phosphorus, or vice versa, the imbalance reduces bone mineral density, leading to soft and weak bones.

Inadequate bone matrix formation

Rickets affects the synthesis of the bone matrix, which provides the scaffold for mineral deposition. This matrix cannot mineralize effectively in the absence of sufficient calcium, phosphorus, or vitamin D, resulting in softened and structurally compromised bones.

Cartilage abnormalities

In young animals, rickets disrupts the normal growth and ossification of cartilage, leading to enlarged joints and weakened structural integrity in weight-bearing bones. This can cause severe deformities, as the growth plates (responsible for bone elongation) fail to calcify properly.

Nutritional management for rickets

Nutrition plays a critical role in the prevention and management of rickets in young animals and birds. Effective dietary strategies focus on ensuring balanced intake and proper bioavailability of calcium, phosphorus and vitamin D.

Vitamin D supplementation

For animals and birds with limited sunlight exposure, dietary vitamin D supplements are essential. Supplements should provide vitamin D_3, as it is the most effective form for raising blood calcium levels. Poultry, in particular, benefit from vitamin D-fortified feeds, as indoor or confined birds often lack adequate UVB exposure. Caution should be taken to avoid excessive vitamin D, as it can lead to toxicity. Veterinary guidance is recommended for dosage.

Balanced calcium and phosphorus intake

Providing a diet with an appropriate calcium-to-phosphorus ratio is essential. This ratio should typically range from 1.2:1 to 2:1 for optimal bone health. Young animals and birds with high growth rates, such as fast-growing poultry breeds, may need additional calcium in their diet. Feed manufacturers often formulate specific growth-stage diets with the correct mineral ratios. Sources of calcium include ground limestone, dicalcium phosphate or calcium carbonate. However, these should be included based on professional nutritional recommendations to avoid imbalances.

Quality protein sources

High-quality protein is necessary to support bone growth and repair, as amino acids contribute to the bone matrix. Deficiency in protein can exacerbate the effects of rickets. Animal-based proteins are particularly beneficial for omnivorous species, providing better digestibility and bioavailability of essential amino acids for bone development.

Supplemental phosphorus for bone mineralization

Dietary phosphorus is available through feeds like grains, which are naturally high in phosphorus, but supplementation may be required if deficiencies are noted. Phosphorus is sometimes bound to compounds like phytate in plant-based diets, which reduces its bioavailability. Adding phytase, an enzyme that breaks down phytates, can improve phosphorus absorption.

Feed quality and consistency

Consistent feed quality is important to avoid fluctuations in nutrient levels that could lead to deficiencies. In environments where forage is limited or unavailable, formulated feeds can provide the necessary nutrients. Feeding practices should minimize environmental contaminants and ensure all animals have access to feed.

Use of UVB lighting (for indoor birds)

Birds, in particular, rely on UVB light to synthesize vitamin D in their skin. For indoor birds that lack natural sunlight exposure, providing artificial UVB lighting can help maintain vitamin D levels and prevent rickets. Lighting should mimic natural sunlight exposure, ideally for 8-12 hours daily, to support healthy vitamin D synthesis.

Rickets is a preventable condition in young animals and birds, primarily resulting from deficiencies in vitamin D, calcium, and phosphorus, along with imbalances in their dietary intake. Recognizing the symptoms early, understanding the mechanisms behind rickets, and implementing appropriate nutritional management can prevent or mitigate this condition effectively. By ensuring proper dietary balance, access to UVB lighting (for birds), and consistent feed quality, caretakers can support optimal bone health and prevent the onset of rickets. Regular veterinary checkups and dietary adjustments are recommended to cater to each animal's unique needs, fostering healthy growth and development in young animals and birds.

b) Osteomalacia

Osteomalacia is a metabolic bone disease characterized by softening of the bones due to improper mineralization, predominantly affecting adult animals and birds. This condition is often linked to deficiencies in vitamin D, calcium, or phosphorus. Unlike rickets, which affects growing animals, osteomalacia occurs after growth plates have closed, leading to weakened bones, pain, and an increased risk of fractures. Here, we'll explore the causes, symptoms, mechanisms and nutritional management of osteomalacia.

Causes of osteomalacia

Osteomalacia in adult animals and birds is primarily due to nutritional deficiencies, environmental factors, and occasionally metabolic disorders. The main causes include:

Vitamin D deficiency

Vitamin D plays a critical role in calcium and phosphorus metabolism, facilitating the absorption of these minerals in the intestines. Adult animals that don't receive enough sunlight or dietary vitamin D are at risk of developing osteomalacia.

Vitamin D deficiencies are more common in animals and birds kept indoors, in regions with limited sunlight, or in environments where UVB exposure is inadequate.

Calcium deficiency

Calcium is a fundamental mineral for bone health. When dietary calcium intake is insufficient, the body compensates by mobilizing calcium from bones, leading to demineralization and bone softening.

This deficiency may be due to low calcium content in feed or imbalances in calcium and phosphorus ratios.

Phosphorus deficiency or imbalance

Phosphorus is essential for proper bone mineralization. Deficiencies or imbalances in calcium-to-phosphorus ratios can disrupt bone structure.

An excessive intake of calcium relative to phosphorus (or vice versa) can interfere with mineral absorption, leading to osteomalacia.

Metabolic disorders and malabsorption

Disorders affecting the absorption of calcium, phosphorus or vitamin D (such as certain gastrointestinal diseases) can contribute to osteomalacia. Liver and kidney diseases that impair vitamin D metabolism also elevate the risk, as these organs are involved in activating vitamin D.

Chronic acidosis

Conditions that lead to metabolic acidosis (high acidity in body fluids) may increase bone demineralization, as the body uses calcium from bones to buffer the blood pH.

Symptoms of osteomalacia

Recognizing the symptoms of osteomalacia is crucial for timely intervention. Symptoms vary but generally include signs of bone weakness and discomfort:

Bone pain: Animals with osteomalacia often experience pain, particularly in weight-bearing bones. Pain may manifest as general discomfort, especially when moving or standing.

Weakness and reluctance to move: Affected animals may show reluctance to walk or exercise due to bone pain and muscular weakness, which often accompanies osteomalacia.

Increased fracture risk: Bones become fragile and more susceptible to fractures, even from minor impacts or normal activity.

Abnormal posture or gait: Animals may develop an abnormal stance or gait to compensate for weakened bones, particularly if the condition affects the spine or legs.

Reduced egg production and soft-shelled eggs (birds): Birds with osteomalacia may lay fewer eggs, and the eggshells may be thin or soft due to poor calcium deposition.

Muscle weakness: Muscle contractions are reliant on calcium, so calcium deficiency can contribute to general muscle weakness.

Skeletal deformities: In severe cases, prolonged osteomalacia can lead to skeletal deformities as bones attempt to adapt to the lack of mineral support.

These symptoms are indicators of osteomalacia and warrant prompt veterinary assessment to avoid further complications.

Mechanism of action in osteomalacia

Osteomalacia develops due to impaired mineralization of the bone matrix, primarily caused by imbalances in vitamin D, calcium and phosphorus.

Role of vitamin D in calcium and phosphorus absorption

Vitamin D is crucial for the intestinal absorption of calcium and phosphorus. Without sufficient vitamin D, calcium absorption decreases, leading to low blood calcium levels (hypocalcemia). To maintain blood calcium levels, the body releases parathyroid hormone (PTH), which mobilizes calcium from bones, leading to demineralization and weakened bones.

Calcium and phosphorus imbalance

Calcium and phosphorus work in tandem for proper bone mineralization. Osteomalacia occurs when the balance between these minerals is disrupted, resulting in poorly mineralized bone matrix. An excess of calcium can limit phosphorus availability, while excess phosphorus can bind calcium and reduce its bioavailability, impairing mineral deposition in the bone matrix.

Failure to mineralize the bone matrix

Osteoblasts in bones produce a collagen matrix, which requires adequate calcium and phosphorus to mineralize properly. In osteomalacia, the lack of these minerals leaves the matrix soft and prone to deformation. This mineralization defect affects adult bone structure, leading to softening and reduced mechanical strength, characteristic of osteomalacia.

Nutritional management for osteomalacia

Proper nutritional management is fundamental for preventing and treating osteomalacia in adult animals and birds. The goal is to provide balanced nutrients that support bone health and restore mineral balance.

Adequate vitamin D supplementation

Vitamin D supplementation is vital, particularly for animals or birds with limited sunlight exposure. Ensuring adequate levels of vitamin D3 is essential, as this form is more bioavailable and effectively supports calcium and phosphorus absorption. Outdoor access, where possible, or UVB lighting for indoor animals and birds can help stimulate natural vitamin D production. Birds, in particular, rely on UVB exposure, so appropriate lighting setups are recommended.

Balanced calcium and phosphorus intake

Feed should be formulated to provide a balanced calcium-to-phosphorus ratio, generally around 1.2:1 to 2:1, depending on the species' specific requirements. Sources of calcium, such as limestone or calcium carbonate, can be added to the diet. For phosphorus, di-calcium phosphate or bone meal is often used. Supplements should be chosen to maintain this balance to prevent imbalances that could exacerbate osteomalacia.

Supplementation with calcium-rich feed

High-calcium feed sources, including legumes and certain mineral supplements, help improve bone density. In birds, calcium supplements like oyster shells or cuttlebone can promote bone health and strengthen eggshells. Animals with higher demands, such as lactating females or egg-laying birds, may require additional calcium to offset losses and support skeletal health.

Ensuring protein quality and digestibility

Quality protein is important, as amino acids provide the foundation for collagen synthesis in the bone matrix. A balanced protein intake supports both muscle and bone strength, reducing stress on the skeletal system. High-quality sources of protein include animal by-products, soy meal, or specific amino acid supplements, which should be included, based on the animal's dietary needs.

Supporting phosphorus bioavailability

Phosphorus in plant-based feeds is often bound to phytate, reducing bioavailability. Supplementing with phytase enzymes can improve phosphorus absorption, especially in diets relying heavily on grains. Animal-based phosphorus sources, such as meat or bone meal, are more bioavailable and may be preferred in diets for carnivorous or omnivorous species.

Corrective measures for metabolic and absorptive issues

For animals with gastrointestinal disorders or metabolic diseases, correcting underlying issues is essential for improving nutrient absorption. Probiotic supplementation or adjustments in diet consistency (such as softened feed) can help improve nutrient uptake. In cases where vitamin D metabolism is impaired (e.g., liver or kidney issues), specialized forms of vitamin D or veterinary-prescribed supplements may be needed.

Monitoring environmental factors

Indoor animals and birds should be exposed to UVB lighting to support natural vitamin D synthesis, especially during winter months or in low-sunlight environments. For herbivores, providing quality forage helps ensure a natural intake of minerals. Regularly testing feed for calcium and phosphorus content helps in fine-tuning dietary management.

Osteomalacia in adult animals and birds is a condition marked by bone softening due to deficiencies or imbalances in vitamin D, calcium, and phosphorus. Recognizing symptoms such as bone pain, weakness, and fractures early can prompt effective interventions. Nutritional management is key in preventing and treating osteomalacia, focusing on providing balanced calcium and phosphorus, adequate vitamin D and quality protein to support bone health. By addressing dietary needs, environmental factors, and any metabolic issues, caretakers can significantly reduce the risk of osteomalacia, ensuring stronger, healthier bones in adult animals and birds.

c) Osteoporosis

Osteoporosis is a metabolic bone disorder characterized by reduced bone mass, decreased bone density, and deterioration of bone microarchitecture, leading to bone fragility and an increased risk of fractures. Unlike other calcium disorders, osteoporosis involves a loss of both bone mineral and organic matrix, making bones brittle and prone to fractures. While commonly associated with aging animals, osteoporosis can also arise from nutritional deficiencies, hormonal imbalances, lack of physical activity, or chronic diseases.

Proper nutritional management plays a vital role in the prevention, management, and improvement of osteoporosis in animals.

Pathophysiology

The development of osteoporosis is attributed to an imbalance in bone remodeling, a dynamic process where bone formation by osteoblasts is outpaced by bone resorption by osteoclasts. This imbalance can result from several factors:

1. ***Calcium deficiency***: Insufficient dietary calcium leads to increased bone resorption to maintain serum calcium levels.
2. ***Vitamin D deficiency***: Reduces calcium absorption from the intestine, exacerbating calcium deficiency.
3. ***Hormonal imbalances***

   Estrogen deficiency (in post-reproductive females) reduces the protective effect on bones.

   Excess glucocorticoids (e.g., in Cushing's disease) promote bone resorption and inhibit bone formation.
4. ***Physical inactivity***: Lack of mechanical loading reduces bone density.
5. ***Chronic diseases***: Conditions such as chronic kidney disease (CKD) disrupt calcium-phosphorus homeostasis and bone metabolism.

Clinical manifestations of osteoporosis

Osteoporosis in animals may remain asymptomatic until significant bone loss occurs. Common clinical signs include:

1. ***Fragility fractures***: Spontaneous or low-trauma fractures, commonly in long bones or the spine.
2. ***Bone pain and lameness***: Resulting from microfractures or structural weakening.
3. ***Skeletal deformities***: In advanced cases, leading to kyphosis or scoliosis.
4. ***Reduced mobility***: Associated with pain, fractures, or joint stiffness.
5. ***Growth retardation***: In young animals due to inadequate bone development.

Diagnosis

Diagnosing osteoporosis involves a combination of clinical history, imaging, and biochemical tests:

1. ***Imaging***:

   Radiographs: Show reduced bone density, cortical thinning, and fractures.

   Dual-energy X-ray absorptiometry (DEXA): Measures bone mineral density (BMD) and is the gold standard for diagnosis.
2. ***Blood tests***: Normal or low serum calcium and phosphorus levels. Elevated parathyroid hormone (PTH) in cases of secondary hyperparathyroidism.
3. ***Histopathology***: Bone biopsy reveals reduced trabecular density and loss of cortical bone.

Treatment goals

The primary objectives in managing osteoporosis are to:

1. Prevent further bone loss.
2. Promote bone remodeling and increase bone mass.
3. Minimize the risk of fractures.
4. Address underlying causes (e.g., hormonal or nutritional imbalances).

Nutritional management of osteoporosis

Nutritional strategies are critical for maintaining bone health and managing osteoporosis. Proper diet and supplementation can restore calcium-phosphorus balance, improve bone mineralization, and support overall skeletal strength.

1. Calcium

Calcium is the primary mineral in bones, and its adequacy is essential for bone strength and structure.

Dietary sources

Dairy products: Milk, cheese, and yogurt.

Calcium-fortified foods: Cereals and plant-based milk.

Leafy greens: Kale, broccoli, and turnip greens.

Calcium supplements

Calcium carbonate or calcium citrate may be used in animals with dietary insufficiency.

Avoid excessive calcium, as it can interfere with phosphorus absorption.

Recommended calcium intake: Varies by species, age, and physiological state. Ensure a balanced calcium-to-phosphorus ratio of 1.2:1 to 2:1.

2. Vitamin D

Vitamin D is crucial for calcium absorption from the gastrointestinal tract and its deposition in bones.

Natural sources

Sunlight exposure: Encourages endogenous vitamin D synthesis in the skin.

Dietary sources: Fatty fish (e.g., salmon, mackerel), liver, and egg yolks.

Vitamin D supplementation

Cholecalciferol (vitamin D_3) is the preferred form for supplementation.

Dosage should be carefully monitored to avoid hypercalcemia.

3. Phosphorus

Phosphorus is a vital component of hydroxyapatite crystals in bone. However, excessive phosphorus can impair calcium absorption and exacerbate osteoporosis.

Balanced phosphorus intake

Avoid phosphorus-rich diets (e.g., organ meats, grains) in animals with calcium deficiency. Maintain an appropriate calcium-to-phosphorus ratio.

Phosphorus binders

Used in animals with CKD to reduce serum phosphorus levels.

4. Protein

Protein is necessary for collagen synthesis, a key component of the bone matrix.

High-quality protein sources

Lean meats, fish, eggs, and dairy products for omnivorous or carnivorous animals.

For herbivores, legumes, and high-quality forage (e.g., alfalfa).

Moderate protein intake

Avoid excessive protein, as it increases urinary calcium excretion.

5. Magnesium

Magnesium supports bone mineralization and calcium metabolism.

Sources:

Nuts, seeds, whole grains, and green leafy vegetables.

Magnesium supplementation may be necessary in cases of deficiency.

6. Trace minerals

Zinc, copper, and manganese are important cofactors in bone matrix synthesis and remodelling.

Dietary sources

Zinc: Found in meat, seafood, and whole grains.

Copper: Present in shellfish, nuts, and legumes.

Manganese: Found in nuts, seeds, and leafy greens.

7. Omega-3 fatty acids

Omega-3 fatty acids have anti-inflammatory properties that may reduce bone resorption.

Sources

Fish oil, flaxseed oil, and chia seeds.

8. Energy requirements

Osteoporotic animals often have increased energy needs to support bone remodeling.

Calorie-dense diets: Provide adequate calories without promoting obesity, which adds stress to weakened bones.

9. Hydration

Hydration supports metabolic functions and kidney health, aiding in calcium and phosphorus balance.

Encourage water intake: Fresh water should always be available.

Supportive therapies

Nutritional management should be complemented by other therapeutic interventions:

1. ***Medications***

   Bisphosphonates: Inhibit bone resorption and increase bone density.

   Hormone therapy: Estrogen replacement in post-reproductive females.

   Calcitonin: Reduces bone resorption and alleviates pain.

2. ***Physical activity***

   Weight-bearing exercises stimulate bone formation and improve bone strength.

3. ***Environmental modifications***

   Prevent falls and fractures by providing non-slippery surfaces and minimizing jumping or high-impact activities.

4. ***Regular monitoring***

   Periodic assessment of blood calcium, phosphorus and vitamin D levels should be monitored.

   Imaging to monitor bone density and healing.

Prognosis

The prognosis for osteoporosis in animals depends on the severity of bone loss and the timeliness of intervention. Early nutritional and medical management can significantly improve bone health and reduce the risk of fractures, while advanced cases may require long-term supportive care.

Osteoporosis is a preventable and manageable condition in animals when addressed through a comprehensive approach, with nutritional management being a cornerstone of therapy. Ensuring adequate calcium, vitamin D, and other bone-healthy nutrients, combined with proper medical care and lifestyle modifications, can improve bone strength, reduce fracture risk, and enhance the quality of life for affected animals.

d) Osteodystrophy fibrosa

Osteodystrophy fibrosa (OF), also known as fibrous osteodystrophy or secondary hyperparathyroidism, is a metabolic bone disorder characterized by bone resorption and replacement of bone tissue with fibrous connective tissue. It results from prolonged disturbances in calcium, phosphorus, and vitamin D metabolism, often secondary to nutritional imbalances or systemic diseases affecting calcium homeostasis. This condition is most commonly seen in growing animals and reptiles, though it can also affect adult animals.

Proper nutritional management, alongside addressing underlying causes, plays a crucial role in both the prevention and treatment of osteodystrophyfibrosa.

Pathophysiology

Osteodystrophy fibrosa is primarily caused by prolonged stimulation of the parathyroid glands, leading to excessive secretion of parathyroid hormone (PTH). This condition is typically classified into:

1. ***Nutritional secondary hyperparathyroidism (NSHP)***: The NSHP is caused by diets deficient in calcium, excessive in phosphorus, or both. It is most common in young, growing animals, exotic pets, and reptiles on improper diets.
2. ***Renal secondary hyperparathyroidism (RSHP)***: The RSHP is associated with chronic kidney disease (CKD), which impairs calcium and phosphorus balance and disrupts vitamin D metabolism.

Key processes in OF development:

1. ***Low blood calcium levels (Hypocalcemia)***

   Inadequate dietary calcium or poor calcium absorption due to vitamin D deficiency leads to hypocalcemia.

   Hypocalcemia triggers the release of PTH, stimulating bone resorption.

2. ***High blood phosphorus levels (Hyperphosphatemia)***: Excess phosphorus in the diet or reduced phosphorus excretion (e.g., in CKD) exacerbates hypocalcemia by suppressing calcium absorption.
3. ***Bone resorption and fibrosis***: Persistent PTH stimulation results in excessive bone resorption and replacement of bone tissue with fibrous connective tissue. This weakens the bones, making them prone to fractures and deformities.

Clinical manifestations

Animals with osteodystrophy fibrosa present with a range of clinical signs depending on the severity of the disease:

1. ***Skeletal deformities***: Bowed limbs, spinal curvature, and an enlarged head observed in reptiles. Mandibular swelling (rubber jaw) is a hallmark sign in dogs, cats, and reptiles.
2. ***Bone weakness and fractures***

   Pathological fractures from minimal trauma.

   Lameness or reluctance to move
3. ***Muscle weakness***: Difficulty standing or walking due to poor bone support and muscle atrophy.
4. ***Dental problems***: Loose teeth or difficulty chewing due to mandibular bone resorption.
5. ***Neurological signs***: Rarely, tetany or seizures from severe hypocalcemia.

Diagnosis

Diagnosing osteodystrophy fibrosa involves a combination of clinical, radiographic, and biochemical assessments:

1. ***Clinical signs***: Observable skeletal deformities and weakness.
2. ***Radiographs***: Generalized osteopenia (reduced bone density), thin cortices and fractures, and ***Floating teeth*** appearance in cases of mandibular resorption.
3. ***Blood tests***:

   Hypocalcemia (low serum calcium),

   Hyperphosphatemia (high serum phosphorus),

   Elevated PTH levels (confirming secondary hyperparathyroidism),
4. ***Dietary history***: A history of imbalanced feeding, such as diets low in calcium or high in phosphorus.

Treatment goals

Treatment focuses on correcting the underlying cause, restoring calcium-phosphorus balance, and promoting bone healing:

1. ***Reduce PTH stimulation***: Restore normal calcium and phosphorus levels to suppress PTH secretion.
2. ***Promote bone remodelling***: Provide sufficient nutrients to support new bone formation.
3. ***Alleviate symptoms***: Address pain and improve mobility.

Nutritional management of osteodystrophy fibrosa

Nutritional management is the cornerstone of treating osteodystrophyfibrosa. Correcting dietary imbalances can halt disease progression and facilitate recovery.

1. Optimize calcium intake

Adequate calcium intake is critical to correcting hypocalcemia and reducing PTH stimulation:

Dietary sources:

High-calcium foods such as dairy products, dark leafy greens (e.g., kale, collards), and fortified feeds. Calcium supplements (e.g., calcium carbonate or calcium gluconate) may be necessary for severe deficiencies.

Avoid calcium deficiency:

Monitor dietary calcium levels and avoid prolonged feeding of calcium-deficient diets.

2. Regulate phosphorus intake

Excessive dietary phosphorus exacerbates hypocalcemia and PTH stimulation:

Low-phosphorus diet: Avoid high-phosphorus foods such as meat, organ meats and grains. Feed diets with a proper calcium-to-phosphorus ratio (1.2:1 to 2:1).

Phosphorus binders: In animals with CKD or severe hyperphosphatemia, phosphorus binders (e.g., aluminum hydroxide) may be used to reduce intestinal absorption of phosphorus.

3. Vitamin D supplementation

Vitamin D is essential for calcium absorption and bone mineralization:

Sources of vitamin D

Sunlight exposure especially in reptiles is to stimulate endogenous vitamin D production. Dietary sources of vitamin D such as fish oil, egg yolks and liver should be incorporated in their diets.

Vitamin D supplements: Calcitriol (active vitamin D) is commonly prescribed in cases of severe deficiency or CKD. Avoid overdosing to prevent vitamin D toxicity (hypercalcemia).

4. Protein management

Protein plays a supporting role in bone health by promoting collagen formation and bone matrix synthesis:

High-quality protein sources: Include lean meats, fish, eggs, and legumes in omnivorous or carnivorous animals.

For herbivores, focus on balanced forage and supplemental protein sources (e.g., alfalfa).

Avoid excessive protein: Overfeeding protein can increase phosphorus levels, worsening hyperphosphatemia.

5. Magnesium and trace minerals

Magnesium and trace minerals are important cofactors for bone health:

Magnesium: Include magnesium-rich foods such as green vegetables, nuts, and seeds. Supplements may be prescribed if magnesium levels are low.

Zinc and copper: Zinc and copper are necessary for collagen synthesis and bone mineralization. They can be supplemented as part of a balanced mineral mix.

6. Energy requirements

Animals with OF often have increased energy needs due to bone remodeling:

Balanced energy intake: Provide a calorie-dense diet to support recovery without causing obesity.

7. Hydration

Maintaining hydration is essential for kidney health and calcium-phosphorus regulation:

Encourage water intake: Ensure animals have access to clean, fresh water. Wet diets or electrolyte solutions may be used if dehydration is a concern.

8. Supplementation with omega-3 fatty acids

Omega-3 fatty acids have anti-inflammatory properties that may help reduce bone resorption:

Sources:

Fish oil, flaxseed or chia seeds.

Monitoring and long-term management

Regular monitoring is essential to evaluate the effectiveness of nutritional and medical interventions:

1. ***Blood tests***: Monitor calcium, phosphorus, and PTH levels periodically.
2. ***Radiographs***: Assess bone density and healing over time.
3. ***Diet adjustments:*** Modify the diet based on progress and any emerging nutritional deficiencies or excesses.

Prognosis

The prognosis for animals with osteodystrophyfibrosa depends on the severity of the disease and how early it is treated. With prompt correction of nutritional imbalances and adequate care, many animals can recover bone strength and function. In chronic or severe cases, skeletal deformities may persist, but pain and mobility can often be managed.

Osteodystrophy fibrosa is a preventable and treatable disorder when addressed through proper nutrition and management. Ensuring balanced calcium and phosphorus intake, adequate vitamin D levels, and overall dietary optimization is critical in restoring bone health and improving the quality of life for affected animals. Combined with medical therapies and regular monitoring, nutritional interventions form the foundation of a successful treatment plan for this condition.

e) Osteogenesis imperfacta

Osteogenesis Imperfecta (OI), also known as brittle bone disease, is a rare congenital disorder characterized by defective collagen synthesis, resulting in fragile bones that are prone to fractures. It is a genetic condition that primarily affects the structural integrity of bone tissue, but it may also involve teeth, joints, and other connective tissues. While the condition is incurable, nutritional management plays a critical role in mitigating symptoms, improving bone health, and enhancing the quality of life for affected animals.

Pathophysiology

Osteogenesis Imperfecta (OI) is caused by mutations in genes responsible for producing type I collagen, which is a critical component of bone matrix and connective tissue. The most common genes implicated are $COL1A_1$ and $COL1A_2$. Mutations in these genes lead to:

1. ***Reduced collagen production (quantitative defect)***: Decreased collagen synthesis results in weak and poorly mineralized bones.
2. ***Abnormal collagen structure (qualitative defect)***: Defective collagen formation compromises bone strength and flexibility, making bones brittle.

Key effects of OI

Fragile bones prone to fractures even under normal physiological stress,

Bone deformities, such as bowing of limbs or spinal curvature,

Joint laxity due to weakened connective tissues,

Dentinogenesis imperfecta, a condition affecting teeth, leading to discoloration and increased susceptibility to dental fractures.

Clinical manifestations

The severity of OI varies widely, ranging from mild forms with occasional fractures to severe cases involving multiple fractures and deformities. Common signs include:

Recurrent fractures: Often occurring with minimal trauma.

Skeletal deformities: Bowed limbs, spinal scoliosis, or kyphosis.

Growth retardation: Short stature due to poor bone development.

Joint laxity: Leading to increased risk of sprains or dislocations.

Blue sclerae: A bluish tinge in the sclera due to thin collagen layers.

Dental abnormalities: Brittle, discolored teeth prone to damage.

Diagnosis of osteogenesis imperfecta

Diagnosis involves a combination of clinical evaluation, genetic testing, and imaging studies:

1. ***Radiographic findings***

   Generalized osteopenia (reduced bone density),

   Thin cortices and widened medullary canals in long bones,

   Multiple healed or healing fractures,
2. ***Genetic testing***: Identification of mutations in the COL1A1 or COL1A2 genes,
3. ***Biochemical tests***:

   Normal or mildly elevated serum calcium and phosphorus levels,

   Normal alkaline phosphatase activity,
4. ***Family history***: OI often follows an autosomal dominant inheritance pattern.

Treatment goals

There is no cure for OI, but treatment focuses on:

Preventing fractures and minimizing deformities,

Enhancing bone strength and mineralization,

Providing supportive care for overall health and mobility.

Nutritional management of osteogenesis imperfecta

Nutritional strategies are critical in supporting bone health and preventing complications associated with OI. While dietary modifications alone cannot cure OI, they play a complementary role alongside medical and physical therapies.

1. Calcium intake

Calcium is essential for bone mineralization and strength. Adequate dietary calcium can improve bone density and reduce the risk of fractures.

Recommended calcium sources

Dairy products: Milk, yogurt, and cheese.

Leafy greens: Kale, broccoli, and turnip greens.

Calcium-fortified foods: Cereals and plant-based milk.

Bone meal supplements (for animals) under veterinary supervision.

Supplementation: Calcium carbonate or calcium citrate supplements may be prescribed in cases of dietary insufficiency.

2. Vitamin D supplementation

Vitamin D is crucial for calcium absorption and utilization in the body. Deficiency can exacerbate bone fragility.

Natural sources:

Sunlight exposure: Encouraging outdoor activities to promote endogenous vitamin D synthesis in the skin.

Foods: Egg yolks, fatty fish (e.g., salmon, mackerel), and liver.

Supplementation

Vitamin D_3 (cholecalciferol) is preferred for its superior efficacy in increasing serum vitamin D levels.

Dosage should be monitored to prevent toxicity (hypercalcemia).

3. Protein

Collagen is a protein-based structure, making protein intake critical for bone and connective tissue health.

High-quality protein sources

Lean meats, fish, eggs, and dairy products.

Plant-based proteins like lentils and soy products for herbivorous animals.

Protein requirements

Ensure adequate but not excessive protein, as high protein intake can increase urinary calcium excretion.

4. Phosphorus balance

Phosphorus, along with calcium, contributes to bone mineralization. An appropriate calcium-to-phosphorus ratio is vital for optimal bone health.

Ideal ratio: Maintain a dietary calcium-to-phosphorus ratio of 1.2:1 to 2:1.

Avoid phosphorus-rich foods like organ meats and grains if phosphorus intake exceeds calcium levels.

5. Magnesium

Magnesium plays a supporting role in bone health by influencing calcium metabolism and collagen synthesis.

Sources

Nuts, seeds, whole grains, and green vegetables.

Magnesium supplements, if necessary, under veterinary guidance.

6. Vitamin C

Vitamin C is essential for collagen synthesis and helps improve the quality of connective tissues.

Sources:

Fresh fruits like oranges, strawberries, and kiwis.

Vegetables such as bell peppers and tomatoes.

7. Omega-3 fatty acids

Omega-3 fatty acids have anti-inflammatory properties and may support bone health by reducing bone resorption.

Sources

Fish oil or algae-based omega-3 supplements.

Flaxseed and chia seeds for herbivorous animals.

8. Zinc and copper

Both zinc and copper are cofactors for enzymes involved in collagen formation and bone mineralization.

Sources

Zinc: Meat, seafood, and whole grains.

Copper: Shellfish, nuts, and legumes.

9. Avoid excess vitamin A

Excessive vitamin A intake can interfere with bone formation and increase fracture risk.

Recommendation

Avoid over-supplementing vitamin A.

Use diets with balanced vitamin levels, especially in commercial pet foods.

10. Weight management

Maintaining an appropriate body weight reduces stress on fragile bones and joints.

Dietary adjustments

Avoid overfeeding to prevent obesity.

Provide nutrient-dense but calorie-controlled diets.

Supportive therapies and monitoring

Nutritional management should be integrated with other therapeutic strategies:

1. ***Medications***: Bisphosphonates to reduce bone resorption and increase bone density.
2. ***Physical therapy***: Exercise and mobility programs to strengthen muscles and prevent joint stiffness.
3. ***Regular monitoring***:

   Periodic blood tests to evaluate calcium, vitamin D, and phosphorus levels.

   Bone density assessments to track progress.

Prognosis

The prognosis for animals with OI depends on the severity of the condition and the effectiveness of management strategies. While fractures and deformities may persist, a comprehensive care plan, including proper nutrition, can significantly improve quality of life and reduce complications.

Osteogenesis imperfecta is a challenging genetic disorder in animals, but nutritional management is a cornerstone of its treatment. By optimizing calcium and vitamin D intake, ensuring adequate protein and micronutrients, and maintaining overall health, affected animals can achieve better bone health and improved quality of life. A tailored nutritional plan, combined with medical and supportive therapies, provides the best approach to managing this complex condition.

f) Pulmonary hyperthyroidism

Pulmonary hyperthyroidism, while not as commonly discussed as other endocrine or metabolic disorders, is an important condition that indirectly influences calcium metabolism in animals. Though the term may sound complex, this condition primarily involves an overproduction of thyroid hormones with secondary effects on calcium homeostasis, often affecting skeletal and systemic health. Nutritional management is an essential component of the overall care plan for animals with pulmonary hyperthyroidism. Proper dietary strategies can help regulate thyroid hormone production, address secondary calcium imbalances, and mitigate systemic complications, enhancing the animal's quality of life.

Pathophysiology

Pulmonary hyperthyroidism arises, when there is excessive production of thyroid hormones, triiodothyronine (T_3) and thyroxine (T_4). These hormones, secreted by the thyroid gland, regulate various physiological processes, including metabolism, cardiovascular function, and skeletal health.

The impact of hyperthyroidism on calcium metabolism occurs through the following mechanisms:

1. ***Increased bone turnover***: Elevated thyroid hormone levels stimulate osteoclast activity, leading to increased bone resorption. This releases calcium into the bloodstream, causing transient or sustained hypercalcemia.
2. ***Reduced bone density***: Chronic stimulation of bone resorption by thyroid hormones results in decreased bone mineral density, predisposing animals to fractures and skeletal deformities.
3. ***Altered vitamin D metabolism***: Hyperthyroidism can disrupt the activation of vitamin D, impairing calcium absorption from the gastrointestinal tract despite elevated serum calcium levels.
4. ***Calcium excretion***: Increased renal excretion of calcium often accompanies hyperthyroidism, further complicating calcium homeostasis and leading to potential deficiencies in bone mineralization.

Causes and risk factors

Pulmonary hyperthyroidism in animals may be linked to various underlying factors, including:

1. ***Thyroid adenomas or hyperplasia***: Benign tumors or glandular hyperplasia often cause excessive thyroid hormone production, especially in cats (common in older felines).
2. ***Thyroid carcinoma***: Though rare, malignant thyroid tumors can result in uncontrolled hormone secretion.
3. ***Iodine imbalance***: Diets deficient or excessive in iodine may disrupt normal thyroid function, contributing to hyperthyroidism.
4. ***Environmental toxins***: Exposure to environmental pollutants, such as flame retardants (polybrominated diphenyl ethers), has been implicated in thyroid dysfunction in domestic animals.
5. ***Genetic predisposition***: Some breeds of cats, dogs, and other species are more prone to thyroid disorders due to inherited susceptibility.

Species affected

Hyperthyroidism primarily affects the following animal species:

1. ***Cats***: Feline hyperthyroidism is the most common endocrine disorder in older cats, with a strong association with thyroid adenomas.
2. ***Dogs***: Canine hyperthyroidism is rare and typically results from thyroid carcinoma rather than benign causes.
3. ***Horses and other livestock***: Thyroid dysfunction in these species is uncommon but may occur secondary to dietary imbalances or exposure to environmental factors.

Clinical signs

The clinical presentation of pulmonary hyperthyroidism in animals varies depending on the species and severity of the condition. Common signs include:

1. ***Skeletal signs***

   Increased bone fragility due to accelerated bone turnover.

   Pain or lameness resulting from microfractures,

2. ***Systemic signs***

   Weight loss despite a normal or increased appetite,

   Hyperactivity or restlessness,

   Vomiting, diarrhea or polyuria (excessive urination),

3. ***Cardiovascular signs:*** Tachycardia (elevated heart rate) and arrhythmias due to heightened metabolic demand.
4. ***Neurological signs***: Neurological signs include weakness or tremors, particularly in advanced cases where calcium homeostasis is severely disrupted.
5. ***Hypercalcemia symptoms:*** Excessive thirst, lethargy and in severe cases renal dysfunction,

Diagnosis

Diagnosing pulmonary hyperthyroidism requires a combination of clinical evaluation, laboratory tests, and imaging studies:

1. ***Thyroid hormone levels***

   Elevated serum T4 or T3 levels confirm hyperthyroidism.

   Free T4 by equilibrium dialysis is often used in borderline cases.
2. ***Calcium and phosphorus levels***: Hypercalcemia with normal or low phosphorus levels may indicate thyroid-induced calcium imbalance.
3. ***Bone density scans***: X-rays or DEXA scans can reveal decreased bone mineral density and fractures.
4. ***Thyroid imaging***: Ultrasonography or scintigraphy can detect thyroid enlargement or nodules.
5. ***Urinalysis and renal function tests***: Assess secondary effects of hypercalcemia on kidney health.

Treatment

The treatment of pulmonary hyperthyroidism focuses on managing excessive thyroid hormone production and correcting calcium imbalances.

Medical therapy

1. ***Anti-thyroid medications***: Drugs like methimazole or carbimazole inhibit thyroid hormone synthesis and are commonly used in cats.
2. ***Calcium-reducing agents***: Bisphosphonates or calcitonin may be prescribed to manage hypercalcemia and prevent bone resorption.

Dietary management

1. ***Low-Iodine Diets***: Restricting iodine intake can reduce thyroid hormone synthesis in some cases.
2. ***Calcium and vitamin D supplementation***: In cases of calcium deficiency or poor bone mineralization, appropriate supplementation is essential.

Surgical and radioactive treatments

1. ***Thyroidectomy***: Surgical removal of the thyroid gland may be indicated in severe or refractory cases.
2. ***Radioactive iodine therapy***: A definitive treatment for hyperthyroidism, particularly in cats, that destroys overactive thyroid tissue without surgery.

Supportive care

1. ***Hydration***: Ensuring adequate hydration helps prevent renal complications associated with hypercalcemia.
2. ***Pain management***: Analgesics may be required to alleviate skeletal discomfort.

Prognosis

The prognosis for animals with pulmonary hyperthyroidism depends on the underlying cause, the extent of systemic involvement, and the success of treatment. Early detection and intervention can improve outcomes significantly, while advanced cases with severe hypercalcemia or bone loss may require long-term management.

Prevention and nutritional management

Preventing pulmonary hyperthyroidism involves addressing known risk factors and maintaining overall health:

1. ***Balanced diet***: Ensure proper iodine levels in the diet to support normal thyroid function.
2. ***Environmental safety***: Minimize exposure to potential endocrine disruptors.
3. ***Regular health check-ups***: Routine veterinary examinations can detect early signs of thyroid dysfunction.

Goals of nutritional management

1. Support optimal thyroid function by regulating iodine intake.
2. Address calcium and phosphorus imbalances caused by excessive thyroid hormone activity.
3. Mitigate weight loss and maintain muscle mass despite increased metabolic demand.
4. Enhance overall health and prevent complications such as bone fragility or renal dysfunction.

Key nutritional strategies

1. Iodine regulation

Iodine is a critical nutrient for thyroid hormone synthesis. Excessive or deficient iodine intake can exacerbate hyperthyroidism:

Low-iodine diets: In hyperthyroid animals, particularly cats, iodine-restricted diets help reduce the production of thyroid hormones. These diets limit iodine intake to levels just sufficient for basic thyroid function without encouraging excessive hormone synthesis.

Sources of iodine: Avoid iodine-rich foods such as fish, shellfish, and iodized salt in cases where a low-iodine diet is recommended.

2. Calcium and phosphorus balance

Hyperthyroidism often leads to increased bone resorption, causing calcium release into the bloodstream and potential hypercalcemia.

Calcium supplementation: In cases where prolonged bone resorption depletes skeletal calcium stores, dietary calcium supplementation may be necessary. However, this should be done cautiously to avoid exacerbating hypercalcemia.

Phosphorus: Maintaining the proper calcium-to-phosphorus ratio (approximately 2:1) is vital for skeletal health.

Vitamin D: This vitamin enhances calcium absorption from the gut. Supplementation may be required if hyperthyroidism disrupts vitamin D metabolism.

3. Protein intake

Hyperthyroid animals often experience muscle wasting due to increased metabolic demand.

High-quality protein: Provide easily digestible, high-quality protein sources, such as eggs, lean meats, or commercial feeds formulated for hyperthyroid animals. This supports muscle maintenance and repair.

4. Energy requirements

Animals with hyperthyroidism typically exhibit increased appetite due to their heightened metabolism, yet they may still lose weight.

Caloric density: A calorie-dense diet can help meet their energy needs without requiring excessive food volume.

Healthy fats: Incorporate healthy fats like omega-3 fatty acids to support energy requirements and reduce inflammation.

5. Antioxidants

Thyroid hormone excess increases oxidative stress, which can damage tissues.

Antioxidant-rich diets: Include foods rich in antioxidants, such as vitamin C (fruits, vegetables) and vitamin E (nuts, seeds, fortified feeds), to neutralize free radicals.

Dietary management plan

1. ***Commercial therapeutic diets***: Prescription diets formulated for hyperthyroid animals, such as iodine-restricted or high-calorie diets, simplify management and ensure balanced nutrition.
2. ***Portion control***: Feed small, frequent meals to manage appetite and digestive efficiency.
3. ***Hydration***: Encourage water intake to support kidney function, which may be ompromised by hypercalcemia or hypermetabolism.

Veterinary monitoring

Nutritional interventions should always be guided by a veterinarian. Regular monitoring of thyroid hormone levels, calcium, phosphorus, and overall nutritional status ensures the diet remains effective and prevents complications. By addressing iodine intake, calcium balance, and metabolic needs, nutritional management plays a crucial role in improving outcomes for animals with pulmonary hyperthyroidism.

Pulmonary hyperthyroidism in animals is a multifaceted disease with significant implications for calcium metabolism and skeletal health. By understanding its pathophysiology, clinical presentation, and treatment options, veterinarians can effectively manage this condition and improve the quality of life for affected animals. Ongoing research into its causes and optimal management strategies will further enhance our ability to address this complex disorder.

g) Secondary hyperthyroidism

Secondary hyperthyroidism is a rare condition in animals, often secondary to systemic disturbances particularly in Ca-P metabolism. Unlike primary hyperthyroidism, where the thyroid gland itself is overactive due to adenomas or carcinomas, secondary hyperthyroidism results from external stimuli causing excessive secretion of thyroid hormones. These hormonal imbalances indirectly influence calcium homeostasis, leading to clinical complications.

Pathophysiology

Secondary hyperthyroidism occurs when external factors, such as chronic iodine imbalances, renal dysfunction, or systemic endocrine disturbances, stimulate the thyroid gland to produce excess thyroxine (T_4) and triiodothyronine (T_3). This overproduction can disrupt calcium metabolism in the following ways:

1. ***Increased bone turnover***: Thyroid hormones accelerate bone remodeling by increasing osteoclast activity, leading to excessive bone resorption and calcium release into the bloodstream.
2. ***Hypercalcemia***: Elevated calcium levels can arise from increased bone breakdown, which in turn can affect neuromuscular, renal, and cardiovascular systems.
3. ***Decreased bone density***: Persistent bone resorption reduces bone mineral density, increasing the risk of fractures and deformities.
4. ***Vitamin D dysregulation***: Hyperthyroidism can impair the activation of vitamin D, reducing calcium absorption from the gut and further complicating calcium balance.

Causes of secondary hyperthyroidism

1. ***Dietary iodine imbalances***: Both excessive and deficient iodine intake can lead to thyroid dysfunction. Diets lacking iodine suppress thyroid hormone production initially, prompting compensatory hyperactivity of the gland over time.
2. ***Chronic kidney disease (CKD)***: Impaired kidney function disrupts phosphorus excretion and vitamin D activation, indirectly stimulating the thyroid gland to overcompensate.
3. ***Endocrine disturbances***: Secondary hyperthyroidism may coexist with hyperparathyroidism, exacerbating calcium-phosphorus imbalances.
4. ***Environmental factors***: Exposure to goitrogens or toxins that interfere with thyroid function can induce secondary hyperthyroidism.
5. ***Breed and species susceptibility***: Some animal breeds and species, particularly cats and dogs, may be predisposed due to genetic or environmental factors.

Clinical signs of secondary hyperthyroidism

The clinical presentation of secondary hyperthyroidism depends on its severity and the duration of the condition. Common signs include:

1. ***Skeletal manifestations***

   Fragile bones, fractures, and skeletal deformities.

   Pain or reluctance to move due to bone weakening.

2. ***Systemic symptoms***

   Weight loss despite a good appetite.

Restlessness, hyperactivity, or irritability.

Polyuria (excessive urination) and polydipsia (excessive thirst) due to hypercalcemia.

3. ***Cardiovascular effects***

Tachycardia (increased heart rate) and arrhythmias.

Hypertension.

4. ***Neurological signs***

Muscle tremors or weakness.

Lethargy in advanced cases.

Diagnosis of secondary hyperthyroidism

1. ***Thyroid function tests***:

Elevated T_4 and T_3 levels confirm hyperthyroidism.

Free T_4 measurement can provide additional accuracy.

2. ***Serum calcium and phosphorus levels***: Hypercalcemia with elevated or normal phosphorus levels is common.
3. ***Radiography and bone density scans***: X-rays or DEXA scans can reveal reduced bone density and fractures.
4. ***Urinalysis and renal function tests***: These assess secondary effects of calcium imbalance on kidney health.
5. ***Dietary history***: Evaluating iodine intake and overall diet composition can help identify nutritional triggers.

Treatment of secondary hyperthyroidism

The primary objective of treatment is to address the underlying cause of thyroid overactivity, restore calcium-phosphorus balance, and prevent systemic complications.

Medical interventions

1. ***Antithyroid medications***: Drugs like methimazole or carbimazole are used to reduce thyroid hormone production.
2. ***Calcium modulators***: Bisphosphonates or calcitonin may be prescribed to manage hypercalcemia and bone loss.
3. ***Vitamin D supplementation***: Activated forms of vitamin D (calcitriol) are often used to support calcium absorption in cases of vitamin D deficiency.
4. ***Iodine regulation***: If iodine imbalance is the cause, dietary adjustments or supplements are necessary.

Surgical and radioactive treatments

Thyroidectomy or radioactive iodine therapy may be indicated in severe or refractory cases.

Nutritional management of secondary hyperthyroidism

Nutritional management plays a crucial role in supporting animals with secondary hyperthyroidism. A tailored diet can help regulate thyroid function, correct calcium imbalances, and enhance overall health.

Key nutritional strategies

1. ***Iodine regulation***

   Low-iodine diets: In hyperthyroid animals, particularly cats, iodine-restricted diets can help reduce thyroid hormone production.

   Avoid iodine excess: Over-supplementation can exacerbate thyroid dysfunction. Limit iodine-rich foods like fish and seaweed unless prescribed by a veterinarian.

2. ***Calcium and phosphorus balance***

   Calcium supplementation: Include calcium-rich foods such as dairy products, calcium-fortified feeds, or calcium carbonate supplements to mitigate bone resorption.

   Phosphorus restriction: Avoid foods high in phosphorus, such as organ meats and grains, especially in animals with renal dysfunction.

   Calcium-phosphorus ratio: Maintain an optimal dietary ratio of approximately 2:1 to support bone health and prevent secondary mineral imbalances.

3. ***Vitamin D optimization***

   Sources: Vitamin D is crucial for calcium absorption. Provide natural sources like fish liver oil or ensure adequate supplementation, particularly in animals with CKD.

   UVB exposure: For reptiles and some exotic pets, UVB lighting facilitates natural vitamin D synthesis.

4. ***High-quality protein:*** Include easily digestible proteins like eggs, lean meats, or commercial feeds to prevent muscle wasting caused by hypermetabolism. However, excessive protein intake should be avoided in animals with kidney disease.

5. ***Antioxidants:*** Include antioxidant-rich foods such as fruits, vegetables, and fortified feeds to combat oxidative stress associated with hyperthyroidism.

6. ***Energy-dense foods:*** Provide calorie-dense diets to compensate for increased metabolic demands. Healthy fats, like omega-3 fatty acids from fish oil, can also reduce inflammation.
7. ***Hydration:*** Encourage water intake to support kidney function and counteract dehydration from polyuria.

Practical dietary approaches

1. ***Prescription diets***: Specialized commercial diets formulated for hyperthyroid animals simplify nutritional management.
2. ***Small, frequent meals***: Feed small portions multiple times a day to maintain stable energy levels and prevent gastrointestinal distress.
3. ***Avoid goitrogens***: Limit foods like soy and cruciferous vegetables (e.g., cabbage, broccoli) that may interfere with thyroid hormone synthesis.

Prognosis and long-term management

With early diagnosis and comprehensive care, many animals with secondary hyperthyroidism respond well to treatment. Nutritional management is particularly effective in addressing systemic effects and improving quality of life. However, animals with concurrent conditions, such as CKD, may require ongoing monitoring and dietary adjustments.

Prevention

Preventing secondary hyperthyroidism involves addressing known risk factors and maintaining overall health:

1. ***Balanced diets***: Ensure appropriate iodine intake and a proper calcium-phosphorus balance in all animal diets.
2. ***Routine Health Checks***: Regular veterinary evaluations can detect early signs of thyroid dysfunction.
3. ***Environmental management***: Provide UVB lighting and avoid exposure to goitrogens or toxins in susceptible species.

Secondary hyperthyroidism is a multifaceted condition that can disrupt calcium homeostasis and systemic health in animals. Proper nutritional management, focusing on iodine regulation, calcium-phosphorus balance, and overall dietary support, is critical for successful treatment and prevention. Collaboration between veterinarians and caregivers ensures the best outcomes for affected animals, restoring their health and quality of life.

h) Primary hyperparathyroidism

Primary hyperparathyroidism (PHPT) is a rare endocrine disorder in animals caused by the overproduction of parathyroid hormone (PTH) by one or more of the parathyroid glands. This leads to disturbances in calcium and phosphorus homeostasis, primarily resulting in hypercalcemia. Effective nutritional management, in conjunction with medical or surgical treatments, is critical to mitigating symptoms and restoring mineral balance.

Pathophysiology

In PHPT, the excessive secretion of PTH is typically caused by:

1. ***Parathyroid adenomas (most common)***: Benign tumors of the parathyroid gland.
2. ***Parathyroid hyperplasia***: Diffuse enlargement of the glands.
3. ***Parathyroid carcinomas (rare)***: Malignant tumors.

PTH plays a central role in calcium regulation by

Increasing bone resorption: Stimulating osteoclasts to release calcium and phosphorus from bones.

Enhancing renal calcium reabsorption: Reducing urinary calcium excretion while promoting phosphorus excretion.

Stimulating vitamin D activation: Increasing intestinal calcium absorption via calcitriol synthesis.

Excessive PTH results in persistent hypercalcemia, often accompanied by hypophosphatemia, due to increased phosphorus excretion.

Clinical manifestations of PHPT

Animals with PHPT may present with subtle or severe clinical signs depending on the degree of hypercalcemia and duration of the disease:

1. ***Renal signs***

   Increased thirst (polydipsia) and urination (polyuria) due to calcium-induced nephrogenic diabetes insipidus.

   Calcium deposition in renal tissues (nephrocalcinosis) can lead to chronic kidney disease (CKD).
2. ***Skeletal signs***: Bone pain, lameness, and fractures due to excessive calcium mobilization from bones.
3. ***Neuromuscular signs***: Weakness, lethargy, and muscle tremors.
4. ***Gastrointestinal signs***: Vomiting, anorexia, and constipation caused by hypercalcemia-induced smooth muscle dysfunction.
5. ***Cardiac signs***: Arrhythmias or bradycardia due to the effects of hypercalcemia on cardiac conduction.

Diagnosis of primary hyperparathyroidism

Diagnosing PHPT involves:

1. ***Serum biochemistry***

   Persistent hypercalcemia

   Low to normal phosphorus levels

   Elevated or inappropriately normal PTH levels (PTH assay)

2. ***Imaging***: Ultrasonography to identify parathyroid gland enlargement
3. ***Exclusion of other causes***: Rule out hypercalcemia of malignancy, chronic kidney disease, and vitamin D toxicosis.

Treatment of primary hyperparathyroidism

The mainstay of treatment for PHPT is surgical removal of the affected parathyroid gland(s). Other options include percutaneous ethanol injection or heat ablation of parathyroid adenomas. Nutritional management is essential for controlling hypercalcemia pre- and post-treatment.

Nutritional management of primary hyperparathyroidism

Nutritional strategies aim to control hypercalcemia, prevent further complications, and support recovery after medical or surgical intervention. Here is a detailed approach:

1. ***Calcium restriction:*** Reducing dietary calcium intake helps lower systemic calcium levels:

   Low-calcium diets: Feed diets specifically formulated to restrict calcium content, such as renal diets for dogs and cats. Avoid high-calcium foods, including dairy products, bone meal, and calcium-rich treats.

   Monitor calcium levels: Over-restriction of calcium can lead to hypocalcemia, especially post-surgery. Balance is key.

2. ***Phosphorus management:*** PHPT often leads to hypophosphatemia due to PTH-induced renal phosphorus excretion. Addressing phosphorus imbalance is crucial:

   Phosphorus supplementation: In cases of significant hypophosphatemia, dietary phosphorus levels may need to be increased cautiously using phosphorus-rich foods or supplements (e.g., monosodium phosphate).

   Balanced calcium-to-phosphorus ratio: Maintain a dietary calcium-to-phosphorus ratio close to 1:1 or slightly higher to support bone health and reduce mineral imbalances.

3. ***Vitamin D supplementation:*** Increased PTH levels can suppress vitamin D activation, reducing calcium absorption from the intestines and contributing to hypercalcemia by increasing bone resorption. Supplementation must be done carefully:

   Calcitriol therapy: Administer calcitriol (active vitamin D) post-surgery to stabilize calcium levels and prevent postoperative hypocalcemia.

   Avoid over-supplementation: Excessive vitamin D can exacerbate hypercalcemia. Regular monitoring is critical.

4. ***Hydration and electrolyte balance:*** Maintaining hydration is vital to enhance renal calcium excretion and prevent kidney damage:

   Encourage water intake: Provide fresh, clean water at all times. Wet diets can increase fluid intake in dogs and cats.

   Electrolyte management: Monitor and correct potassium and magnesium imbalances to avoid complications like muscle weakness or arrhythmias.

5. ***Protein management:*** Protein intake influences phosphorus levels and overall metabolic balance:

   Moderate protein levels: Use diets with moderate protein content to avoid excessive phosphorus intake while meeting the animal's nutritional needs.

   High-quality protein sources: Include easily digestible protein sources, such as lean meats or eggs.

6. ***Essential fatty acids (EFAs):*** Supplementing omega-3 fatty acids can reduce inflammation and support kidney function:

   Sources of EFAs: Include fish oil or diets enriched with omega-3 fatty acids to benefit animals with nephrocalcinosis or CKD.

7. ***Avoid high-oxalate foods:*** High oxalate foods can increase the risk of calcium oxalate stone formation, which is a potential complication of hypercalcemia:

   Avoid foods such as: Spinach, beet greens, sweet potatoes, and rhubarb.

8. ***Post-surgical nutritional support:*** Post-surgery, animals are at risk of hypocalcemia due to the sudden drop in PTH levels. Nutritional support should focus on:

   Calcium supplementation: Provide oral calcium (calcium carbonate or calcium gluconate) if hypocalcemia develops.

   Vitamin D therapy: Maintain calcitriol supplementation as needed to support calcium homeostasis.

Frequent monitoring: Adjust dietary and supplemental calcium based on serum calcium levels.

Monitoring and adjustments

Nutritional management of PHPT requires regular monitoring and adjustments:

Biochemical monitoring

Regularly measure serum calcium, phosphorus, PTH, and renal function markers

Bone health assessment

Evaluate skeletal recovery through radiographs or physical exams.

Body condition and appetite

Monitor weight and overall health to ensure adequate nutritional intake.

Prognosis

The prognosis for animals with PHPT is generally favorable if treated early and effectively. Surgical removal of the affected gland(s) typically resolves hypercalcemia. Long-term dietary and nutritional management support recovery and prevent complications such as kidney damage or bone demineralization.

Primary hyperparathyroidism is a rare but significant calcium disorder in animals, leading to persistent hypercalcemia and associated complications. Nutritional management, focusing on calcium restriction, phosphorus balancing, vitamin D supplementation, and overall dietary optimization, is a cornerstone of treatment. When combined with appropriate medical or surgical interventions, nutritional strategies can greatly improve outcomes, enhance recovery, and ensure long-term health. Regular monitoring and individualized care are essential for managing this condition effectively.

i) Secondary hyperparathyroidism

Secondary hyperparathyroidism (SHPT) is a common calcium-related disorder in animals, often associated with imbalances in calcium, phosphorus, and vitamin D. Unlike primary hyperparathyroidism, which arises from a functional abnormality in the parathyroid glands, SHPT occurs as a compensatory response to chronic conditions that disrupt calcium homeostasis. Nutritional management plays a critical role in treating SHPT, especially when addressing the underlying causes such as renal disease, dietary imbalances, or intestinal malabsorption.

Pathophysiology

SHPT occurs due to persistent stimulation of the parathyroid glands in response to hypocalcemia or hyperphosphatemia. Common underlying causes include:

1. ***Renal secondary hyperparathyroidism (Renal SHPT)***:

   Chronic kidney disease (CKD) impairs the kidneys' ability to excrete phosphorus and activate vitamin D (1,25-dihydroxyvitamin D or calcitriol).

   Hyperphosphatemia and decreased calcitriol levels lead to reduced intestinal calcium absorption and systemic hypocalcemia, which stimulates excessive secretion of parathyroid hormone (PTH).

2. ***Nutritional secondary hyperparathyroidism (nutritional SHPT)***:

   Results from diets deficient in calcium, excessive in phosphorus, or imbalanced in the calcium-to-phosphorus ratio.

   Common in young, growing animals fed inappropriate diets, such as meat-only diets for dogs and cats or high-phosphorus diets in herbivores.

3. ***Intestinal secondary hyperparathyroidism***: Chronic gastrointestinal diseases (e.g., inflammatory bowel disease) impair calcium and vitamin D absorption, leading to hypocalcemia and compensatory PTH secretion.

4. ***Pseudohyperparathyroidism***: Rarely, certain tumors produce PTH-like substances, mimicking SHPT but unrelated to true parathyroid activity.

Clinical manifestations

SHPT primarily affects the skeletal system and may also cause systemic complications:

Skeletal abnormalities

Bone pain, deformities, and fractures due to osteoclastic bone resorption.

In young animals, growth deformities such as rickets or limb bowing.

Soft tissue mineralization

Hyperphosphatemia and high PTH levels can lead to calcification of soft tissues, including blood vessels and kidneys.

Neuromuscular signs

Weakness, lethargy, and, in severe cases, tetany or seizures due to hypocalcemia.

Facial Swelling (Rubber Jaw):

Common in severe cases of nutritional SHPT in dogs and cats, where fibrous tissue replaces demineralized jawbones.

Diagnosis

Diagnosis of SHPT involves a combination of history, clinical signs, laboratory testing, and imaging:

1. ***Laboratory tests***: Elevated serum phosphorus (especially in renal SHPT) or normal serum calcium.

   Elevated PTH levels (via PTH assay).

   Low calcitriol levels in renal and intestinal SHPT.
2. ***Imaging***: Radiographs may show osteopenia, skeletal deformities, and fractures.
3. ***Dietary history***: Identifying imbalanced or inappropriate diets is crucial for diagnosing nutritional SHPT.

Nutritional management

Nutritional therapy is a cornerstone of SHPT management, aiming to restore calcium-phosphorus balance, normalize PTH secretion, and address the underlying cause.

1. ***Phosphorus restriction***

   Controlling dietary phosphorus intake is critical in managing hyperphosphatemia, particularly in renal SHPT.

   Low-phosphorus diets: Commercial renal diets for dogs and cats (e.g., Hill's k/d, Royal Canin Renal) are formulated to reduce phosphorus intake while meeting overall nutritional needs.

   For herbivores, avoid phosphorus-rich feeds such as grains and bran.

   Phosphate binders: If dietary phosphorus restriction is insufficient, use phosphate binders (e.g., aluminum hydroxide, calcium carbonate) to reduce phosphorus absorption in the gastrointestinal tract. Administer these with meals.
2. ***Calcium supplementation***

   Calcium supplementation is necessary to correct hypocalcemia and support bone health.

   Calcium sources: Use calcium carbonate or calcium gluconate as supplements.

   For herbivores, provide calcium-rich forages (e.g., alfalfa) and avoid calcium-deficient feeds.

   Calcium-to-phosphorus ratio: Maintain a dietary calcium-to-phosphorus ratio of 1.2:1 to 2:1 to promote proper mineral balance and reduce PTH secretion.

3. ***Vitamin D supplementation***

Vitamin D is essential for enhancing calcium absorption and suppressing PTH secretion.

Calcitriol therapy: Administer calcitriol (active vitamin D) in cases of renal or intestinal SHPT, where vitamin D metabolism is impaired.

Dosage must be carefully monitored to prevent hypercalcemia or soft tissue mineralization.

Natural sources: For herbivores, ensure adequate sunlight exposure to support natural vitamin D synthesis.

4. ***Protein management***

Protein intake influences phosphorus levels and overall metabolic health:

Moderate Protein Levels: For animals with renal SHPT, feed moderate levels of high-quality protein to reduce phosphorus load while meeting nutritional needs. Avoid excessive protein restriction in young, growing animals with nutritional SHPT.

5. ***Essential fatty acids (EFAs)***

Supplementing omega-3 fatty acids can help reduce inflammation and slow CKD progression, indirectly benefiting SHPT.

Sources of EFAs: Include fish oil or diets enriched with omega-3 fatty acids.

6. ***Balanced energy intake***

Animals with SHPT often experience weight loss or poor growth. Ensure adequate caloric intake to meet energy demands.

Nutrient-dense foods: Use energy-dense diets or add healthy fats (e.g., fish oil) to increase caloric intake without increasing phosphorus levels.

Palatability enhancers: Use flavour additives or warm food to encourage eating in animals with poor appetite.

7. ***Electrolyte and micronutrient balance***

Potassium: Supplement potassium if hypokalemia is present, particularly in cats with CKD.

Magnesium: Monitor and correct magnesium levels as imbalances can worsen calcium disorders.

Antioxidants: Include antioxidants like vitamin E and selenium to reduce oxidative stress, particularly in CKD.

8. ***Avoiding harmful practices***

   Avoid excessive phosphorus in treats or supplements: Eliminate high-phosphorus treats or chews from the diet.

   Minimize vitamin A excess: Avoid excessive vitamin A, which can exacerbate bone resorption and counteract calcitriol therapy.

Monitoring and adjustments

Nutritional management requires ongoing evaluation and adjustments to ensure efficacy and prevent complications:

Biochemical monitoring

Regularly measure serum calcium, phosphorus, PTH, and kidney function markers.

Skeletal assessments

Periodic imaging evaluates bone density and deformities.

Weight and growth tracking

Monitor body condition and growth in young animals.

Prognosis

The prognosis for animals with SHPT depends on the underlying cause and the timeliness of intervention. Nutritional management, when combined with medical therapies, can significantly improve outcomes and enhance quality of life. However, advanced skeletal deformities or concurrent severe CKD may limit recovery potential.

Secondary hyperparathyroidism in animals is a multifaceted disorder with profound effects on calcium metabolism and skeletal health. Nutritional management is a cornerstone of treatment, focusing on phosphorus restriction, calcium and vitamin D supplementation, and overall dietary balance. Tailored dietary interventions, regular monitoring, and collaboration with veterinarians and nutritionists are essential for successful long-term management and improved animal welfare.

j) Renal osteodystrophy

Renal osteodystrophy (ROD) is a metabolic bone disease associated with chronic kidney disease (CKD). It results from disruptions in calcium, phosphorus, and vitamin D metabolism due to declining renal function. Animals with ROD suffer from skeletal abnormalities, pain, and systemic complications that compromise their quality of life. Nutritional management is a cornerstone of treatment, complementing medical therapies to slow disease progression and alleviate symptoms.

Pathophysiology

ROD arises from a combination of factors linked to CKD:

1. ***Hyperphosphatemia***: Diseased kidneys lose the ability to excrete phosphorus efficiently, leading to its accumulation in the bloodstream. Elevated phosphorus levels stimulate fibroblast growth factor-23 (FGF-23), which inhibits renal production of 1,25-dihydroxyvitamin D (calcitriol), the active form of vitamin D.
2. ***Hypocalcemia***: Reduced calcitriol impairs calcium absorption from the intestines, causing systemic hypocalcemia. Hypocalcemia triggers the release of parathyroid hormone (PTH), leading to secondary hyperparathyroidism.
3. ***Secondary hyperparathyroidism***: Prolonged elevation of PTH results in excessive calcium mobilization from bones, leading to demineralization, osteomalacia, and fibrous tissue deposition within the skeletal system.
4. ***Bone remodelling abnormalities***: The combined effects of high PTH, hyperphosphatemia, and calcitriol deficiency disrupt normal bone remodelling, causing weakened bone structure, pain, and deformities.

Clinical Manifestations in Animals

Animals with ROD may display

Skeletal abnormalities: Bone pain, fractures, and deformities.

Growth retardation: In young animals, impaired bone development due to mineral imbalances.

Facial swelling: Enlargement of the maxilla or mandible ("rubber jaw") due to fibrous tissue replacement.

Lethargy and weakness: Associated with pain and muscle dysfunction.

Neurological signs: Hypocalcemia-induced tetany or seizures in severe cases.

Diagnosis of renal osteodystrophy

1. ***Laboratory testing***:

   Elevated serum phosphorus.

   Low or normal serum calcium.

   Elevated PTH (secondary hyperparathyroidism).

   Decreased calcitriol levels.

   Indicators of CKD: Increased creatinine and blood urea nitrogen (BUN).
2. ***Imaging***: Radiographs reveal osteopenia, fractures, and skeletal deformities.

3. ***Bone Biopsy (rarely performed)***: Confirms histopathological changes in bone tissue.

Nutritional management of renal osteodystrophy

Nutrition plays a vital role in addressing the underlying metabolic imbalances contributing to ROD. The goals are to control phosphorus levels, support calcium homeostasis, and prevent further skeletal damage while meeting the animal's overall dietary needs.

1. Phosphorus control

Reducing dietary phosphorus intake is critical to managing hyperphosphatemia and preventing further activation of secondary hyperparathyroidism.

Low-phosphorus diet: Feed a renal diet specifically formulated with reduced phosphorus levels. Commercial renal diets for dogs and cats (e.g., Hill's k/d, Royal Canin Renal) are balanced to meet this requirement.

Phosphate binders: If dietary restriction is insufficient, use phosphate binders (e.g., aluminum hydroxide, calcium carbonate) to reduce phosphorus absorption in the gastrointestinal tract. Administer these with meals for maximum efficacy.

2. Calcium supplementation

To counteract hypocalcemia and support bone health, dietary calcium may be supplemented cautiously.

Calcium carbonate or calcium acetate: These can serve dual roles as calcium supplements and phosphate binders. However, calcium levels must be carefully monitored to avoid hypercalcemia.

Calcium-to-phosphorus ratio: Maintain a dietary calcium-to-phosphorus ratio between 1.2:1 and 2:1 to promote skeletal health and minimize complications.

3. Vitamin D supplementation

Active vitamin D (calcitriol) supplementation is essential to enhance intestinal calcium absorption, suppress PTH secretion, and mitigate secondary hyperparathyroidism.

Calcitriol therapy: Administer in low doses under veterinary supervision. Over-supplementation can lead to hypercalcemia and soft tissue mineralization.

Monitoring: Regularly check calcium, phosphorus, and PTH levels to ensure safe and effective calcitriol dosing.

4. Protein management

Protein intake should be carefully controlled to balance the nutritional needs of the animal while minimizing phosphorus load.

Moderate protein levels: Renal diets typically include high-quality, easily digestible protein sources in restricted amounts. This helps limit phosphorus intake while meeting the animal's energy requirements.

Avoid excessive restriction in young animals: Growing animals with ROD require sufficient protein for development. In these cases, moderate restriction rather than severe limitation is appropriate.

5. Essential fatty acids (EFAs)

Supplementing omega-3 fatty acids can help reduce inflammation and slow the progression of CKD, indirectly benefiting ROD.

Sources of EFAs: Fish oil is a rich source of omega-3 fatty acids. Include supplements or diets enriched with EPA and DHA.

6. Electrolyte balance

Electrolyte disturbances, especially hypokalemia, are common in CKD and can exacerbate muscle weakness and lethargy.

Potassium supplementation: Use potassium gluconate or potassium citrate if hypokalemia is present, particularly in cats with CKD.

Sodium restriction: Limit sodium intake to prevent hypertension, which is common in CKD.

7. Energy and caloric density

Animals with CKD and ROD often experience reduced appetite and weight loss. A nutrient-dense diet can help maintain body condition.

High-calorie formulations: Use energy-dense renal diets or add healthy fats (e.g., fish oil) to increase calorie intake without increasing phosphorus levels.

Palatability Enhancers: Warming the food, adding broths, or using flavor additives can encourage fussy eaters.

8. Avoiding toxins and excesses

Avoid excess vitamin A: High vitamin A levels can exacerbate bone resorption and counteract calcitriol therapy.

Limit oxalate-rich foods: Avoid foods like spinach and beet greens, which may contribute to calcium oxalate deposition.

Monitoring and adjustments

Nutritional management requires ongoing monitoring to ensure efficacy and avoid complications:

Biochemical monitoring

Regularly check serum calcium, phosphorus, PTH, and kidney function markers to guide dietary adjustments.

Bone health assessment

Periodic imaging or physical exams assess the progression of skeletal changes.

Weight and appetite tracking

Monitor body condition and caloric intake to prevent malnutrition.

Prognosis

The prognosis for animals with ROD varies based on the severity of CKD, the degree of bone involvement, and the timeliness of intervention. While nutritional management cannot reverse bone damage, it significantly slows disease progression, alleviates symptoms, and improves the animal's quality of life.

Renal osteodystrophy is a serious complication of chronic kidney disease that requires a comprehensive approach to management. Nutritional interventions, including phosphorus restriction, calcium and vitamin D supplementation, and controlled protein intake, are essential for addressing the underlying metabolic imbalances. When tailored to the individual animal's needs and monitored closely, these strategies can improve outcomes and support a better quality of life. Collaboration between veterinarians, nutritionists, and pet owners is key to achieving long-term success in managing this challenging condition.

k) Excess adrenal corticoid secretion

Excessive adrenal corticoid secretion, also known as hyperadrenocorticism or Cushing's disease, is a disorder characterized by the overproduction of cortisol by the adrenal glands. Cortisol, a glucocorticoid hormone, plays a crucial role in regulating metabolism, stress response, and immune function. However, when produced in excess, cortisol can disrupt calcium metabolism and contribute to secondary calcium disorders in animals. Understanding the effects of hyperadrenocorticism on calcium homeostasis and implementing appropriate nutritional strategies are key for managing the condition and improving animal health.

Pathophysiology

The adrenal glands, located above the kidneys, produce cortisol under the regulation of the hypothalamus and pituitary gland. In normal physiology, cortisol helps regulate glucose metabolism, immune responses, and stress adaptation. However, in hyperadrenocorticism, excessive cortisol production leads to various metabolic disturbances, including calcium imbalances. This can occur through several mechanisms:

Increased bone resorption

Cortisol stimulates osteoclast activity, leading to increased bone resorption and reduced bone mineralization. As a result, calcium is released from the bones into the bloodstream, contributing to altered calcium levels.

Reduced calcium absorption

High cortisol levels impair the gastrointestinal tract's ability to absorb calcium from the diet, exacerbating calcium deficiency.

Altered parathyroid hormone (PTH) regulation

Cortisol can interfere with parathyroid hormone (PTH) secretion, which plays a role in regulating blood calcium levels. This can lead to imbalances in calcium and phosphorus metabolism.

Renal calcium loss

Excessive cortisol increases renal calcium excretion, further depleting calcium levels in the body.

Causes of hyperadrenocorticism

Hyperadrenocorticism can be classified into two main types based on its origin:

Pituitary-dependent hyperadrenocorticism (PDH): The most common cause, where a pituitary tumor leads to excessive production of adrenocorticotropic hormone (ACTH), stimulating the adrenal glands to produce excess cortisol.

Adrenal-dependent hyperadrenocorticism (ADH): Caused by a tumor in one of the adrenal glands that directly secretes excess cortisol, bypassing the pituitary regulation.

Iatrogenic hyperadrenocorticism

This condition occurs due to the prolonged use of corticosteroid medications such as prednisone which can suppress the hypothalamic-pituitary-adrenal axis and lead to excessive cortisol production. Pituitary-dependent hyperadrenocorticism is most common in dogs, whereas, adrenal-dependent hyperadrenocorticism is more frequently seen in cats and horses. In all species, hyperadrenocorticism can lead to disturbances in calcium metabolism and bone health.

Clinical signs of hyperadrenocorticism and calcium disorders

Animals with hyperadrenocorticism may exhibit a range of symptoms, many of which are related to calcium imbalances:

General symptoms

Increased thirst and urination (polydipsia and polyuria)

Increased appetite (polyphagia) and weight gain

Thin, fragile skin and hair loss

Abdominal bloating and muscle weakness

Calcium imbalance symptoms

Osteoporosis or osteopenia, leading to fractures, bone pain, and weakness

Hypocalcemia (low calcium) may result in muscle tremors, seizures, or tetany

Hypercalcemia (less common but possible) can lead to lethargy, anorexia, and kidney dysfunction

Other symptoms

Elevated blood pressure (hypertension)

Increased susceptibility to infections due to immune suppression

Diagnosis

The diagnosis of hyperadrenocorticism and its associated calcium imbalances involves several steps:

1. ***Clinical history and examination***: Identification of classic signs such as polyuria, polydipsia, and muscle weakness,
2. ***Blood tests***: Measure cortisol levels through tests such as the ACTH stimulation test or low-dose dexamethasone suppression test. Assess calcium, phosphorus, and alkaline phosphatase (ALP) levels, as elevated ALP may indicate bone turnover.
3. ***Urine tests***: Measure urine cortisol/creatinine ratio to assess for excessive cortisol excretion,
4. ***Imaging***: Use ultrasound or CT scans to detect adrenal tumors or pituitary abnormalities.

Nutritional management of hyperadrenocorticism

Managing hyperadrenocorticism and its associated calcium imbalances requires dietary interventions aimed at supporting calcium homeostasis and mitigating the effects of excessive cortisol secretion. The following nutritional strategies can be used:

A. Calcium and phosphorus balance

1. ***Calcium supplementation***: Provide calcium supplements, especially in cases of hypocalcemia, to restore normal blood calcium levels. The use of calcium carbonate or calcium citrate supplements can help ensure proper bone health.
2. ***Phosphorus regulation***: Maintain a proper calcium-to-phosphorus (Ca:P) ratio, ideally 1.5:1 to 2:1, as excess phosphorus can worsen calcium imbalance and contribute to bone demineralization. Consider phosphorus binders in cases of hyperphosphatemia (high phosphorus levels).

B. Vitamin D supplementation

1. ***Role of vitamin D***: Vitamin D is crucial for calcium absorption from the gastrointestinal tract and helps regulate bone metabolism. Supplementation with vitamin D (e.g., cholecalciferol) may be necessary to enhance calcium absorption, particularly in animals with impaired gut absorption due to excess cortisol.
2. ***Caution***: Ensure that vitamin D supplementation does not lead to hypercalcemia, as excessive vitamin D can exacerbate bone resorption and lead to calcification of soft tissues.

C. Magnesium supplementation

1. ***Magnesium's role***: Magnesium is involved in bone mineralization and works synergistically with calcium. Inadequate magnesium levels can worsen calcium imbalances. Supplement magnesium, especially in animals with muscle tremors or seizures associated with hypocalcemia.

D. Protein and energy management

1. ***High-quality protein***: Provide high-quality, easily digestible protein to support muscle mass and bone health. Low-protein diets may exacerbate muscle wasting and osteopenia. Protein sources such as chicken, turkey, or fish are ideal for managing weight gain while maintaining muscle integrity.
2. ***Energy intake***: Monitor caloric intake to prevent obesity, a common complication of hyperadrenocorticism. Use balanced energy-dense feeds that are rich in essential nutrients but low in empty calories.

E. Avoidance of goitrogenic foods

1. ***Goitrogenic foods*:** Certain foods, such as soy and cabbage, contain goitrogens that can interfere with thyroid and adrenal function. While not directly related to cortisol production, these foods should be avoided in animals with adrenal dysfunction.

F. Omega-3 fatty acids

1. ***Anti-inflammatory effects:*** Omega-3 fatty acids, particularly from fish oils, help reduce the inflammatory effects of excessive cortisol and promote overall metabolic health. Supplementing omega-3s may also help manage hypertension, a common symptom of hyperadrenocorticism.

Long-term management and prevention

1. ***Regular monitoring:*** Regular blood tests to monitor cortisol, calcium, phosphorus, and other metabolic markers are crucial for managing the condition and adjusting nutritional support.
2. ***Balanced diet*:** Ongoing provision of a balanced diet, rich in essential nutrients like calcium, magnesium, and protein, is vital for maintaining bone health and overall well-being.
3. ***Medication*:** In addition to nutritional management, pharmacological treatment (e.g., trilostane or mitotane) may be used to control cortisol production.

Prognosis

The prognosis for animals with hyperadrenocorticism depends on the severity of the condition and the effectiveness of treatment. With appropriate medical and nutritional management, many animals can live a normal life. However, untreated or poorly managed cases can lead to complications such as osteoporosis, fractures, kidney failure, and hypertension, which can significantly reduce life expectancy.

Excessive adrenal corticoid secretion (hyperadrenocorticism) can lead to serious calcium imbalances, with implications for bone health and metabolic function. Nutritional management plays a vital role in supporting calcium homeostasis, improving bone health, and mitigating the effects of excessive cortisol secretion. By providing balanced calcium and phosphorus levels, vitamin D, magnesium, and high-quality protein, along with monitoring and medical intervention, caretakers can significantly improve the quality of life for animals suffering from this condition.

l) Hypothyroidism

Hypothyroidism, a condition characterized by an underactive thyroid gland, affects various metabolic processes, including calcium homeostasis. The thyroid gland, located in the neck, plays a critical role in regulating metabolism, growth, and mineral balance through the secretion of thyroid hormones, thyroxine (T_4) and triiodothyronine (T_3). In animals, hypothyroidism can disrupt calcium metabolism, potentially leading to secondary calcium disorders. Addressing this complex condition requires a comprehensive approach that includes nutritional management to alleviate symptoms and support overall health.

Pathophysiology of hypothyroidism and calcium metabolism

Thyroid hormones influence calcium homeostasis through their regulatory effects on bone turnover and calcium absorption:

Bone metabolism: Thyroid hormones stimulate osteoblast and osteoclast activity, maintaining bone remodeling. Hypothyroidism reduces this activity, leading to decreased calcium mobilization from bones.

Intestinal calcium absorption: Reduced thyroid hormone levels can impair calcium absorption in the gut, contributing to hypocalcemia (low blood calcium levels).

Parathyroid gland interaction: Hypothyroidism often coexists with parathyroid gland dysfunction, further disrupting calcium-phosphorus balance.

Mechanisms in calcium disorders

1. ***Hypocalcemia***: Prolonged hypothyroidism can lead to low serum calcium levels, causing neuromuscular and skeletal symptoms.
2. ***Hypercalcemia (rare cases)***: Thyroid dysfunction may sometimes cause elevated calcium levels due to altered bone metabolism or parathyroid hormone activity.

Causes of hypothyroidism

Hypothyroidism in animals can be classified as primary (affecting the thyroid gland) or secondary (due to dysfunction in the hypothalamus or pituitary). Key causes include:

1. ***Primary causes***

   Autoimmune thyroiditis (common in dogs)

   Iodine deficiency or toxicity

   Congenital hypothyroidism due to genetic mutations

   Thyroid neoplasia

2. ***Secondary causes***: Pituitary gland dysfunction leading to insufficient thyroid-stimulating hormone (TSH) production. Hypothalamic disorders affecting thyrotropin-releasing hormone (TRH) secretion.
3. ***Nutritional factors***: Inadequate iodine intake is a significant cause of hypothyroidism-related calcium disorders. Goitrogenic substances in feed, such as certain brassica plants, interfere with iodine utilization.

Clinical signs of hypothyroidism

Hypothyroidism manifests through a range of symptoms, some of which are directly or indirectly related to calcium metabolism:

General symptoms

Lethargy, weight gain, and intolerance to cold,

Calcium disorder symptoms

Hypocalcemia: Muscle tremors, tetany, or seizures.

Skeletal: Weakness, delayed bone growth, or osteopenia.

Other Symptoms

Dry skin, hair loss, and poor coat condition

Bradycardia (slow heart rate) and reduced appetite

Diagnosis of hypothyroidism

1. ***Clinical examination***

   Assess symptoms such as lethargy, poor growth, and neurological signs.
2. ***Laboratory tests***

   Measure serum T_4, T_3, and TSH levels to confirm hypothyroidism.

   Assess calcium, phosphorus, and parathyroid hormone (PTH) levels to identify related disorders.
3. ***Dietary analysis***: Evaluate iodine and mineral intake in the animal's diet.
4. ***Imaging***: Ultrasound or radiography to identify thyroid gland abnormalities.

Nutritional management of hypothyroidism

Proper nutritional management is essential for managing hypothyroidism and its impact on calcium homeostasis. Dietary strategies aim to support thyroid function, correct calcium imbalances, and enhance overall health.

A. Iodine supplementation

1. ***Role of iodine***: Iodine is a key component of thyroid hormones, and its deficiency directly leads to hypothyroidism.
2. ***Sources***: Include iodized salt, seaweed, or commercial iodine supplements in the diet.
3. ***Caution***: Excessive iodine can also cause thyroid dysfunction, so supplementation should be carefully regulated.

B. Calcium and phosphorus balance

1. ***Calcium supplementation***: Provide calcium carbonate or gluconate in cases of hypocalcemia to restore normal serum calcium levels. Maintain the ideal calcium-to-phosphorus (Ca:P) ratio (1.5:1 to 2:1).
2. ***Phosphorus***: Supplement phosphorus using dicalcium phosphate if phosphorus levels are low.

C. Vitamin D supplementation

1. ***Role of vitamin D***: Vitamin D enhances calcium absorption and utilization, mitigating hypocalcemia associated with hypothyroidism.
2. ***Sources***: Provide dietary supplements or ensure adequate sunlight exposure.
3. ***Dosage***: Avoid excessive vitamin D, as it can lead to hypercalcemia and soft tissue calcification.

D. Selenium supplementation

1. ***Role of selenium***: Selenium supports thyroid hormone synthesis and protects against oxidative stress in the thyroid gland.
2. ***Sources***: Include selenium-rich feeds or supplements like sodium selenite or selenomethionine.

E. Protein and energy balance

1. ***Protein***: Adequate dietary protein is essential for thyroid hormone production and overall metabolic health. Include high-quality protein sources such as soybean meal, fish meal, or legumes.
2. ***Energy***: Provide balanced energy levels to prevent weight gain, which is common in hypothyroid animals.

F. Avoid goitrogenic substances

1. ***Common goitrogens***: Brassica plants (e.g., cabbage, kale) and soy products contain compounds that interfere with iodine uptake.
2. ***Dietary adjustments***: Limit goitrogenic feeds or balance their inclusion with adequate iodine supplementation.

G. Omega-3 fatty acids

1. *Role of omega-3 fatty acids*: Omega-3 fatty acids reduce inflammation and support thyroid function.
2. *Sources*: Add flaxseed, fish oil, or chia seeds to the diet.

H. Trace minerals

1. *Zinc and iron*: Zinc supports thyroid hormone activity, while iron deficiency can impair thyroid function.
2. *Magnesium*: Magnesium supplementation may help regulate calcium metabolism and prevent muscle cramps or tetany.

Long-term management and prevention

1. ***Regular monitoring***: Periodic assessment of thyroid hormone levels, calcium-phosphorus balance, and dietary adequacy.
2. ***Customized diets***: Formulate diets based on the animal's physiological needs (e.g., lactating animals, growing young).
3. ***Environmental enrichment***: Reduce stress and provide adequate exercise to enhance metabolic health.

Prognosis

The prognosis for animals with hypothyroidism depends on the severity of the condition and the timeliness of intervention:

Mild cases

Nutritional and medical management can effectively restore normal thyroid function and correct calcium imbalances.

Severe or chronic cases

Long-term management, including thyroid hormone replacement therapy, may be necessary.

Hypothyroidism in animals disrupts calcium metabolism, leading to secondary calcium disorders such as hypocalcemia and skeletal abnormalities. Effective nutritional management is crucial for addressing iodine deficiencies, restoring calcium-phosphorus balance, and supporting overall thyroid function. By providing a balanced diet rich in iodine, vitamin D, selenium, and other essential nutrients, caretakers can improve the health and productivity of affected animals. Early diagnosis and intervention, combined with ongoing monitoring, are key to successful management of hypothyroidism and its associated calcium disorders.

m) Osteopetrosis

Osteopetrosis is a rare, inherited or acquired skeletal condition characterized by excessive bone density and abnormal bone remodelling. Unlike osteoporosis, this involves bone resorption and decreased density, osteopetrosis results in an accumulation of dense but brittle bone due to defective osteoclast function. This disorder disrupts normal calcium homeostasis, as bones become excessively mineralized and fail to maintain normal calcium turnover. Effective nutritional management is vital to mitigate complications, improve quality of life, and manage underlying calcium imbalances.

Pathophysiology

Osteopetrosis also known as marble bone disease arises from defective osteoclast activity:

Osteoclast dysfunction

Impaired bone resorption leads to an inability to remodel bones and maintain calcium-phosphorus balance.

Abnormal bone accumulation

Bones become abnormally dense, losing their structural integrity, leading to fractures and reduced marrow space.

Calcium dysregulation

Bone mineral deposition reduces calcium release into the bloodstream, potentially causing hypocalcemia. Excessive calcium within the bone may lead to secondary calcium metabolic issues.

Etiology

1. ***Genetic***: Congenital osteopetrosis is caused by genetic mutations affecting osteoclast differentiation or function. This is seen in various animal species, including cattle, sheep, and dogs.
2. ***Acquired***: Rare cases in animals may arise due to exposure to toxins, infections, or metabolic disturbances.

Clinical signs of osteopetrosis

Affected animals exhibit a range of clinical manifestations depending on the severity and species:

Musculoskeletal abnormalities

Increased bone density and thickness,

Susceptibility to fractures due to brittle bones,

Skeletal deformities, such as short limbs or cranial malformations,

Neurological symptoms

Neurological symptoms include compression of cranial nerves which causes abnormal bone growth, leading to blindness, deafness or facial paralysis.

Bone marrow suppression

Anemia, leukopenia and thrombocytopenia due to reduced marrow space, which causes increased susceptibility to infections.

Systemic effects:

Poor growth rates, weakness and hypocalcemia-related tetany in severe cases.

Nutritional management of osteopetrosis

Nutritional strategies for managing osteopetrosis aim to optimize calcium homeostasis, support bone health, and alleviate symptoms associated with the disorder.

a) Calcium and phosphorus balance

1. ***Maintain optimal ratios***: An appropriate calcium-to-phosphorus (Ca:P) ratio (1.5:1 to 2:1) is essential to prevent further disruption of bone mineral metabolism. Avoid excess calcium supplementation, which could exacerbate bone density issues and hypocalcemia.
2. ***Phosphorus supplementation***: Phosphorus is critical for maintaining bone integrity and aiding in calcium metabolism. Supplement with di-calcium phosphate or bone meal if dietary phosphorus is insufficient.

b) Vitamin D supplementation

1. ***Role of vitamin D***: Vitamin D is essential for calcium absorption and mobilization. In osteopetrosis, vitamin D supplementation may help improve calcium utilization and reduce hypocalcemia.
2. ***Sources***: Provide cholecalciferol (vitamin D_3) through fortified feeds, supplements, or natural sunlight exposure.
3. ***Caution***: Avoid excessive vitamin D, as it may promote metastatic calcification or worsen mineral imbalances.

c) Magnesium supplementation

1. ***Role of magnesium***: Magnesium plays a crucial role in regulating calcium metabolism and supporting bone remodelling. Magnesium deficiency can exacerbate hypocalcemia and neuromuscular symptoms.
2. ***Sources***: Supplement diets with magnesium oxide or magnesium sulfate if deficiency is suspected.

d) Protein and energy requirements

1. ***Protein for bone growth*:** High-quality protein is necessary for collagen synthesis and bone matrix formation. Provide balanced rations with adequate protein levels to support skeletal and overall health.
2. ***Energy needs*:** Ensure sufficient caloric intake to prevent weight loss and support growth in young animals.

e) Trace mineral supplementation

1. ***Zinc*:** Zinc is critical for bone development and osteoclast function. Supplementation may enhance bone remodeling and improve skeletal health.
2. ***Copper*:** Copper is involved in collagen cross-linking and bone strength. Deficiency should be corrected with dietary supplementation.
3. ***Manganese*:** Manganese supports cartilage and bone matrix formation. Provide balanced mineral mixes containing manganese.

f) Avoiding bone-hardening agents

Excess calcium: Avoid feeding high-calcium forages or supplements, as they can worsen the condition.

Acid-base balance: Diets that induce metabolic alkalosis (e.g., high-anion diets) should be avoided, as they reduce calcium solubility and availability.

Supporting bone health through nutrition

1. ***Encouraging bone remodelling***: Provide nutrients that support balanced bone formation and resorption. Include natural sources of vitamin K (e.g., leafy greens), which promote bone remodelling.
2. ***Omega-3 fatty acids***: Anti-inflammatory properties of omega-3 fatty acids can help mitigate bone inflammation and improve overall skeletal health. Provide flaxseed or fish oil as a dietary supplement.

Prevention of secondary complications

Nutritional management also aims to prevent complications associated with osteopetrosis:

1. ***Prevent fractures***: Ensure animals receive adequate bedding and avoid excessive exercise to reduce fracture risk.
2. ***Bone marrow support***: Address anaemia and immune deficiencies caused by bone marrow suppression with iron, copper, and vitamin B_{12} supplementation.
3. ***Neurological protection***: Manage compression-related symptoms through dietary and environmental support.

Monitoring and long-term management

1. ***Regular blood testing***: Monitor serum calcium, phosphorus, magnesium, and alkaline phosphatase levels to assess bone metabolism.
2. ***Dietary adjustments***: Regularly evaluate and adjust the diet based on clinical progress and blood test results.
3. ***Collaboration with veterinarians***: Work closely with a veterinary professional to monitor and manage the condition effectively.

Prognosis and outcomes

The prognosis for animals with osteopetrosis depends on the severity and whether the condition is congenital or acquired:

Mild cases

Nutritional management can improve symptoms and quality of life.

Severe cases

Animals with significant skeletal deformities or bone marrow suppression may have a poor prognosis.

Osteopetrosis in animals is a complex condition resulting in abnormal calcium metabolism and skeletal abnormalities. Nutritional management plays a critical role in mitigating symptoms, supporting bone health, and preventing secondary complications. By maintaining an optimal calcium-phosphorus balance, ensuring adequate vitamin D and magnesium levels, and addressing overall dietary requirements, caretakers can enhance the quality of life for affected animals. Early intervention and tailored nutritional strategies, combined with veterinary care, a key to managing this rare but impactful condition.

n) Pituitary gland disorders/ Pituitary gland abnormalities

Calcium disorders in animals can arise from various systemic conditions, including pituitary gland dysfunction. The pituitary gland, as the master endocrine organ, regulates multiple hormones that influence calcium metabolism, primarily through its effects on the parathyroid glands, kidneys, bones, and vitamin D metabolism. Disorders of the pituitary gland, such as hyperadrenocorticism (Cushing's disease) and hypopituitarism, can indirectly cause disturbances in calcium homeostasis. Effective nutritional management is crucial for mitigating these imbalances and maintaining optimal health in affected animals.

Role of pituitary gland in calcium metabolism

The pituitary gland influences calcium balance indirectly through hormone regulation:

Growth hormone (GH): Stimulates bone growth and remodelling, indirectly affecting calcium storage in the skeleton.

Adrenocorticotropic hormone (ACTH): Regulates cortisol secretion, which impacts calcium absorption and bone resorption.

Thyroid-stimulating hormone (TSH): Influences thyroid function, indirectly affecting calcium levels through calcitonin secretion.

Prolactin: Plays a role in calcium mobilization during lactation.

Indirect parathyroid regulation: The pituitary impacts parathyroid hormone (PTH) activity, which governs calcium mobilization from bones, renal reabsorption, and intestinal absorption.

Common pituitary disorders associated with calcium dysregulation

1. **Hyperadrenocorticism (Cushing's disease)**

   Pathophysiology: Excess cortisol production caused by pituitary tumors (pituitary-dependent hyperadrenocorticism) or adrenal tumors disrupts calcium balance. Chronic cortisol excess leads to increased bone resorption reduced intestinal calcium absorption, and renal calcium excretion. Animals may develop osteoporosis, hypocalcemia, or soft tissue calcification (metastatic calcification).

   Clinical signs: Weakness, muscle atrophy, polyuria/polydipsia (PU/PD), and fragile bones, Poor wound healing and recurrent infections,

2. **Hypopituitarism**

   Pathophysiology: Insufficient secretion of pituitary hormones, including GH and ACTH, affects calcium metabolism. Reduced growth hormone levels impair bone formation and calcium storage. Hypothyroidism secondary to hypopituitarism can alter calcium homeostasis through decreased calcitonin secretion.

   Clinical signs: Stunted growth (in juveniles), lethargy, hypocalcemia and impaired bone mineral density.

3. **Pituitary tumors:** Hormone-secreting tumors can cause hypercalcemia or hypocalcemia by disrupting systemic hormone levels. Non-secreting tumors may indirectly affect calcium metabolism by damaging pituitary tissue.

Nutritional management of pituitary gland dysfunction

Effective nutritional strategies are vital to managing calcium imbalances in animals with pituitary gland disorders. These strategies aim to correct deficiencies or excesses, prevent complications, and support overall health.

A. Dietary calcium-phosphorus balance

1. ***Calcium and phosphorus ratios***: Maintain a dietary calcium-to-phosphorus ratio of 1.5:1 to 2:1. Avoid high-phosphorus diets, which exacerbate hypocalcemia by stimulating parathyroid hormone (PTH) secretion.
2. ***Calcium supplementation***: Hypocalcemic animals may require dietary calcium supplementation using calcium carbonate or calcium gluconate. Avoid over-supplementation to prevent hypercalcemia, which can lead to metastatic calcification.

B. Vitamin D regulation

1. ***Role of vitamin D***: Vitamin D enhances calcium absorption from the intestines and mobilization from bones. Supplementation is crucial in animals with hypocalcemia due to reduced calcium absorption.
2. ***Sources***: Provide cholecalciferol (vitamin D_3) through fortified feeds or supplements.

   Ensure sunlight exposure for natural vitamin D synthesis in grazing animals.
3. ***Monitoring***: Avoid excessive vitamin D, as it can lead to hypercalcemia and calcification of soft tissues.

C. Protein intake

1. ***Adequate protein levels***: Protein is essential for bone matrix formation and maintaining calcium-binding proteins in the blood. Provide high-quality protein from animal or plant sources in balanced amounts.
2. ***Excess protein***: Avoid diets excessively high in protein, as it may increase urinary calcium excretion and worsen calcium loss.

D. Magnesium supplementation

1. ***Role in calcium homeostasis***: Magnesium supports calcium metabolism by enhancing PTH activity and maintaining bone integrity. Supplement diets with magnesium oxide or magnesium sulfate to correct deficiencies.
2. ***Magnesium deficiency***: Hypomagnesemia can exacerbate hypocalcemia, leading to tetany and neuromuscular dysfunction.

E. Phytate and oxalate considerations

1. ***Phytates***: Found in grains and legumes, phytates bind calcium and reduce its bioavailability.

   Minimize phytate-rich feeds in animals with hypocalcemia.

2. ***Oxalates***: Oxalates in some forages (spinach and beet greens) and in dry poor quality roughages (wheat straw and paddy straw) form insoluble calcium oxalate, reducing calcium absorption. Limit oxalate-containing plants in the diet of at-risk animals.

F. Electrolyte balance

1. ***Sodium and potassium***: Sodium and potassium influence renal calcium excretion. Provide balanced electrolyte supplementation to support calcium metabolism.
2. ***Chloride***: Chloride influences the acid-base balance, affecting calcium solubility and excretion.

G. Energy management

1. ***Energy requirements***: Provide adequate energy to support normal metabolic functions and prevent mobilization of calcium from bones. High-energy diets may be needed in hyperadrenocorticism to counteract muscle wasting and weight loss.
2. ***Avoid obesity***: Excessive weight can exacerbate skeletal stress and calcium imbalances.

Preventative measures

1. ***Routine monitoring***: Regular blood tests should be done to monitor calcium, phosphorus, and magnesium levels. Hormone assays to evaluate pituitary function.
2. ***Balanced rations***: Ensure diets meet the specific nutritional needs of the species and physiological state (e.g., lactation, growth).
3. ***Environmental management***: Provide proper housing, exercise, and stress reduction to support overall endocrine health.

Prognosis and outcomes

With proper nutritional management and veterinary intervention, many animals with pituitary-related calcium disorders can achieve improved quality of life.

Long-term outcomes depend on the underlying cause, the severity of calcium imbalance, and adherence to dietary recommendations.

Calcium disorders in animals with pituitary gland dysfunction are complex and multifactorial. Nutritional management is essential for addressing these imbalances, supporting recovery, and preventing complications. By ensuring a balanced calcium-phosphorus ratio, regulating vitamin D and magnesium levels, and tailoring dietary energy and protein intake, caretakers can mitigate the effects of these disorders and promote optimal health in affected animals.

o) Paget's disease

Paget's disease, also known as osteitis deformans, is a chronic condition that affects the skeletal system by disrupting the normal cycle of bone remodelling. While it is predominantly reported in humans, Paget's disease or similar calcium-related disorders have also been observed in animals. These disorders alter the balance between bone resorption and formation, leading to structural abnormalities, pain, and compromised skeletal integrity.

Nutritional management plays a supportive role in addressing Paget's disease by promoting optimal bone health and maintaining calcium balance. Although medical treatments like bisphosphonates are the primary therapy, a well-balanced diet can enhance treatment outcomes and improve the overall health of affected animals.

This document provides an in-depth exploration of Paget's disease and its relevance as a calcium metabolism disorder in animals, covering its pathophysiology, clinical presentation, diagnosis, treatment, and potential impact on animal health.

Pathophysiology

The hallmark of Paget's disease is abnormal bone remodelling. Bone is a dynamic tissue that undergoes continuous turnover through the actions of osteoclasts (bone-resorbing cells) and osteoblasts (bone-forming cells). In Paget's disease, this tightly regulated process is disrupted:

1. ***Excessive bone resorption***: Osteoclasts become hyperactive, leading to accelerated bone degradation.
2. ***Compensatory bone formation***: Osteoblasts attempt to compensate by laying down new bone rapidly. However, the new bone is often structurally weak and disorganized.
3. ***Calcium imbalance***: This dysregulated remodelling causes a local and systemic disruption in calcium homeostasis, leading to secondary complications like hypercalcemia in severe cases.

In animals, as in humans, these changes result in skeletal deformities, fragility, and an increased risk of fractures.

Causes and risk factors

While the exact cause of Paget's disease in animals is not fully understood, several factors are believed to contribute:

1. ***Genetic predisposition***: Genetic mutations affecting osteoclast function have been implicated in some cases of Paget-like disorders in animals.

2. ***Viral infections***: In humans, paramyxo viruses have been linked to Paget's disease. Although evidence in animals is limited, similar viral aetiologies are under investigation.
3. ***Dietary imbalances***: Deficiencies or excesses of dietary calcium, phosphorus or vitamin D can exacerbate calcium disorders, potentially triggering or worsening skeletal abnormalities.
4. ***Environmental factors***: Chronic exposure to toxins or environmental stressors may disrupt normal bone metabolism in animals.

Species affected

Paget's disease or similar calcium-related skeletal disorders have been observed in various species:

1. ***Dogs***: Certain breeds may develop conditions resembling Paget's disease, particularly those with genetic predispositions to skeletal abnormalities.
2. ***Cats***: Feline cases are rare, but calcium imbalance disorders affecting bone remodeling have been reported.
3. ***Horses***: In horses, calcium-related bone disorders often manifest as lameness, fractures, or bone deformities, though they are not always classified as Paget's disease.
4. ***Livestock***: Ruminants like cattle and sheep can experience calcium metabolic disorders due to dietary imbalances, which may lead to Paget-like skeletal changes.

Clinical signs

The clinical presentation of Paget's disease in animals varies depending on the species, severity, and bones affected. Common signs include:

1. ***Skeletal deformities***: Bowed limbs, spinal curvature, or skull enlargement due to abnormal bone growth.
2. ***Pain and lameness***: Increased bone fragility and microfractures lead to discomfort and impaired mobility.
3. ***Neurological symptoms***: Compression of nerves due to bone overgrowth, particularly in the skull or spine, can cause neurological deficits.
4. ***Fractures***: Weak, brittle bones are prone to fractures even with minimal trauma.
5. ***Systemic signs***: In some cases, hypercalcemia may cause lethargy, anorexia, vomiting, or kidney dysfunction.

Diagnosis

Diagnosing Paget's disease or similar disorders in animals involves a combination of clinical examination, imaging, and laboratory tests:

1. ***Radiographic imaging***: X-rays reveal characteristic bone changes, including thickening, deformities, and areas of increased or decreased bone density.
2. ***Bone scans***: Nuclear scintigraphy can help identify active areas of abnormal bone remodelling. (Scintigraphy is a diagnostic procedure that uses a radioactive chemical to create images of the inside of the body. It's also known as a gamma scan or radionuclide).
3. ***Biochemical tests:*** Elevated levels of alkaline phosphatase (ALP) suggest increased osteoblastic activity. Calcium and phosphorus levels may indicate systemic imbalances.
4. ***Bone biopsy***: Histopathological examination of affected bone tissue confirms the diagnosis by showing disorganized bone architecture and increased remodelling activity.

Treatment

There is no definitive cure for Paget's disease, but treatment aims to manage symptoms, slow disease progression, and improve the animal's quality of life.

1. ***Medications***:

   Bisphosphonates: These drugs inhibit osteoclast activity, reducing excessive bone resorption and normalizing bone remodelling.

   Calcitonin: A hormone that regulates calcium levels and reduces bone turnover, providing symptomatic relief.

   Pain management: Non-steroidal anti-inflammatory drugs (NSAIDs) or opioids may be used to control pain and inflammation.
2. ***Dietary management***: Balanced levels of calcium and phosphorus are essential to support healthy bone metabolism. Supplementation with vitamin D may be necessary in cases of deficiency.
3. ***Surgery***: In severe cases, surgical intervention may be required to stabilize fractures or correct deformities.
4. ***Physical therapy***: Controlled exercise and physiotherapy can help maintain mobility and prevent muscle atrophy.

Prognosis

The prognosis for animals with Paget-like disease depends on the extent of skeletal involvement, the presence of complications, and the effectiveness of treatment. Early detection and intervention can significantly improve outcomes, although chronic cases may require ongoing management.

Prevention and nutritional management of Paget's disease

Preventive measures focus on minimizing risk factors and maintaining optimal skeletal health:

1. ***Proper nutrition***: Ensuring a diet with appropriate levels of calcium, phosphorus, and vitamin D is crucial for preventing calcium metabolic disorders.
2. ***Genetic screening***: Breeding programs in domestic animals can help reduce the prevalence of genetic predispositions to bone disorders.
3. ***Regular veterinary care***: Routine check-ups and early detection of abnormalities can prevent disease progression.

Key nutritional considerations

1. Calcium

Role: Calcium is essential for bone mineralization and strength. In Paget's disease, abnormal bone remodelling increases the need for calcium to support healthy bone formation.

Sources: Include calcium-rich foods such as dairy products, leafy greens, and calcium-fortified feeds in the animal's diet. Supplements may be necessary if dietary intake is inadequate.

Caution: Excess calcium supplementation should be avoided, as it can contribute to hypercalcemia and kidney issues.

2. Phosphorus

Role: Phosphorus works in tandem with calcium for bone development. Maintaining the proper calcium-to-phosphorus ratio (usually 2:1) is critical to prevent imbalances that could exacerbate skeletal deformities.

Sources: Meat, fish, grains, and bone meal are common sources of phosphorus.

3. Vitamin D

Role: Vitamin D enhances calcium absorption from the gastrointestinal tract and regulates bone mineralization. Deficiency can impair calcium metabolism, worsening Paget's symptoms.

Sources: Sunlight exposure is a natural source of vitamin D, and dietary sources include fish oil, liver, and fortified feeds. Supplementation may be necessary in cases of deficiency.

4. Protein

Role: Adequate protein intake is essential for maintaining muscle mass and providing the amino acids required for bone matrix synthesis.

Sources: Meat, eggs, legumes, and high-quality animal feeds provide sufficient protein.

5. Antioxidants

Role: Oxidative stress may play a role in bone degeneration. Antioxidant-rich diets, including vitamins C and E, help combat free radical damage and support overall health.

Sources: Fruits, vegetables, and fortified feeds are good sources of antioxidants.

Dietary management strategies

Balanced diet: Tailor the diet to the animal's specific needs, considering age, species, and disease severity.

Weight management: Maintain a healthy weight to reduce stress on weakened bones.

Supplements: Consult a veterinarian before introducing calcium, vitamin D, or other supplements to avoid potential complications.

By ensuring optimal nutrition, animals with Paget's disease can achieve better bone health and improved quality of life.

Research and future directions

Research into Paget's disease and related conditions in animals is ongoing. Advances in genetic studies, biomarker identification, and imaging technologies may enhance our understanding of the disease and improve diagnostic and therapeutic approaches. Comparative studies between human and animal cases also offer valuable insights into the pathophysiology of Paget's disease and its management.

Paget's disease is a complex skeletal disorder that disrupts calcium metabolism, leading to significant health challenges in affected animals. Although less commonly reported in animals than in humans, it highlights the importance of understanding bone remodelling disorders and their impact on animal welfare. By improving diagnostic techniques, treatment options, and preventive strategies, veterinarians and researchers can enhance the quality of life for animals suffering from these debilitating conditions.

29

Dwarfism Due to High-level/ Long-duration Feeding of *Leucaena leucocephala*

Leucaena leucocephala, commonly known as Subabool or Koobabool, is a fast-growing leguminous shrub often used as fodder for livestock due to its high protein content and palatability. Despite its nutritional benefits, long-term or excessive feeding of Subabool can lead to adverse health effects, including dwarfism in livestock. This condition is primarily caused by the presence of toxic compounds in the plant, notably mimosine and its breakdown products, which interfere with normal growth and metabolic functions.

Nutritional profile of subabool

Subabool is a valuable fodder crop due to its high protein content (20–30%) and digestibility. It also contains significant levels of vitamins and minerals, making it a preferred choice for supplementing livestock diets, particularly in regions with limited forage availability. However, its nutritional benefits are overshadowed by the presence of anti-nutritional factors, particularly mimosine and its derivatives.

Mimosine and its toxic effects

Mimosine, a non-protein amino acid, is the primary toxic compound in Subabool. It constitutes about 5–12% of the dry weight of the plant and can exert toxic effects on animals through:

Inhibition of cell division

Disruption of protein synthesis

Chelation of essential minerals like zinc and iron, and impaired metabolic and enzymatic functions.

When ingested, mimosine is metabolized in the rumen by rumen microbes into 3-hydroxy-4(1H)-pyridone (DHP), which is even more toxic. The accumulation of mimosine and DHP in the animal's system leads to adverse health effects, including dwarfism.

Mechanism of dwarfism due to subabool feeding

Dwarfism caused by Subabool feeding is the result of several interconnected factors that impair normal growth:

1. ***Disruption of protein metabolism***: Mimosine interferes with protein synthesis by inhibiting enzymes involved in DNA replication and cell division, leading to reduced growth and development in young animals.
2. ***Mineral Imbalance***: The chelation of essential minerals likc iron and zinc by mimosine disrupts metabolic processes necessary for bone growth and hormone function.
3. ***Thyroid dysfunction***: Mimosine and DHP have goitrogenic properties, meaning they interfere with thyroid hormone production. Hypothyroidism slows metabolic rate and growth, contributing to stunted development and dwarfism.
4. ***Reduced feed intake and digestive efficiency***: Subabool feeding can cause anorexia or reduced appetite due to the bitter taste of mimosine, further limiting nutrient intake and growth. The toxic effects on rumen microbes impair fermentation and digestion, reducing nutrient absorption.
5. ***Cumulative toxicity***: Prolonged exposure to mimosine and DHP leads to systemic toxicity, manifesting as poor body condition, hair loss, and growth retardation.

Clinical signs of subabool-induced dwarfism

Animals fed Subabool for long durations or at high levels exhibit a range of symptoms associated with mimosine toxicity and growth impairment:

1. ***Growth retardation***: Stunted growth or dwarfism particularly observed in young animals. Failure to achieve expected body size and body weight in young ones.
2. ***Thyroid enlargement***: Goiter due to disruption of thyroid hormone synthesis.
3. ***Hair and skin issues***

   Hair loss (alopecia) and rough coat texture are common.

   Skin lesions or depigmentation may occur in severe cases.
4. ***Anemia***: Chelation of iron leads to reduced hemoglobin levels and anemia, contributing to lethargy and weakness.
5. ***Poor body condition***: Weight loss, muscle wasting, and reduced body fat reserves.

6. ***Reduced reproductive performance***: In mature animals, prolonged Subabool feeding can affect fertility and milk production.

Nutritional management of subabool-induced dwarfism

Effective management of dwarfism caused by Subabool involves reducing exposure to its toxic compounds, detoxifying its effects, and providing balanced nutrition to support recovery and growth.

1. **Controlled feeding of subabool**

   Limit subabool proportion in diet: Subabool should constitute no more than 30% of the total dry matter intake to minimize toxicity risks. Mix subabool with other forages like grasses or leguminous fodder without anti-nutritional factors to dilute its toxic effects.

   Avoid feeding to vulnerable groups: Young, pregnant, or lactating animals are more susceptible to mimosine toxicity and should receive minimal Subabool in their diets.

2. **Processing and detoxification**

   Processing Subabool can significantly reduce its mimosine content and toxicity:

   Drying and chopping: Sun-drying or wilting Subabool reduces mimosine levels by up to 30%.

   Ensiling: Fermentation during ensiling can degrade mimosine and improve palatability.

   Soaking in water: Soaking Subabool leaves in water for 24–48 hours leaches out mimosine.

   Chemical treatments: Calcium hydroxide or lime treatments can neutralize mimosine in Subabool.

3. **Supplementation to counteract toxicity:** Providing supplements to counteract the adverse effects of mimosine is crucial for animals already affected by subabool toxicity:

   1. ***Minerals***: Supplement with iron, zinc, and copper to counteract deficiencies caused by mimosine chelation. Use mineral blocks or fortified feeds to ensure consistent intake.

   2. ***Protein and energy***: Feed high-quality protein sources (e.g., soybean meal, fishmeal) to compensate for reduced protein synthesis. Energy-dense feeds like molasses or grains can support recovery.

   3. ***Iodine supplementation***: Administer iodine supplements or iodized salt to mitigate goitrogenic effects and support thyroid function.

4. ***Vitamin support***: Provide vitamins A, D, and E to support immune function, bone growth, and recovery.

4. ***Probiotic and rumen modifiers:*** Introduce probiotics or rumen additives to restore rumen function and detoxify DHP:

   Probiotics: Strains like *Lactobacillus* and *Saccharomyces* help improve digestion and reduce toxin accumulation.

   Rumen microbial adaptation: Some rumen microbes (e.g., *Synergistes jonesii*) can degrade mimosine and DHP. Inoculating livestock with these microbes can enhance tolerance to subabool.

5. ***Alternative forages:*** To reduce reliance on subabool, incorporate alternative high-protein forages like:

   Lucerne (alfalfa)

   Desmanthus or *Stylosanthes species*

   Grasses like Napier or Rhodes grass.

Recovery and long-term management

1. ***Monitor growth and body condition***: Regularly track weight and body size to assess recovery in affected animals.
2. ***Gradual dietary adjustments***: Transition affected animals to a balanced diet with minimal subabool content.
3. ***Avoid prolonged feeding***: Rotate Subabool with other forage crops to prevent cumulative toxicity.

Subabool is a nutritious fodder source but its excessive or prolonged use can lead to significant health issues, including dwarfism, due to the toxic effects of mimosine and its derivatives. Proper feeding practices, detoxification techniques, and dietary supplementation are essential to mitigate these risks and support animal growth and productivity. By managing subabool feeding responsibly and providing a balanced diet, livestock farmers can maximize its benefits while minimizing its adverse effects.

30

Calcinosis in Grazing Animals

Calcinosis is a condition characterized by the abnormal deposition of calcium salts in soft tissues, often leading to significant morbidity in affected animals. This condition is particularly relevant in grazing animals, as it can be associated with dietary imbalances, exposure to toxic plants, and metabolic disorders. Depending on the etiology, calcinosis can present as dystrophic, metastatic, or calcinosis circumscripta. Understanding the pathogenesis, clinical signs, diagnosis, and management is crucial for prevention and control in grazing animals.

Types of calcinosis

1. ***Dystrophic calcinosis:*** This type of calcinosis occurs when calcium deposits in damaged or necrotic tissues despite normal blood calcium levels. Associated with trauma, infections, or localized inflammation.
2. ***Metastatic calcinosis:*** This type of calcinosis results from hypercalcemia, which leads to calcium deposition in normal tissues.

   Common causes include:

   Nutritional imbalances (excessive calcium or vitamin D intake),

   Toxic plants (e.g., Cestrum diurnum),

   Chronic renal failure, causing disturbances in calcium-phosphorus metabolism,

   Paraneoplastic syndromes associated with tumors.
3. ***Calcinosis circumscripta***

   A localized form of calcification seen in subcutaneous tissues or near joints,

   Rare in grazing animals but may be associated with trauma or idiopathic causes.

Etiology of calcinosis in grazing animals

A. Plant-induced calcinosis

Cestrum diurnum (day-blooming jessamine)

This plant contains glycosides that mimic vitamin D activity, causing increased intestinal calcium absorption and hypercalcemia. It is common in tropical and subtropical regions where *Cestrum diurnum* grows abundantly.

Solanum malacoxylon

Found in South America, this plant produces substances similar to vitamin D. The grazing animals consume this plant often develop metastatic calcinosis.

B. Dietary imbalances

Excess dietary calcium or vitamin D

Occurs in improperly formulated feed supplements,

High phosphorus diets can also disrupt calcium-phosphorus balance, contributing to calcification.

Deficient magnesium levels

Hypomagnesemia can interfere with calcium homeostasis, indirectly contributing to calcinosis.

C. Chronic disease

Renal failure

Impaired renal function disrupts calcium and phosphorus metabolism, leading to hyperphosphatemia and soft tissue calcification.

Endocrine disorders

Parathyroid hormone imbalances (hyperparathyroidism) can result in metastatic calcification.

Pathophysiology

Calcinosis arises when there is an imbalance between calcium and phosphorus homeostasis, leading to precipitation of calcium salts (calcium phosphate or calcium carbonate) in tissues. The mechanism varies by type:

In metastatic calcinosis, hypercalcemia or hyperphosphatemia leads to the deposition of calcium in normal tissues.

In dystrophic calcinosis, calcium salts accumulate in areas of tissue necrosis without systemic calcium disturbances.

Plant-induced calcinosis involves compounds that mimic vitamin D, causing hypercalcemia and calcification of soft tissues such as arteries, kidneys, and tendons.

Key tissues affected include

Muscles: Stiffness and reduced mobility.

Tendons and ligaments: Calcification leads to lameness.

Arteries: Loss of elasticity, predisposing to cardiovascular complications.

Kidneys: Impaired renal function due to calcification.

Clinical signs

The clinical presentation of calcinosis varies with the severity and extent of calcium deposition:

Early signs

Reduced feed intake and weight loss,

Generalized stiffness or reluctance to move,

Musculoskeletal symptoms

Lameness and swollen joints,

Muscle hardening due to calcification,

Systemic symptoms

Depression and lethargy

Decreased milk production in lactating animals

Other signs

Calcified arteries may cause poor circulation.

Respiratory distress if the lungs are involved.

In cases of severe renal involvement, signs of kidney failure may occur, such as polyuria and polydipsia.

Diagnosis

Clinical examination

Observation of stiffness, lameness, and hard nodules under the skin

History of exposure to toxic plants or dietary imbalances

Laboratory tests

1. ***Blood chemistry:***

   Elevated serum calcium and phosphorus levels in metastatic calcinosis

   Normal calcium levels in dystrophic calcinosis.

2. ***Urine analysis***: Increased calcium excretion (hypercalciuria) in cases of plant-induced calcinosis

3. ***Enzyme activity***: Elevated alkaline phosphatase, indicating bone metabolism disturbances,

Imaging

Radiography: Reveals calcification in soft tissues, tendons, and joints

Ultrasound: Useful for detecting calcification in internal organs like kidneys

Histopathology

Tissue biopsy showing calcium salt deposits confirms the diagnosis

Treatment

1. ***Immediate management***

   Remove the source of calcium or vitamin D overload: Stop access to toxic plants.

   Adjust dietary formulations to correct calcium-phosphorus ratios.

   Provide supportive care to maintain hydration and electrolyte balance.

2. ***Medical therapy***

   Chelation therapy: Sodium EDTA may be used to bind excess calcium and promote excretion.

   Diuretics: Furosemide increases calcium excretion via the kidneys.

   Bisphosphonates: Inhibit bone resorption and reduce hypercalcemia.

   Glucocorticoids: May reduce calcium absorption and mobilization but are rarely used long-term in grazing animals.

3. ***Surgical intervention***

   In cases of calcinosis circumscripta, surgical removal of calcified nodules may be necessary.

4. ***Supportive care***

   Pain management for animals with lameness or musculoskeletal discomfort,

   Nutritional support to address weight loss and deficiencies,

Prevention

A. Pasture management

Identify and remove toxic plants like *Cestrum diurnum* and *Solanum malacoxylon* from grazing areas.

Rotate pastures to prevent overgrazing and minimize exposure to risky flora.

B. Dietary control

Maintain balanced rations with appropriate calcium and phosphorus levels.

Avoid excessive supplementation of vitamin D.

Ensure magnesium levels are adequate to support calcium metabolism.

C. Regular monitoring

Perform routine blood tests in herds grazing in high-risk areas.

Monitor feed composition, especially in intensively managed systems.

D. Education

Train farmers and herders to recognize toxic plants and early signs of calcinosis.

Prognosis

The prognosis for animals with calcinosis depends on the extent of calcium deposition and the timeliness of intervention:

Mild cases

Animals recover well with early diagnosis and removal of the causative factor.

Severe cases

Chronic calcification in vital organs or the musculoskeletal system may lead to permanent impairment or death.

Nutritional management of calcinosis

Calcinosis in grazing animals, characterized by abnormal calcium deposition in soft tissues, often stems from dietary imbalances, exposure to toxic plants, or metabolic disorders. Effective nutritional management plays a pivotal role in preventing, controlling, and mitigating this condition.

1. Correcting calcium-phosphorus balance

Maintaining an appropriate calcium-to-phosphorus (Ca:P) ratio in the diet is crucial:

Ideal Ratio: For most grazing animals, the optimal Ca:P ratio is between 1.5:1 and 2:1.

Excess calcium: Over-supplementation of calcium, often through mineral licks or improperly formulated feeds, must be avoided to prevent hypercalcemia.

Phosphorus supplementation: Phosphorus deficiency can exacerbate calcinosis. Providing phosphorus-rich supplements like di-calcium phosphate helps restore balance.

2. Vitamin D regulation

Excessive vitamin D intake is a major factor in plant-induced calcinosis:

Avoid over-supplementation of vitamin D in feed, as it can mimic the effects of toxic plants like *Cestrum diurnum* or *Solanum malacoxylon*.

For animals grazing in areas with high exposure to these plants, dietary vitamin D levels should be minimized to reduce the additive effect.

3. Magnesium supplementation

Magnesium plays a critical role in calcium metabolism:

Deficiency: Hypomagnesemia can exacerbate calcium deposition by disrupting calcium-phosphorus homeostasis.

Solution: Supplementation with magnesium oxide or magnesium sulfate in feed or mineral mixes can help prevent calcinosis.

4. Use of alternative feeds

In areas where toxic plants like *Cestrum diurnum* are prevalent:

Restrict access: Prevent grazing on affected pastures by fencing or rotational grazing.

Alternative forages: Provide non-toxic fodder such as hay or silage to reduce reliance on natural pastures.

Supplementary feeding: Use balanced concentrates to meet nutritional requirements without exposing animals to risky forage.

5. Electrolyte and water management

Proper hydration and electrolyte balance support metabolic processes:

Clean water: Ensure access to fresh, clean water to facilitate the excretion of excess calcium and toxins.

Electrolyte supplementation: Provide balanced mineral supplements containing sodium, potassium, and chloride to support renal and systemic homeostasis.

6. Monitoring and adjustment

Regular assessment of dietary composition and animal health is essential:

Blood testing: Monitor serum calcium, phosphorus, and magnesium levels to detect imbalances early.

Dietary audits: Periodically evaluate feed and forage quality for nutrient content and adjust rations accordingly.

Proactive prevention

Educate herders about the risks of toxic plants and the importance of balanced nutrition.

Rotate pastures to prevent overgrazing and minimize exposure to harmful plants.

Economic impact

Calcinosis can have significant economic repercussions for livestock producers:

Reduced productivity due to weight loss, lameness, and decreased milk yield,

Increased veterinary costs for treatment and management,

Loss of affected animals in severe cases,

Preventative strategies are, therefore, essential to minimize economic losses in affected regions.

Calcinosis in grazing animals is a multifactorial condition with potentially severe consequences. Early recognition, effective management and preventative measures are crucial for controlling this disease. Proper pasture management, balanced nutrition, and regular health monitoring are essential for protecting the health and productivity of grazing animals.

Nutritional management is integral to preventing and managing calcinosis in grazing animals. By maintaining a balanced diet, regulating vitamin D intake, and addressing mineral imbalances, farmers can safeguard animal health and productivity while reducing the economic impact of this condition.

31

Diseases of Phosphorus Disorders

Phosphorus is a critical mineral for animal health, playing a vital role in bone development, energy metabolism, cell signalling, and overall physiological functions. Disorders related to phosphorus imbalance, whether due to deficiency or excess, can lead to significant health issues in animals.

a) Pica

Pica is a condition characterized by the persistent ingestion of non-nutritive substances such as soil, rocks, plastic, cloth, or other inappropriate materials. While commonly observed in humans, pica is also prevalent in animals, including livestock, domestic pets, and wildlife. It is often a sign of underlying health, environmental, or nutritional issues. Understanding the causes, implications, and management of pica is crucial for animal welfare and productivity.

Causes of pica in animals

The aetiology of pica in animals is multifaceted and can be broadly categorized into nutritional, medical, behavioural, and environmental factors.

Nutritional deficiencies

Nutritional imbalances are the most common cause of pica. Deficiencies in key minerals like phosphorus, sodium, calcium, or trace elements such as iron, zinc, and cobalt often lead animals to consume non-nutritive substances in an attempt to compensate. For example:

***Phosphorus defic*iency**: Animals may consume bones or wood.

Salt deficiency: They may lick soil or other salty objects.

Ruminants are particularly susceptible to pica when their diets lack balanced mineral supplementation.

Medical issues

Gastrointestinal disorders

Conditions like gastritis, gastric ulcers, or intestinal parasites can cause animals to eat non-food items.

Neurological or metabolic disorders

Issues like liver disease or conditions affecting the nervous system can lead to abnormal behaviours, including pica.

Behavioural factors

Pica can also stem from stress, boredom, or anxiety. This is particularly observed in confined or socially isolated animals, such as dogs chewing on inappropriate items or confined livestock ingesting bedding materials.

Environmental factors

Poor housing, lack of access to natural grazing, or exposure to contaminated environments can trigger pica. For example, animals housed on barren or contaminated grounds may ingest inappropriate materials due to lack of options.

Consequences of pica

Pica can have significant health implications for animals. These include:

1. ***Physical injuries and obstructions:*** Ingested objects can cause damage to the digestive tract, leading to perforations, blockages, or impactions. For example, plastic ingestion often causes severe intestinal obstructions.
2. ***Toxicity:*** Non-nutritive materials may contain toxins or chemicals, leading to poisoning. Lead toxicity from ingesting paint chips is a notable example in dogs and livestock.
3. ***Parasitic infections:*** Consuming contaminated soil or faeces may expose animals to parasites such as roundworms, hookworms, or protozoa.
4. ***Reduced productivity:*** In livestock, pica can lead to reduced growth rates, decreased milk production, and overall poor body condition, affecting farm profitability.

Nutritional management of pica

Addressing pica requires a holistic approach, with nutritional management playing a central role. Key strategies include:

1. ***Identifying and correcting nutritional deficiencies***

   Proper diagnosis is crucial to identify deficiencies causing pica. This involves:

   Feed analysis: Testing the nutrient profile of feed to identify any deficiencies.

   Mineral supplementation: Providing balanced mineral mixes tailored to the species and physiological needs of the animal. For instance:

Phosphorus and calcium can be supplemented through di-calcium phosphate or bone meal. Salt licks containing sodium chloride can prevent salt cravings. Trace minerals like zinc and copper should be included in mineral premixes.

2. ***Balancing diets:*** Diets should be balanced to meet the animal's energy, protein, and mineral needs. Ensuring a proper forage-to-concentrate ratio in ruminants and providing varied, high-quality diets in pets are effective measures.
3. ***Improving forage quality:*** Low-quality forages lacking in nutrients may drive animals to seek alternative sources. Ensuring access to nutrient-rich pastures or supplementing with high-quality hay can prevent pica.
4. ***Supplementing vitamins and trace elements:*** Trace element deficiencies (e.g., cobalt, selenium, iodine) are common in pica cases. Injectable or oral mineral supplements may be required for quick correction. Similarly, vitamin deficiencies, such as vitamin A, can be addressed through fortified feeds.

Environmental and behavioural management

In addition to nutritional strategies, managing environmental and behavioural factors is vital to preventing pica:

1. ***Improving housing and enrichment:*** Providing adequate space, social interaction, and environmental enrichment can reduce stress and boredom, which often contribute to pica.

   For pets: Toys, exercise, and training can prevent destructive chewing behaviours.

   For livestock: Access to clean, spacious pens and grazing opportunities can mitigate pica.
2. ***Access to clean water and salt:*** Dehydration or lack of access to clean water may drive animals to consume inappropriate materials. Similarly, providing free-choice access to salt blocks or electrolyte solutions can address sodium deficiencies.
3. ***Avoiding contaminated environments:*** Ensuring animals are housed in clean, uncontaminated environments prevents accidental ingestion of harmful substances. Proper waste management and soil testing for heavy metals are particularly important for grazing animals.

Monitoring and long-term management

1. ***Regular health checks:*** Routine veterinary check-ups are essential to detect and address underlying medical conditions that may predispose animals to pica.
2. ***Continuous nutritional assessment:*** Regular evaluation of dietary components and periodic blood tests to monitor mineral and vitamin levels ensure the diet remains balanced over time.
3. ***Behavioural training:*** In pets, addressing pica through behavioural interventions, such as training and positive reinforcement, can be effective in preventing recurrence.

Case studies and examples

1. ***Livestock***

   Cattle and buffaloes: Pica is frequently observed in cattle and buffaloes grazing on phosphorus-deficient pastures. Supplementing with phosphorus-rich mineral mixes and improving pasture quality effectively curbs the behaviour.

   Goats and sheep: Salt deficiency in small ruminants often leads to soil consumption. Salt licks are a simple and effective remedy.

2. ***Pet animals***

   Dogs and cats: Behavioural pica in dogs, such as chewing on fabric or eating plastic, is often resolved by addressing boredom and providing appropriate chew toys. Medical pica caused by gastrointestinal upset or anaemia is managed by treating the underlying condition and improving the diet.

3. ***Zoo and captive animals:*** Animals in captivity, such as primates or elephants, may develop pica due to stress or boredom. Enrichment activities and balanced diets help reduce such behaviours.

Pica is a multifaceted condition in animals with significant implications for health and productivity. Nutritional deficiencies are often at its core, but environmental, medical, and behavioural factors also play critical roles. Proper diagnosis, nutritional intervention, and environmental management are essential for effective prevention and treatment. For livestock, providing balanced diets and access to mineral supplements is key, while for pets, a combination of diet optimization and behavioural training is often required. Ultimately, addressing pica involves a comprehensive approach that prioritizes both the physical and psychological well-being of the animal.

b) Hypophosphatemia

Hypophosphatemia is a condition characterized by abnormally low levels of phosphate in the blood. Phosphorus is a critical element in animal physiology, playing a central role in energy metabolism, bone health, and cellular function. It is a key component of ATP (adenosine triphosphate), phospholipids, and nucleic acids, making it essential for cellular function, tissue repair, and growth. Hypophosphatemia can affect a wide range of animal species, including livestock, pets, and wildlife. The causes, clinical signs, diagnosis, treatment, and prevention strategies of hypophosphatemia vary depending on the underlying condition, species, and management practices.

Causes of hypophosphatemia

Several factors can contribute to hypophosphatemia, including dietary deficiencies, metabolic disorders, and certain diseases. These factors can lead to reduced phosphorus intake, impaired absorption, increased excretion, or abnormal distribution of phosphate within the body.

Dietary causes

Phosphorus deficiency in diet: A primary cause of hypophosphatemia in animals is insufficient phosphorus in the diet. This is particularly common in herbivores that graze on pasture with low phosphorus content or consume poorly balanced rations. Phosphorus deficiencies can occur if the feed is low in phosphorus, such as certain forages (e.g., tropical grasses), or if the phosphorus in the diet is poorly bioavailable due to high calcium or magnesium levels.

Inadequate phosphorus supplements: Animals that are not provided with adequate phosphorus supplements in their diet, especially in dairy, poultry, and swine production systems, are at risk of hypophosphatemia. This is particularly significant in lactating animals that have higher phosphorus requirements due to milk production.

2. Metabolic causes

Renal dysfunction: The kidneys play a major role in regulating phosphate balance. Conditions that impair kidney function, such as renal failure or nephritis, can lead to phosphate wasting and resultant hypophosphatemia.

Hyperparathyroidism: Excessive parathyroid hormone (PTH) secretion can increase phosphate excretion by the kidneys, resulting in lower phosphate levels in the blood.

Vitamin D Deficiency: Vitamin D is essential for phosphorus absorption from the gastrointestinal tract. A deficiency of vitamin D, which may occur due to poor sunlight exposure or dietary insufficiency, can reduce phosphorus absorption and contribute to hypophosphatemia.

Endocrine Disorders

Insulin therapy in diabetic animals: The administration of insulin in diabetic animals can cause an intracellular shift of phosphorus, leading to a decrease in blood phosphate levels. This is particularly seen during the treatment of diabetic ketoacidosis.

Refeeding syndrome: Animals that have been severely malnourished and then begin to receive nutritional support too quickly can experience a rapid intracellular shift of phosphorus, leading to severe hypophosphatemia. This is commonly seen in animals recovering from prolonged starvation or anorexia.

Acute and chronic diseases

Sepsis: Severe infection can lead to systemic inflammation and a shift in phosphate from the extracellular to the intracellular space. This can contribute to low blood phosphorus levels, especially if the animal has inadequate nutritional support during recovery.

Trauma or burns: Injuries such as trauma or severe burns can cause a shift of phosphate from the blood into tissues during the healing process, leading to hypophosphatemia.

Other causes

Toxins and poisons: Certain toxins, such as those from plants (e.g., bracken fern) or industrial chemicals, can interfere with phosphorus metabolism and contribute to hypophosphatemia.

Chronic diarrhea: Chronic gastrointestinal diseases, such as diarrhea, can lead to a loss of phosphorus, resulting in hypophosphatemia.

Clinical signs of hypophosphatemia

Hypophosphatemia can present with a wide range of clinical signs, depending on the severity of the deficiency and the underlying causes. Clinical manifestations generally affect the nervous, musculoskeletal, and cardiovascular systems.

Neurological signs

Weakness and lethargy: One of the earliest signs of hypophosphatemia is generalized weakness and lethargy, as phosphate is essential for energy production in all body cells.

Ataxia and tremors: Muscle weakness can progress to ataxia (lack of coordination) and muscle tremors, particularly in severe cases.

Seizures: Low phosphate levels may lead to seizures or convulsions in severe hypophosphatemia due to impaired cellular function and neural excitability.

Musculoskeletal signs

Muscle weakness: Phosphate plays an essential role in muscle contraction and metabolism. Deficiency leads to muscle weakness, particularly in large skeletal muscles, which can affect an animal's ability to stand or move.

Bone deformities and osteomalacia: Chronic phosphorus deficiency can result in weakened bones (osteomalacia) and in severe cases, bone deformities, especially in growing animals or those with higher calcium demands.

Cardiovascular and Respiratory Signs

Heart failure: Phosphate is important for maintaining normal heart function. Hypophosphatemia can lead to cardiovascular dysfunction, including arrhythmias and congestive heart failure.

Respiratory depression: In severe cases, respiratory failure may occur due to impaired muscle function, including diaphragm weakness.

Other signs

Decreased milk production: In lactating animals, hypophosphatemia may result in reduced milk yield.

Reproductive failure: Phosphorus deficiency can impair reproductive function, leading to infertility, delayed parturition, or poor reproductive performance in livestock.

Diagnosis of hypophosphatemia

The diagnosis of hypophosphatemia involves a combination of clinical signs, history, and laboratory tests. The most common diagnostic tool is blood testing to measure phosphate levels.

1. ***Serum phosphate levels:*** The most reliable diagnostic test for hypophosphatemia is the measurement of serum phosphorus. Phosphorus concentration in serum lower than the normal range (typically 2.5-6 mg/dL in most species) confirms the diagnosis of hypophosphatemia.
2. ***Urinary phosphate excretion:*** Urinary phosphorus levels can be measured to assess renal phosphate handling and help differentiate between causes of hypophosphatemia, particularly renal vs. non-renal causes.
3. ***Other tests:*** *Calcium and vitamin D levels*: Since calcium and vitamin D are involved in phosphorus metabolism, measurements of serum calcium and vitamin D levels may provide additional diagnostic information.

Blood gas analysis: In cases where phosphate shifts into cells, blood gas analysis may reveal abnormalities in the pH or bicarbonate levels, indicating systemic metabolic disturbances.

Treatment of hypophosphatemia

The treatment of hypophosphatemia depends on the severity of the condition and its underlying cause. Immediate intervention is critical, especially in severe cases.

1. ***Phosphate supplementation***

   Oral phosphate: For mild cases, oral phosphate supplements (e.g., monosodium phosphate or potassium phosphate) can be administered. These can be provided in the form of powders or solutions added to feed or water.

   Intravenous or subcutaneous phosphate: In severe cases or where oral supplementation is not feasible, phosphate can be administered intravenously or subcutaneously. Care should be taken to avoid overly rapid administration to prevent adverse effects such as tissue damage or cardiac arrhythmias.

2. ***Addressing underlying causes***

   Dietary correction: Adjusting the animal's diet to include more phosphorus-rich foods or providing appropriate mineral supplements is crucial in cases of dietary deficiency.

   Treating renal disease: In cases where renal dysfunction is the cause, appropriate treatment (e.g., fluid therapy, renal support) may be necessary to manage the underlying condition.

   Insulin therapy: For diabetic animals undergoing insulin therapy, careful monitoring of phosphorus levels is necessary, and adjustments to insulin dosage may be required.

3. ***Supportive care***

   Fluid therapy: Rehydration with intravenous fluids may be needed to correct dehydration, especially in animals with concurrent gastrointestinal disease or sepsis.

   Electrolyte management: Other electrolytes, including calcium, potassium, and sodium, may also require correction, particularly in cases of refeeding syndrome.

Prevention of hypophosphatemia

Preventing hypophosphatemia focuses on ensuring that animals receive an adequate and balanced diet, as well as minimizing conditions that predispose them to phosphate depletion.

1. ***Balanced diets:*** Ensure that the diet provides adequate levels of phosphorus, especially for lactating animals, young animals, and those under stress. Phosphorus supplements may be needed in animals fed poor-quality forages or on high-calcium diets.
2. ***Regular monitoring:*** Monitor phosphorus levels in high-risk animals, particularly those with kidney disease, diabetes, or those undergoing refeeding after malnutrition.
3. ***Avoid rapid changes in diet:*** Gradual reintroduction of food after periods of malnutrition can help avoid shifts in phosphate levels and reduce the risk of refeeding syndrome.

Hypophosphatemia is a serious metabolic disorder in animals that can have wide-ranging effects on health, including weakness, bone deformities, and even death if left untreated. Early detection through clinical signs and laboratory testing are very-2 important tools.

c) Hyperphosphatemia (excess phosphorus)

Hyperphosphatemia refers to elevated phosphorus levels in the blood. This condition can result from excessive dietary intake, reduced excretion, or certain diseases.

Causes

Dietary excess: Feeding diets with high phosphorus content, such as excessive grain or protein meals.

Kidney disease: Impaired renal function reduces phosphorus excretion, leading to accumulation.

Bone disorders: Increased bone resorption releases phosphorus into the bloodstream.

Hypoparathyroidism: Reduces the regulation of phosphorus levels.

Symptoms

Soft tissue mineralization (calcification),

Impaired kidney function,

Secondary hyperparathyroidism (compensatory increase in parathyroid hormone due to chronic imbalances),

Reduced appetite and growth retardation,

Diseases associated

Nutritional secondary hyperparathyroidism (NSH): Also known as big head disease in horses, caused by a high phosphorus-to-calcium ratio.

Urolithiasis: High phosphorus levels can contribute to the formation of urinary calculi in ruminants.

Soft tissue calcification: Chronic hyperphosphatemia can lead to mineral deposits in organs, impairing their function.

Management

Reducing dietary phosphorus and ensuring proper calcium-to-phosphorus ratios,

Treating underlying conditions such as kidney disease,

Using phosphate binders in severe cases to reduce absorption,

d) Nutritional secondary hyperparathyroidism (NSH)

The NSH occurs due to chronic imbalances in calcium and phosphorus, typically characterized by high dietary phosphorus and low calcium intake. It is common in horses and exotic animals fed diets with excessive grains or unbalanced ratios.

Symptoms

Facial swelling (big head disease in horses),

Lameness and difficulty walking,

Painful and deformed bones,

Poor growth and reduced performance,

Diagnosis

Blood tests to measure calcium and phosphorus levels.

Radiographs showing bone demineralization and fractures.

Management

Correcting dietary imbalances by adding calcium-rich feeds like alfalfa or limestone,

Limiting phosphorus-rich feeds such as grains and bran,

Providing vitamin D supplementation to improve calcium absorption,

e) Postparturient haemoglobinuria (PPH)

This condition is primarily seen in high-yielding dairy cows after calving. It is linked to severe phosphorus deficiency, leading to red blood cell destruction and hemoglobin release into the urine.

Causes

Inadequate dietary phosphorus intake during the dry period,

Stress of lactation increases phosphorus demand,

Symptoms

Red to dark-colored urine,

Anemia and pale mucous membranes,

Weakness and reduced milk production.

Jaundice in severe cases.

Management

Immediate dietary correction with phosphorus supplements,

Intravenous administration of phosphorus in severe cases,

Proper dry-period nutrition to prevent recurrence,

f) Urolithiasis

Urolithiasis, or urinary stone formation, is a common condition in ruminants such as sheep, goats, and cattle. It often results from imbalanced phosphorus levels, particularly a high dietary phosphorus-to-calcium ratio.

Causes

High-grain diets or excessive use of phosphorus-rich concentrates.

Low water intake leading to concentrated urine.

Inadequate calcium levels to balance dietary phosphorus.

Symptoms

Difficulty urinating (straining),

Swollen abdomen due to bladder rupture,

Blood in the urine (hematuria),

Restlessness and pain,

Management

Immediate surgical intervention in severe cases,

Dietary adjustments to include low-phosphorus and high-fiber feeds,

Increasing water intake to dilute urine,

Adding urinary acidifiers to prevent stone formation,

g) Phosphorus imbalance in poultry

Poultry are particularly sensitive to phosphorus disorders due to their rapid growth and high calcium and phosphorus requirements for eggshell formation.

Deficiency symptoms

Poor growth and reduced feed efficiency,

Weak bones and fractures,

Reduced egg production and soft-shelled eggs,

Excess symptoms

Increased risk of kidney damage,

Reduced feed intake and poor performance,

Management

Providing balanced diets with the correct calcium-to-phosphorus ratio (typically 2:1),

Supplementing with phytase enzymes in plant-based poultry feeds to improve phosphorus availability.

h) Phosphorus imbalance in exotic animals

Exotic animals including reptiles and zoo animals are also prone to phosphorus disorders. Many captive animals develop nutritional secondary hyperparathyroidism due to unbalanced diets.

Symptoms

Bone deformities and fractures,

Reduced activity and poor growth,

Soft shells in turtles,

Management

Feeding species-specific, balanced diets,

Ensuring access to ultraviolet light to synthesize vitamin D is crucial for calcium and phosphorus metabolism.

Nutritional management of phosphorus disorders

Phosphorus is an essential mineral in animal nutrition, playing a crucial role in bone development, energy metabolism, and various cellular functions. Imbalances in phosphorus intake can lead to disorders such as hypophosphatemia (deficiency) and hyperphosphatemia (excess), resulting in significant health problems. Effective nutritional management is critical in preventing and treating these disorders, ensuring optimal animal health and productivity. This article explores the nutritional strategies for managing phosphorus-related diseases in animals.

Nutritional management strategies

A. Balancing calcium-to-phosphorus ratios

A proper calcium-to-phosphorus (Ca:P) ratio is critical for preventing both deficiency and excess-related disorders.

Ideal ratios

Ruminants and Horses: 1.5:1 to 2:1.

Poultry and Swine: 2:1 for layers, 1.2:1 for broilers.

Companion animals: 1:1 to 2:1 depending on species and life stage.

Imbalances: High phosphorus relative to calcium reduces calcium absorption, leading to secondary hyperparathyroidism.

Practical steps

Use phosphorus-rich feeds (e.g., grains, oilseed meals) cautiously in combination with calcium sources like limestone or dicalcium phosphate.

Monitor forage and feed analyses to assess mineral content and adjust accordingly.

B. Dietary phosphorus supplementation

Phosphorus deficiency is managed by ensuring adequate dietary phosphorus intake through high-quality supplements and feeds.

Common Phosphorus Supplements

1. *Di-calcium phosphate (DCP)*: Supplies both calcium and phosphorus, ideal for balancing ratios.
2. *Monosodium phosphate*: A water-soluble source for rapid correction of deficiencies.
3. *Bone meal*: A natural source of phosphorus and calcium.

Feeding guidelines

Adjust phosphorus levels based on the animal's physiological stage:

Growing animals: Require higher levels for skeletal development.

Lactating animals: Need additional phosphorus for milk production.

Pregnant animals: Require more phosphorus for foetal development.

C. Managing excess phosphorus intake

Excess dietary phosphorus contributes to disorders like hyperphosphatemia, urolithiasis, and secondary hyperparathyroidism.

Strategies to reduce excess

1. ***Avoid overfeeding concentrates***: Limit phosphorus-rich grains like corn and soybean meal.
2. ***Incorporate roughages***: For ruminants, high-fiber forages dilute dietary phosphorus.
3. ***Use low-phosphorus diets***: Particularly for animals with kidney disease or predisposition to urinary calculi.

Phosphate binders

In cases of severe hyperphosphatemia, phosphate binders (e.g., aluminum hydroxide) can reduce phosphorus absorption.

D. Role of phytase in non-ruminants

In non-ruminants such as poultry and swine, phosphorus from plant-based feeds is often bound to phytate, making it less bioavailable.

Phytase enzymes: Break down phytate, improving phosphorus availability and reducing dietary supplementation needs.

Benefits

Enhances phosphorus absorption,

Reduces environmental phosphorus excretion,

Phytase supplementation is particularly important in intensive farming systems where feed efficiency and waste management are priorities.

E. Nutritional management of specific disorders

1. Rickets and osteomalacia

Supplement phosphorus with bone meal, DCP, or monosodium phosphate.

Ensure adequate vitamin D to facilitate phosphorus absorption.

Maintain balanced Ca:P ratios.

2. Postparturient haemoglobinuria (PPH)

Increase dietary phosphorus during the dry period to meet lactation demands.

Use easily absorbable sources like monosodium phosphate.

Provide a balanced diet with adequate energy and protein to prevent metabolic stress.

3. Nutritional secondary hyperparathyroidism (NSH)

Correct imbalances by reducing phosphorus-rich feeds (bran and grains) and supplementing calcium in the diet.

For horses, avoid feeding excess wheat bran and provide calcium-rich roughages like alfalfa hay.

Introduce vitamin D supplementation if needed.

4. Urolithiasis

Prevention: Balance dietary calcium and phosphorus, and ensure ample water intake to dilute urine.

Management

Reduce phosphorus in the diet.

Add urinary acidifiers (e.g., ammonium chloride) to prevent struvite stone formation.

F. Kidney disease in companion animals

Limit phosphorus intake to reduce renal load.

Use prescription renal diets with low phosphorus and high-quality protein.

G. Addressing environmental concerns

Excess phosphorus excretion in intensive livestock farming can contribute to environmental pollution, particularly eutrophication of water bodies.

Nutritional interventions

Optimize phosphorus levels in feeds to minimize waste.

Use phytase enzymes in poultry and swine diets to reduce undigested phosphorus excretion.

Promote precision feeding practices to align nutrient supply with animal needs.

H. Monitoring and diagnostics

Regular monitoring of dietary phosphorus intake and animal health is essential to prevent and manage disorders.

Key steps

1. ***Feed analysis***: Periodic testing of feed ingredients for phosphorus content.
2. ***Blood testing***: Measuring serum phosphorus levels to detect deficiencies or excess.
3. ***Bone health assessment***: Radiographs and physical exams to identify skeletal abnormalities.

Veterinary and nutritional expertise should guide diet formulation and supplementation decisions.

Phosphorus disorders in animals can lead to significant health issues, affecting growth, reproduction, and overall productivity. Prevention and management depend on providing balanced diets with appropriate calcium-to-phosphorus ratios, addressing underlying health conditions, and tailoring nutritional strategies to the species' needs. Early diagnosis and intervention are critical to minimizing the impact of these disorders and ensuring the well-being of animals.

Nutritional management is the cornerstone of prevention and treatment, focusing on balanced diets, appropriate supplementation, and tailored feeding strategies. By addressing species-specific needs and physiological demands, and by incorporating modern tools like phytase enzymes, livestock and companion animal health can be optimized. Additionally, careful management reduces the environmental impact of phosphorus excretion, contributing to sustainable animal agriculture. Proactive monitoring and collaboration between veterinarians, nutritionists, and producers ensure effective control of phosphorus-related diseases.

32

Diseases of Magnesium Disorders

Magnesium is a vital mineral for animals, playing critical roles in enzyme activation, muscle contraction, nerve transmission, and bone development. Disorders related to magnesium imbalance can lead to severe health issues, affecting productivity, reproduction, and overall well-being. This chapter explores diseases associated with magnesium disorders in animals, their causes, symptoms, and management strategies.

Importance of magnesium in animal health

Magnesium (Mg) is involved in several physiological processes:

Activation of over 300 enzymes, including those in energy metabolism,

Regulation of muscle and nerve function,

Maintenance of normal heart rhythm,

Bone structure and calcium-phosphorus metabolism,

Magnesium disorders primarily manifest as hypomagnesemia (deficiency) or hypermagnesemia (excess), with hypomagnesemia being more common and significant.

a) Hypomagnesemia (Grass /Lactation tetany)

Hypomagnesemia, commonly known as grass tetany or lactation tetany, is a metabolic disorder in ruminants characterized by abnormally low levels of magnesium (Mg) in the blood. This condition primarily affects lactating animals, particularly cattle, sheep, and goats, grazing on lush pastures low in magnesium or high in antagonistic elements. Grass tetany is a potentially fatal condition if not promptly recognized and treated.

Magnesium and its role in animal physiology

Magnesium is a vital mineral involved in numerous physiological functions, including:

1. Neuromuscular activity: Magnesium acts as a cofactor in enzyme systems that regulate nerve impulse transmission and muscle contraction.

2. Enzyme activation: It is essential for the activity of enzymes involved in energy metabolism, DNA synthesis, and protein synthesis.
3. Calcium and potassium balance: Magnesium influences the balance of calcium and potassium, which are critical for cellular function.

Ruminants rely on dietary magnesium absorption, primarily in the rumen, as they cannot store large amounts of this mineral. Consequently, a deficiency can quickly lead to clinical signs.

Causes of hypomagnesemia

Several factors contribute to magnesium deficiency, often in combination:

1. Dietary Causes

Low magnesium content in pastures: Pastures rich in rapidly growing grasses (e.g., ryegrass, wheatgrass) during spring often have inadequate magnesium levels.

Excess potassium and nitrogen: High potassium (K) and nitrogen (N) levels in the soil inhibit magnesium absorption in plants and reduce magnesium availability to grazing animals.

High calcium and phosphorus: Imbalanced dietary calcium (Ca) or phosphorus (P) ratios may impair magnesium absorption in the rumen.

2. Physiological stressors

Lactation: Lactating animals have higher magnesium requirements due to milk production. Milk contains significant magnesium, leading to rapid depletion if intake is insufficient.

Pregnancy: Late-pregnant animals, particularly ewes, are at risk due to the demands of fetal growth and reduced rumen capacity.

3. Environmental factors

Cold weather: Low temperatures can increase stress, reduce grazing time, and alter magnesium metabolism.

Rainy or overcast conditions: Wet weather can lead to reduced pasture intake, exacerbating magnesium deficiency.

4. Absorption inhibitors

Oxalates and organic acids: Some forages contain oxalates or organic acids that bind magnesium and reduce its bioavailability.

Rumen pH: High concentrate diets can lower rumen pH, inhibiting magnesium absorption.

Clinical signs of hypomagnesemia

The clinical presentation varies from subclinical to severe forms:

1. Subclinical hypomagnesemia

Reduced milk yield,

Poor body condition,

Decreased appetite and weight loss,

Subtle behavioral changes such as irritability,

2. Acute grass tetany

Neuromuscular signs: Tremors, muscle stiffness, and hyperexcitability.

Behavioral signs: Aggressiveness, restlessness, and vocalization.

Seizures: Animals may collapse and exhibit paddling movements due to tetanic spasms.

Death: Sudden death is common in untreated cases, often attributed to respiratory failure or cardiac arrest.

3. Chronic hypomagnesemia

Chronic cases may present with lethargy, reduced fertility, and poor growth.

Diagnosis of hypomagnesemia

Diagnosis is based on clinical signs, history, and laboratory tests:

1. Clinical history and examination

Recent access to lush pastures or high-stress periods (e.g., lactation),

Observable neuromuscular symptoms and behavioral changes,

2. Laboratory tests

Serum magnesium levels: Hypomagnesemia is confirmed if serum Mg concentration is below 1.2 mg/dL.

Urine magnesium: Low urinary magnesium levels indicate inadequate dietary intake.

Postmortem findings: Vitreous humor magnesium concentration can be used as a diagnostic indicator in deceased animals.

Treatment of hypomagnesemia

Immediate intervention is critical for acute cases, as untreated animals may die rapidly.

1. ***Intravenous therapy***

   Magnesium sulfate ($MgSO_4$): Administered intravenously to restore serum magnesium levels. Care must be taken to avoid rapid administration, which can cause cardiac arrest.

   Combination solutions: Calcium-magnesium combinations (e.g., calcium borogluconate with magnesium) are often used to simultaneously address hypocalcemia, which commonly co-occurs with hypomagnesemia.

2. ***Subcutaneous therapy:*** For mild cases, magnesium sulfate can be administered subcutaneously. This method provides a slower release of magnesium into the bloodstream.
3. ***Oral magnesium:*** Oral magnesium supplements (e.g., magnesium oxide or magnesium chloride) can be given to replenish magnesium stores and prevent recurrence.

Prevention of hypomagnesemia

Prevention strategies focus on managing diet, soil conditions, and environmental stressors:

1. ***Dietary management***

   Magnesium supplementation: Offer free-choice mineral supplements or include magnesium salts in concentrates to ensure adequate intake.

   Pasture management: Rotate pastures and avoid grazing animals on lush, magnesium-deficient grass for extended periods.

   High-magnesium feeds: Provide hay or silage containing higher levels of magnesium.

2. ***Soil and fertilizer management***

   Reduce potassium fertilizer: Excessive potassium application inhibits magnesium uptake by plants.

   Lime application: Liming acidic soils can improve magnesium availability.

3. ***Stress reduction:*** Minimize stressors such as cold weather and abrupt changes in diet to reduce magnesium depletion. Provide shelter and supplemental feed during adverse weather conditions.
4. ***Monitoring and early detection:*** Regularly monitor serum magnesium levels in high-risk animals, especially during early lactation or after dietary changes. Observe for early signs of hypomagnesemia to enable prompt intervention.

Economic and welfare implications

Hypomagnesemia has significant economic and welfare impacts:

Economic losses: Reduced milk production, poor growth, reproductive inefficiencies, and livestock deaths result in financial losses for farmers.

Animal welfare: Severe cases cause distress and pain, making it crucial to adopt effective prevention and treatment strategies.

Hypomagnesemia, or grass tetany, is a critical metabolic disorder affecting ruminants, especially during high-risk periods such as lactation or grazing on lush pastures. The condition arises from a complex interplay of dietary, environmental, and physiological factors. Effective prevention through dietary management, magnesium supplementation, and careful pasture management is essential to avoid significant economic losses and ensure animal welfare. Early diagnosis and prompt treatment are vital for mitigating the effects of hypomagnesemia and improving outcomes in affected animals.

b) Hypermagnesemia (Magnesium excess)

Hypermagnesemia is less common than hypomagnesemia but can occur under certain conditions, particularly in animals with impaired kidney function or excessive magnesium intake.

Causes of hypermagnesemia

1. ***Excessive magnesium supplementation:*** Overuse of magnesium-containing feed additives, mineral supplements or laxatives (e.g., magnesium sulfate).
2. ***Kidney disease***: Reduced renal excretion of magnesium can lead to accumulation.
3. ***Ingestion of high-Magnesium Water:*** Water sources with high magnesium concentrations can contribute to hypermagnesemia

Muscle weakness and reduced reflexes.

Depressed respiratory function,

Cardiac arrhythmias or slowed heart rate,

Lethargy and coma in severe cases,

Management of hypermagnesemia

1. ***Immediate intervention:*** Intravenous administration of calcium gluconate to antagonize magnesium's effects,
2. ***Dietary modifications***:

   Reduce or eliminate magnesium supplements.

   Provide low-magnesium water sources.

3. ***Address underlying conditions***: Treat kidney dysfunction to restore magnesium excretion.
4. ***Magnesium imbalances and other metabolic diseases:*** Magnesium disorders often interact with other metabolic conditions, exacerbating their effects or complicating management.
   A. *Hypocalcemia (Milk fever):* Magnesium influences calcium metabolism by modulating parathyroid hormone (PTH) secretion. Magnesium deficiency can impair PTH activity, exacerbating hypocalcemia.

      Prevention involves providing adequate magnesium alongside calcium supplements.
   B. *Ketosis:* Magnesium deficiency reduces energy metabolism efficiency, increasing the risk of ketosis in lactating cows.
   C. *Reproductive disorders:* Magnesium imbalances can lead to infertility, poor conception rates, or weak offspring due to impaired energy metabolism and muscle function.
5. ***Monitoring and diagnostic tools:*** Effective management of magnesium disorders requires regular monitoring of animal health and dietary magnesium levels.

Key diagnostic approaches

1. ***Blood tests***: Serum magnesium levels (normal range: 0.7–1.2 mmol/L in cattle).
2. ***Urine analysis***: Assess urinary magnesium excretion to evaluate dietary intake and absorption.
3. ***Forage and feed analysis***: Determine magnesium content in pastures and feeds to ensure adequacy.

Magnesium disorders, particularly hypermagnesemia, pose significant challenges in animal health management, especially in ruminants. Preventive strategies such as dietary supplementation, pasture management, and regular monitoring are essential for maintaining optimal magnesium balance. By understanding the causes, symptoms, and management of magnesium-related diseases, livestock producers and veterinarians can ensure the health, productivity, and welfare of animals, reducing economic losses and improving overall farm performance.

33

Diseases of Sodium, Potassium and Chloride Disorders

Sodium (Na), potassium (K), and chloride (Cl) are essential electrolytes that play critical roles in maintaining fluid balance, acid-base homeostasis, nerve transmission, and muscle function in animals. Imbalances in these electrolytes, whether due to deficiency or excess, can lead to serious health issues that affect productivity and well-being. This chapter discusses the disorders associated with sodium, potassium, and chloride imbalances in animals, their causes, symptoms, and nutritional management strategies.

1. Sodium disorders

Sodium is the primary extracellular cation and is crucial for maintaining osmotic pressure, fluid balance, and nerve signal transmission.

a) Hyponatremia (Sodium deficiency)

Causes

1. ***Dietary deficiency***: Inadequate sodium in feed or water.
2. ***Excessive loss***: Through sweat, diarrhea, vomiting, or prolonged lactation.
3. ***Dilutional hyponatremia***: Excessive water intake diluting sodium levels.

Symptoms

Weakness and lethargy,

Reduced feed intake and growth retardation,

Muscle twitching and tremors,

In severe cases, neurological signs like incoordination and seizures.

Diseases associated

Salt deficiency syndrome: Common in ruminants grazing on sodium-deficient soils or fed low-salt diets. Symptoms include pica (eating non-food items), poor growth, and reduced milk production.

Management

Supplement diets with common salt (sodium chloride) at appropriate levels:

Ruminants: 0.25–0.5% of the diet.

Poultry and swine: 0.1–0.2% of the diet.

Provide free access to salt licks or mineral blocks.

Ensure adequate sodium in drinking water (minimum 0.2 g/L for ruminants).

b) Hypernatremia (Sodium Excess)

Causes

1. ***Excessive salt intake***: Overconsumption of salted feeds or water.
2. ***Dehydration***: Lack of water exacerbates sodium concentration.
3. ***Kidney dysfunction***: Impaired sodium excretion.

Symptoms

Thirst, lethargy, and reduced appetite,

Neurological signs like twitching, ataxia, and seizures,

Dehydration and dry mucous membranes,

Management

Provide immediate access to fresh water to dilute sodium levels.

Avoid over-supplementing salt in diets or mineral blocks.

Treat underlying dehydration or kidney dysfunction.

Potassium disorders

Potassium is the primary intracellular cation and is essential for nerve impulse transmission, muscle contraction, and enzymatic processes.

a) Hypokalemia (Potassium deficiency)

Causes

1. ***Inadequate dietary potassium***: Seen in diets based on poor-quality forages or grains.
2. ***Excessive loss***: Due to diarrhea, prolonged sweating, or excessive use of diuretics.
3. ***Shift of potassium into cells***: Occurs during metabolic alkalosis or insulin therapy.

Symptoms

Weakness and lethargy,

Muscle tremors and reduced reflexes,

In severe cases, cardiac arrhythmias and paralysis,

Diseases associated

Hypokalemic myopathy: Muscle weakness and stiffness in animals with chronic potassium deficiency.

Management

Supplement diets with potassium sources like potassium chloride (KCl) or potassium carbonate.

For cattle, ensure adequate potassium in forages (1.0–1.5% of dry matter).

Use oral or intravenous potassium supplementation in severe cases.

b) Hyperkalemia (Potassium excess)

Causes

1. ***Excessive potassium intake***: High-potassium forages (e.g., alfalfa, silage) or supplements.
2. ***Kidney dysfunction***: Impaired excretion of potassium.
3. ***Tissue damage***: Release of intracellular potassium during trauma or severe acidosis.

Symptoms

Muscle weakness and flaccid paralysis,

Cardiac abnormalities (arrhythmias, bradycardia),

Depression and reduced feed intake,

Management

Reduce potassium intake by modifying diets.

Treat underlying kidney disease or acidosis.

Administer calcium gluconate to stabilize heart function in severe cases.

3. Chloride disorders

Chloride is the primary extracellular anion and is essential for acid-base balance, fluid regulation, and digestion (as a component of gastric hydrochloric acid).

a) Hypochloremia (Chloride Deficiency)

Causes

1. ***Inadequate dietary chloride***: Rare, but can occur in animals on low-salt diets.
2. ***Excessive loss***: Through vomiting, diarrhea, or excessive sweating.
3. ***Metabolic alkalosis***: Associated with increased bicarbonate levels and decreased chloride.

Symptoms

Reduced growth and feed efficiency,

Weakness and lethargy,

Alkalosis symptoms, including shallow breathing and muscle twitching,

Diseases associated

Abomasal displacement in cattle: Chloride loss in abomasal fluid leads to metabolic alkalosis and hypochloremia.

Management

Supplement chloride using sodium chloride or potassium chloride in diets.

Treat underlying causes like diarrhea or alkalosis.

For metabolic alkalosis, administer intravenous isotonic saline (NaCl) to restore chloride levels.

b) Hyperchloremia (Chloride excess)

Causes

1. ***Excessive chloride intake***: Overuse of salt or chloride-containing supplements.
2. ***Dehydration***: Concentrates chloride in the blood.
3. ***Kidney disease***: Impaired chloride excretion.

Symptoms

Dehydration and excessive thirst,

Acidosis symptoms, such as rapid breathing,

Neurological signs like weakness and depression,

Management

Provide fresh water to dilute chloride levels.

Modify diets to reduce chloride intake.

Treat underlying kidney disease or dehydration.

Sodium, potassium and chloride interactions

These electrolytes are interconnected in maintaining fluid balance and acid-base homeostasis. Imbalances in one electrolyte often affect the others.

Electrolyte balance in diets

The dietary cation-anion difference (DCAD) is a useful tool for balancing diets in ruminants, particularly during the transition period:

Positive DCAD diets: Promote milk production and energy balance.

Negative DCAD diets: Used prepartum to prevent hypocalcemia.

Key points for balancing electrolytes

Ensure adequate intake of sodium, potassium, and chloride based on species-specific requirements.

Monitor feed and water quality for electrolyte content.

Avoid excessive use of supplements that can disrupt the balance.

Monitoring and diagnostics

Regular monitoring of electrolyte levels and animal health is critical for managing imbalances.

Diagnostic tools

1. *Blood tests*: Measure serum levels of sodium, potassium and chloride.
2. *Urine analysis*: Evaluate electrolyte excretion.
3. *Feed and water analysis*: Check mineral content in diets and water sources.

A. Nutritional management of sodium deficiency (hyponatremia)

Causes

Diets low in sodium.

Excessive sodium losses through sweat, milk production, diarrhea, or heat stress.

Symptoms

Reduced feed intake, weight loss, poor growth,

Pica (abnormal eating behavior, such as licking soil or wood),

Weakness and dehydration,

Management

1. ***Salt supplementation***: Sodium is commonly provided as sodium chloride (common salt). Incorporating salt into animal feed is a practical and cost-effective solution.

   Recommended dietary inclusion rates

   Cattle and Sheep: 0.25–0.5% of dry matter.

   Poultry: 0.15–0.20% of feed.

2. ***Salt licks or blocks***: Free access to salt blocks ensures animals can self-regulate their sodium intake.
3. ***Water sodium content***: Ensure the water contains at least 0.2 g/L sodium, especially for grazing livestock.
4. ***Prevention during stressful Periods***: Increase dietary sodium levels during heat stress, high milk production, or heavy lactation periods.

Nutritional management of sodium excess (hypernatremia)

Causes

Over-supplementation of salt or ingestion of saline water,

Dehydration or reduced water intake.

Symptoms

Thirst, reduced feed intake, weakness.

Severe cases: neurological signs such as twitching or seizures.

Management

1. ***Reduce sodium intake***:

   Identify and eliminate high-sodium feed or water sources.

   Adjust the formulation of feeds containing excessive salt.

2. ***Provide fresh water***: Ensure animals have constant access to clean, fresh water to dilute sodium concentrations in the body.
3. ***Correct electrolyte imbalance***: Offer electrolyte solutions containing balanced sodium, potassium, and chloride levels to restore homeostasis.

Potassium disorders

Nutritional Management of Potassium Deficiency (Hypokalemia)

Causes

Low-potassium diets or potassium-depleted forages,

Diarrhea, heavy sweating, or kidney dysfunction leading to potassium loss,

Symptoms

Muscle weakness, lethargy, poor growth,

Cardiac arrhythmias and impaired feed intake in severe cases,

Management

1. ***Dietary potassium supplementation***:

   Add potassium chloride (KCl) or potassium carbonate to the feed. Recommended dietary potassium levels:

   Dairy cattle: 1.2–1.8% of dry matter.

   Poultry: 0.4–0.6% of feed.

2. ***Quality forages:*** Legumes such as alfalfa and clover are naturally rich in potassium and can be included in the diet.
3. ***Electrolyte supplementation***: During periods of heat stress or illness, provide oral rehydration solutions containing potassium.
4. ***Monitor and address losses***: Reduce potassium losses by managing diarrhea and dehydration effectively.

Nutritional management of potassium excess (hyperkalemia)

Causes

Diets rich in high-potassium feeds like silage, alfalfa, or concentrates,

Impaired potassium excretion due to kidney dysfunction,

Symptoms

Weakness, flaccid paralysis, and reduced feed intake,

Cardiac abnormalities such as bradycardia or arrhythmias,

Management

1. ***Reduce potassium intake***:

   Limit high-potassium forages in the diet.

   Replace high-potassium feed ingredients with alternatives.

2. ***Balance dietary cation-anion difference (DCAD)***: In ruminants, maintain DCAD levels to ensure optimal potassium utilization and prevent excess accumulation.
3. ***Treat underlying causes***: Address kidney dysfunction or metabolic acidosis contributing to hyperkalemia.
4. ***Calcium supplementation***: Intravenous calcium gluconate may help stabilize cardiac function in severe cases.

Nutritional management of chloride deficiency (hypochloremia)

Causes

Low-chloride diets or excessive losses through sweat, vomiting, or diarrhea,

High dietary bicarbonate levels leading to metabolic alkalosis.

Symptoms

Poor growth, reduced feed efficiency,

Lethargy, alkalosis symptoms (e.g., slow breathing, muscle tremors),

Management

1. ***Salt supplementation:*** Sodium chloride or potassium chloride can restore chloride levels.
2. ***Adjust DCAD levels:*** Decrease bicarbonate-rich feeds and include anionic salts like ammonium chloride or calcium chloride to correct alkalosis.
3. ***Rehydration therapy:*** Oral or intravenous fluids containing chloride (e.g., saline solution) help restore balance in severe cases.

Nutritional management of chloride excess (hyperchloremia)

Causes

Excessive chloride intake from over-supplementation or saline water,

Dehydration or kidney dysfunction leading to impaired chloride excretion,

Symptoms

Rapid breathing (acidosis), lethargy, poor performance,

Severe cases: neurological signs and dehydration,

Management

1. *Reduce chloride intake*: Identify and remove high-chloride feed or water sources.
2. *Provide balanced diets*: Ensure sodium, potassium, and chloride levels are balanced according to species-specific requirements.
3. *Rehydration*: Administer fresh water or balanced oral rehydration solutions to dilute excess chloride.
4. *Integrative nutritional strategies*

   Dietary cation-anion difference (DCAD)

   DCAD is critical for ruminants, particularly during the transition period around calving.

Positive DCAD diets: Promote milk production and metabolic efficiency (e.g., +250 to +400 mEq/kg diet).

Negative DCAD diets: Used in prepartum cows to prevent milk fever and improve calcium metabolism,

Electrolyte monitoring

Regular monitoring of feed and water electrolyte content is essential.

Adjust feed formulations to ensure optimal Na, K and Cl levels.

5. ***Prevention and monitoring***

   1. ***Nutritional balancing***

      Use formulated diets to meet the electrolyte requirements of livestock and poultry.

      Ensure adequate salt, potassium, and chloride in feeds, especially during high-stress periods.

   2. ***Environmental management***: Minimize heat stress through proper housing and ventilation, as electrolyte imbalances are often exacerbated during high temperatures.

   3. ***Regular health checks***: Monitor serum and urine electrolyte levels to detect early signs of imbalances.

Sodium, potassium and chloride are critical for livestock and poultry health, and their imbalances can lead to severe metabolic, muscular, and neurological disorders and thus severely impact productivity and performance. Nutritional management strategies, including balanced diets, proper supplementation, and regular monitoring, are key to preventing and addressing these disorders. By maintaining optimal electrolyte balance, farmers and veterinarians can ensure the health and productivity of their animals, contributing to sustainable livestock and poultry production. Collaboration between veterinarians, nutritionists, and producers is the key to ensuring optimal health, productivity, and welfare in animals.

34

Diseases of Sulphur Disorders

Sulphur (S) is a vital mineral for livestock and poultry, playing a key role in the synthesis of amino acids (methionine, cysteine), vitamins (biotin, thiamine), and other critical metabolites. While sulphur deficiencies or excesses are less common than imbalances in other minerals, they can lead to significant health and production challenges. Effective nutritional management is crucial to prevent and address sulphur-related disorders.

Importance of sulfur in livestock and poultry

Sulfur is essential for:

1. ***Protein synthesis***: It is a component of sulfur-containing amino acids like methionine and cysteine, which are critical for growth and tissue repair.
2. ***Structural integrity***: Sulfur contributes to the structure of keratin, essential for wool, feathers, and hooves.
3. ***Metabolic functions***: It is involved in the synthesis of vitamins (biotin and thiamine), enzymes, and coenzymes.
4. ***Detoxification***: Sulfates aid in detoxifying harmful substances in the liver.

Sulphur deficiency (hypo-sulfurosis)

Causes

1. ***Low dietary sulphur***: Diets low in sulfur-containing ingredients, such as grains or poor-quality forages.
2. ***Imbalanced nitrogen-sulphur ratio***: Excessive non-protein nitrogen (e.g., urea) in ruminants' diets without adequate sulfur supplementation.
3. ***Poor feed quality***: Grazing on sulfur-deficient soils or consuming sulfur-depleted feeds.

Symptoms

Poor growth and reduced feed efficiency,

Decreased milk and egg production,

Rough coat, poor feather development or brittle hooves,

Pica (abnormal eating behavior),

In ruminants: Reduced microbial protein synthesis in the rumen, leading to poor fiber digestion.

Diseases associated

Polioencephalomalacia (PEM): In ruminants, excessive dietary sulfur can paradoxically cause a deficiency at the tissue level due to thiamine destruction.

Wool break and reduced feather quality: In sheep and poultry, low sulfur intake affects keratin synthesis.

Nutritional management of sulphur deficiency

Dietary sulphur supplementation

Include sulfur-rich feed ingredients like oilseed meals, alfalfa, and legumes. Use sulphur supplements such as calcium sulphate, magnesium sulphate, or ammonium sulphate.

Ensure proper nitrogen-sulphur ratio

Maintain a nitrogen-to-sulphur ratio of 10:1 to 15:1 in ruminant diets to optimize microbial protein synthesis in the rumen.

Monitor forage and soil quality

Test soils and forages for sulphur content and supplement accordingly if deficiencies are identified.

Incorporate sulphur-containing amino acids

For poultry and monogastric animals, supplement with methionine or methionine analogs to meet sulfur amino acid requirements.

Water analysis

Assess the sulfur content in water sources to ensure that animals are not receiving suboptimal amounts.

Vitamin supplementation

Provide adequate biotin and thiamine to compensate for reduced synthesis during deficiency.

Sulfur toxicity (hyper-sulphurosis)

Causes

Excessive dietary sulphur: Feeding high-sulphur feeds (e.g., distillers grains, molasses, or cruciferous plants).

High Sulphates in water: Water sources with >500 ppm sulphates can contribute to toxicity.

Over-supplementation: Overuse of sulphur-containing supplements or feeds.

Symptoms

Polioencephalomalacia (PEM)

Neurological signs like ataxia, blindness, seizures, and head pressing,

More common in cattle, sheep, and goats consuming excessive sulphur,

Reduced performance

Lower feed intake, poor growth, and reduced productivity.

Respiratory and digestive issues

In poultry, excess sulphur can lead to poor gut health and increased ammonia production.

Interference with mineral absorption

High sulphur levels can inhibit copper and selenium absorption, causing secondary deficiencies.

Nutritional management of sulphur toxicity

1. ***Reduce high-sulphur feed ingredients***

   Limit the inclusion of distillers grains, molasses, and cruciferous forages in the diet.

   Replace these with low-sulphur alternatives.

2. ***Monitor and adjust water quality***: Test water sources for sulphate levels. If high, provide alternative water sources or use filtration systems.

3. ***Dietary adjustments***

   Incorporate rumen-protective agents like thiamine to prevent sulfur-induced PEM.

   Balance the diet to avoid excessive sulfur intake (<0.4% sulfur in total diet for ruminants).

4. ***Supplement copper and selenium***: Address secondary deficiencies by supplementing with copper sulfate or selenium selenite, as appropriate.

5. ***Early intervention in cases of PEM***: Administer intravenous thiamine or supportive therapies for animals exhibiting PEM symptoms.

Sulphur requirements in livestock and poultry

Livestock

1. ***Cattle and Sheep***

   Total dietary sulphur content: 0.2–0.4% of dry matter.

   Higher levels are required for high-producing dairy cows or during periods of high nitrogen supplementation.

2. ***Swine***: Sulphur requirements are generally met through methionine supplementation, as they cannot utilize inorganic sulphur effectively.
3. ***Goats***: Similar requirements to sheep but may need adjustments based on breed and production stage.

Poultry

1. ***Broilers and layers***

   Sulphur is primarily supplied via methionine and cysteine in poultry diets.

   Methionine should comprise ~0.35–0.5% of the diet, depending on the bird's age and production stage.

2. ***Sulfur deficiency symptoms in poultry***: Feather abnormalities, poor growth, and reduced egg production.
3. ***Supplementation***: Use DL-methionine or methionine hydroxyanalogs to correct deficiencies.

Monitoring and prevention

For livestock

Regularly test feed, forage, and water for sulphur content.

Monitor nitrogen-to-sulphur ratios in diets to avoid imbalances.

Ensure adequate supplementation of sulphur-containing amino acids and vitamins.

For poultry

Evaluate dietary methionine and cysteine levels regularly.

Use high-quality protein sources like soybean meal to meet amino acid requirements.

General practices

Educate farmers on the importance of balanced sulphur nutrition and potential sources of excess or deficiency.

Periodic health checks to detect early signs of sulphur-related disorders.

Sulphur is a critical mineral for livestock and poultry, essential for growth, production, and metabolic processes. Imbalances in sulphur intake—whether deficiency or excess—can lead to significant health and production challenges. Proper nutritional management, including balanced diets, careful supplementation and regular monitoring of feed and water sources, is essential to prevent and address sulphur-related disorders. Collaboration between farmers, veterinarians and nutritionists is key to maintaining optimal animal health and productivity.

35

Diseases of Common Salt Disorders

Common salt, or sodium chloride (NaCl), is an essential mineral that plays a critical role in the health and productivity of livestock and poultry. It is a major source of sodium (Na) and chloride (Cl), both of which are vital for maintaining osmotic balance, acid-base regulation, nerve conduction, and muscle function. However, deficiencies or excesses of common salt can lead to severe health issues, reduced productivity, and even mortality in animals. This chapter discusses common salt disorders, their causes, symptoms, and the nutritional management strategies necessary to prevent and mitigate these problems.

Importance of common salt in animal health

1. ***Electrolyte balance***: Sodium and chloride maintain osmotic pressure and fluid balance within the body.
2. ***Acid-base regulation***: Sodium and chloride are crucial for maintaining acid-base homeostasis in the blood.
3. ***Nerve and muscle function***: Sodium is essential for nerve impulse transmission and muscle contractions.
4. ***Digestive health***: Chloride is a major component of gastric acid (hydrochloric acid), which aids in digestion.
5. ***Milk and egg Pproduction***: Adequate salt levels are required for optimal production in dairy animals and laying poultry.

Nutritional management of salt deficiency

1. ***Inclusion of salt in feed***:

   Add common salt to animal diets to meet species-specific requirements.

   Recommended levels:

   Cattle and sheep: 0.25–0.5% of dry matter intake.

   Poultry: 0.15–0.2% of total feed.
2. ***Free access to salt licks or blocks***: Provide free-choice salt blocks to allow animals to regulate their sodium intake.

3. ***Supplementation in drinking water***: Dissolve salt in drinking water, especially during high stress or lactation periods.
4. ***Use of balanced electrolytes***: In case of dehydration or diarrhoea, administer oral rehydration solutions containing sodium, chloride, and potassium.
5. ***Monitor pasture quality***: For grazing animals, ensure that pastures are not excessively rich in potassium (which can inhibit sodium absorption).

Nutritional management of salt excess

1. ***Reduce salt intake***

   Identify and limit high-salt feed ingredients or saline water sources.

   Reformulate diets to reduce sodium and chloride levels.
2. ***Provide fresh, clean water***: Ensure animals have constant access to clean, non-saline water to dilute excess sodium and chloride in the body.
3. ***Electrolyte balancing***: Supplement potassium and other electrolytes to restore balance in the event of salt toxicity.
4. ***Monitor salt levels in feed and water***: Regularly test feed and water for sodium and chloride content to prevent accidental over-supplementation.
5. ***Gradual transition***: Introduce any changes in salt supplementation or feed composition gradually to avoid sudden imbalances.

4. ***Species-specific considerations***

   Ruminants (Cattle, Sheep, Goats)

   Sodium and chloride are crucial for rumen microbial activity and fiber digestion.

   Salt deficiency can lead to poor rumen function and decreased productivity.

   High salt intake combined with limited water can cause salt poisoning, especially in grazing animals.

Poultry

Sodium and chloride are important for maintaining egg production and quality.

Excessive salt can lead to wet litter conditions, promoting diseases like footpad dermatitis and ammonia toxicity.

Swine

Salt deficiencies impair growth rates and feed efficiency in pigs.

Excessive salt intake can lead to salt poisoning, particularly if water is restricted.

Preventative strategies

Balanced diet formulation

Use precise formulations to meet sodium and chloride requirements without over-supplementing.

Regular monitoring

Test feed and water for salt content regularly.

Observe animal behavior and performance for early signs of deficiencies or toxicities.

Provide adequate water

Ensure free access to fresh water to support hydration and maintain electrolyte balance.

Environmental management

In poultry, manage housing conditions to prevent wet litter caused by excess salt in the diet.

Education and training

Train farmers and caretakers on the importance of proper salt supplementation and monitoring practices.

Common salt is a critical nutrient for livestock and poultry, essential for maintaining health, productivity, and overall performance. Both deficiencies and excesses of salt can lead to serious health problems and economic losses. Nutritional management strategies, including proper supplementation, regular monitoring, and species-specific dietary adjustments, are key to preventing and managing salt-related disorders. By ensuring a balanced intake of sodium and chloride, farmers can optimize animal health and productivity while minimizing the risk of salt toxicity or deficiency.

36

Dehydration

Dehydration is a critical condition in animals characterized by an imbalance between fluid intake and loss, leading to a deficit in total body water. It is a common problem across all species and can result from a variety of causes, including disease, environmental factors, and inadequate water intake. Understanding the causes, clinical signs, diagnosis, and treatment of dehydration is vital for ensuring the health and well-being of animals.

Physiology of fluid balance

Water is essential for physiological processes, including nutrient transport, waste elimination, temperature regulation, and cellular metabolism. Animals maintain fluid balance through mechanisms that regulate intake (via drinking and food consumption) and output (via urine, feces, respiration, and sweat in some species). Dehydration occurs when water loss exceeds water intake.

Body's water compartments are divided into

Intracellular fluid (ICF): About two-thirds of total body water: One-third of total body water, including plasma and interstitial fluid. Dehydration affects these compartments differently based on severity and cause, potentially leading to hypovolemia (decreased blood volume) and compromised organ function.

Causes of dehydration

Insufficient water intake

Neglect or inaccessibility: Lack of access to clean, fresh water.

Behavioural issues: Reluctance to drink due to stress or environmental changes.

Illness: Conditions like oral pain or esophageal obstruction can reduce water consumption.

Excessive fluid loss

Gastrointestinal loss: Vomiting, diarrhoea, and excessive salivation (e.g., in heatstroke or poisoning) are major contributors.

Renal loss: Diseases like chronic kidney disease or diabetes mellitus can lead to excessive urination.

Skin loss: Sweating (in horses) or fluid loss through burns or wounds.

Respiratory loss: Increased evaporation due to panting, especially in dogs, or hyperventilation.

Blood loss: Hemorrhage resulting from trauma, surgery, or coagulopathies.

Lactation: High-producing animals, such as dairy cows, can experience dehydration due to fluid loss in milk.

Systemic conditions

Fever, sepsis, or systemic inflammation can increase metabolic demand for water and contribute to fluid loss.

Clinical signs of dehydration

The clinical presentation of dehydration varies with its severity. It can be categorized as mild, moderate, or severe:

a) ***Mild dehydration*** (5 - 6% fluid deficit): Subtle decrease in skin elasticity dry or tacky mucous membranes and normal vital signs,

b) ***Moderate dehydration (7 - 9% fluid deficit)*:** Pronounced loss of skin turgor (slow skin tenting), dry, sticky mucous membranes and sunken eyes, weakness or lethargy and increased heart rate (tachycardia).

c) ***Severe dehydration (>10% fluid deficit)*:** Skin remains tented after pinching, Very dry mucous membranes, cold extremities due to poor perfusion, weak pulse and signs of shock, collapse or unconsciousness.

Species-specific signs may also be observed

Horses: Prolonged capillary refill time (CRT), dry gums and sunken eyes.

Cattle: Reduced rumen motility, sunken eyes, and lethargy.

Dogs and Cats: Decreased skin elasticity, dry gums, and tachycardia.

Diagnosis

Clinical assessment

Skin turgor test: Pinching the skin to observe its elasticity.

Mucous membranes: Examining colour, moisture, and CRT.

Weight monitoring: Acute weight loss can indicate fluid loss.

Heart rate and pulse: Increased heart rate and weak pulses suggest hypovolemia.

Laboratory tests

Packed cell volume (PCV) and total solids (TS): Increased PCV and TS indicate hemoconcentration.

Serum chemistry: Elevated blood urea nitrogen (BUN) and creatinine can indicate dehydration-related renal compromise.

Electrolyte imbalance: Sodium, potassium and chloride levels help determine the type and severity of dehydration.

Urine analysis: Concentrated urine (high specific gravity) reflects dehydration.

Additional diagnostics

Imaging or advanced testing may be necessary to identify underlying causes, such as gastrointestinal disease or renal failure.

Management and treatment

a) Assessing severity

Treatment strategies depend on the degree of dehydration, underlying cause, and the species involved.

b) Fluid replacement

Routes of administration

Oral rehydration

Suitable for mild dehydration when the animal is not vomiting or anorexic,

Electrolyte solutions (oral rehydration salts) are preferred over plain water to replenish lost electrolytes.

Subcutaneous fluids

Used in small animals for mild to moderate dehydration,

Easily administered in outpatient settings,

Intravenous (IV) fluids

Gold standard for moderate to severe dehydration,

Rapid rehydration and correction of electrolyte imbalances,

Intraperitoneal or Intraosseous fluids: Used in neonates or animals where IV access is challenging.

Fluid types

Crystalloids: Balanced electrolyte solutions like ringer's lactate or normal saline.

Colloids: Plasma expanders (e.g., hetastarch) for severe hypovolemia or shock.

Hypertonic saline: Useful for rapid volume expansion in large animals.

Volume and rate

Calculate based on percentage of dehydration and ongoing losses.

Rehydrate over 12-24 hours unless the animal is in shock, which requires faster correction.

Treat underlying causes

Antiemetics for vomiting,

Antidiarrheal agents or probiotics for gastrointestinal issues,

Antibiotics for infections,

Management of systemic conditions like diabetes or kidney disease,

Supportive care

Monitoring urine output, body weight and hydration parameters during recovery,

Providing free access to clean, fresh drinking water,

Nutritional support to promote recovery,

Prevention

Adequate water access

Ensure continuous access to clean, fresh water in sufficient quantities.

Monitor water intake during extreme weather conditions.

Dietary management

High-quality diets with adequate water content (especially for cats and herbivores).

Include electrolyte supplementation during stress condition or higher activity.

Environmental considerations

Avoid prolonged exposure to heat.

Provide shade and ventilation in hot climates.

Regular veterinary care

Early diagnosis and treatment of conditions that predispose to dehydration,

Vaccination and parasite control to minimize disease risk,

Special considerations

Species-specific factors

Horses: Dehydration during intense exercise (e.g., endurance races) can lead to colic or exhaustion. Monitor hydration status using CRT, heart rate, and skin turgor.

Ruminants: Dehydration due to ruminal disorders like bloat or diarrhea requires careful correction of both fluid and electrolyte imbalances.

Dogs and cats: Heatstroke is a common cause of dehydration, especially in brachycephalic breeds.

Neonates and geriatric animals

Neonates are more prone to dehydration due to their higher water turnover and immature renal function.

Geriatric animals may have reduced thirst perception and are at increased risk due to chronic diseases.

Prognosis

The prognosis for dehydrated animals depends on the severity of dehydration and the speed of treatment. Mild cases usually resolve quickly with appropriate fluid therapy, while severe cases with delayed intervention can lead to organ failure or death. Addressing underlying causes is critical to prevent recurrence.

Dehydration is a potentially life-threatening condition in animals that requires prompt recognition and intervention. By understanding the causes, recognizing clinical signs, and implementing effective treatment strategies, caregivers and veterinarians can manage dehydration effectively and prevent long-term complications. Proactive measures such as ensuring access to clean water, addressing environmental stressors and managing underlying conditions are the key for prevention of dehydration.

37

Hemorrhagic Disorders in Animals

Hemorrhagic disorders refer to conditions that impair the ability of an animal's blood to clot properly, resulting in prolonged or excessive bleeding. These disorders can be broadly categorized into congenital coagulation defects and acquired coagulation defects, each with distinct etiologies, pathophysiologies, and clinical manifestations.

a) Congenital coagulation defects

Congenital coagulation defects are inherited conditions caused by genetic mutations affecting one or more components of the coagulation cascade. These defects are often breed-specific and can be observed in a variety of domestic animals.

Common types of congenital coagulation defects

Hemophilia- A

Cause: Deficiency of factor VIII.

Affected animals: Common in dogs, especially German Shepherds, and less common in cats and horses,

Pathophysiology: Factor VIII is essential for forming the intrinsic tenase complex, which amplifies thrombin generation. Its deficiency results in inadequate fibrin formation and poor clot stability.

Clinical signs: Spontaneous bleeding, hematomas, hemarthrosis (bleeding into joints), and prolonged bleeding after trauma or surgery.

Diagnosis: Prolonged activated partial thromboplastin time (aPTT), normal prothrombin time (PT), and reduced Factor VIII activity.

Hemophilia-B

Cause: Deficiency of factor IX.

Affected Animals: Predominantly seen in dogs, especially Labrador Retrievers, and less frequently in cats.

Pathophysiology: Similar to haemophilia-A, factor IX deficiency disrupts the intrinsic pathway of coagulation.

Clinical signs: Similar to Hemophilia-A, but may vary in severity depending on the residual activity of factor IX.

Diagnosis: Prolonged aPTT, normal PT, and reduced factor IX activity.

von Willebrand disease (vWD)

Cause: Deficiency or dysfunction of von Willebrand factor (vWF), a protein necessary for platelet adhesion.

Types

Type 1: Partial quantitative deficiency (common in Doberman Pinschers)

Type 2: Qualitative defect (rare),

Type 3: Complete absence of vWF (severe, seen in Scottish Terriers and Shetland Sheepdogs),

Clinical signs: Mucosal bleeding (epistaxis, gingival bleeding), prolonged bleeding from wounds, and excessive haemorrhage during surgery or oestrus.

Diagnosis: Reduced vWF concentration (measured by vWF antigen assay) and prolonged buccal mucosal bleeding time.

Other factor deficiencies

Factor VII deficiency: Common in Beagle dogs; mild bleeding tendency,

Factor XI deficiency: Observed in Holstein cattle and Kerry Blue Terriers ,

Factor XII deficiency: Seen in cats but typically asymptomatic as factor-XII is not essential for hemostasis.

Management

Plasma transfusion: Fresh or fresh-frozen plasma provides the deficient coagulation factors.

Cryoprecipitate: Used for von Willebrand disease and Hemophilia-A as it is rich in factor VIII and vWF.

Supportive care: Avoidance of trauma and careful monitoring during surgeries or invasive procedures.

b) Acquired coagulation defects

Acquired coagulation defects result from external or internal factors that disrupt normal clotting mechanisms. These defects are more common than congenital defects and can arise due to various diseases, toxins, or deficiencies.

Common causes of acquired coagulation defects

Vitamin K deficiency or antagonism

Cause

Ingestion of anticoagulant such as rodenticides (warfarin, brodifacoum),

Hepatic disease impairing vitamin K metabolism,

Dietary deficiencies (rare),

Pathophysiology: Vitamin K is essential for the activation of clotting factors II, VII, IX, and X. Deficiency or antagonism results in non-functional factors, leading to impaired clot formation.

Clinical signs: Hematuria, melena, epistaxis, petechiae, ecchymoses, and prolonged bleeding from minor wounds.

Diagnosis: Prolonged PT (early indicator), prolonged aPTT, and response to vitamin K therapy.

Treatment

Vitamin K_1 (phytonadione) administered subcutaneously or orally,

Blood transfusion in severe cases to replace clotting factors,

Disseminated intravascular coagulation (DIC)

Cause: Secondary to severe systemic diseases like sepsis, neoplasia, trauma, or heatstroke.

Pathophysiology: Widespread activation of coagulation leads to the consumption of platelets and clotting factors, resulting in microvascular thrombosis and subsequent bleeding (consumptive coagulopathy).

Clinical signs: Multifocal hemorrhages, petechiae, organ dysfunction, and signs of the primary disease.

Diagnosis: Prolonged PT and aPTT, decreased fibrinogen, thrombocytopenia, and increased D-dimers or fibrin degradation products.

Treatment

Address the underlying cause.

Supportive care with plasma transfusions and heparin should be undertaken in specific cases.

Liver disease

Cause: Hepatic insufficiency due to chronic hepatitis, cirrhosis, or hepatic neoplasia.

Pathophysiology: The liver is responsible for producing most clotting factors and regulating fibrinolysis. Liver dysfunction impairs these processes, leading to coagulopathies.

Clinical Signs: Variable bleeding tendencies depending on the severity of liver disease.

Diagnosis: Abnormal PT and aPTT, low albumin, and elevated liver enzymes.

Treatment: Management of liver disease and plasma transfusions.

Immune-mediated thrombocytopenia (IMTP)

Cause: Autoimmune destruction of platelets.

Pathophysiology: Reduced platelet count leads to impaired primary hemostasis and increased bleeding risk.

Clinical signs: Petechiae, ecchymoses, mucosal bleeding, and anemia from blood loss.

Diagnosis: Severe thrombocytopenia on a complete blood count (CBC), with no evidence of underlying bone marrow disorders.

Treatment

Immunosuppressive therapy such as corticosteroids and cyclosporine should be applied.

Blood transfusions should be done in severe cases.

Toxicities

Snake venom: Venoms from certain snakes contain anticoagulant proteins or induce DIC.

Toxins: Mycotoxins and heavy metals may interfere with clotting mechanisms.

Uremia

Cause: Chronic kidney disease leading to the accumulation of uremic toxins.

Pathophysiology: Uremic toxins impair platelet function and disrupt vascular integrity.

Clinical signs: Mucosal bleeding and bruising.

Diagnosis: Azotemia, thrombocytopathy on platelet function assays.

Treatment: Dialysis and supportive care.

Hypothermia and shock

Pathophysiology: Severe hypothermia or hypovolemic shock can impair clotting enzyme activity and platelet function, resulting in coagulopathies.

Treatment: Correction of temperature and perfusion abnormalities.

Diagnostic approaches for hemorrhagic disorders

History and physical examination: Determine the onset and nature of bleeding, possible exposure to toxins, or family history of bleeding disorders.

Laboratory tests

PT and aPTT to assess the extrinsic and intrinsic coagulation pathways,

Platelet count and function tests,

Fibrinogen levels and D-dimer for fibrinolysis evaluation,

Factor assays for specific deficiencies,

Advanced diagnostics: Genetic testing for congenital defects and imaging for bleeding localization are important advanced techniques.

Hemorrhagic disorders in animals can range from mild to life-threatening. Early recognition and accurate diagnosis are critical for effective management. While congenital defects often require lifelong supportive care, many acquired disorders can be treated successfully by addressing the underlying cause. Advances in veterinary medicine, including genetic testing and targeted therapies, continue to improve outcomes for animals with these challenging conditions.

38

Metabolic Problems Related to Acid-base Balance

Maintaining a proper acid-base balance is crucial for the health, productivity, and survival of livestock and poultry. The physiological processes that regulate this balance are sensitive to various factors such as diet, environmental conditions, and metabolic disorders. Any disruption can lead to metabolic problems that negatively impact growth, reproduction, immunity, and overall performance. The acid-base balance in animals refers to the maintenance of the blood's pH within a narrow range, typically 7.35–7.45. It is regulated by three main systems:

Buffer systems

The bicarbonate buffer system is the most significant, comprising bicarbonate ions (HCO_3^-) and carbonic acid (H_2CO_3). It resists rapid pH changes by neutralizing excess acids or bases.

Respiratory regulation

Animals adjust blood pH by modulating the rate and depth of respiration, which affects carbon dioxide (CO_2) levels. Increased CO_2 leads to acidosis, while decreased CO_2 causes alkalosis.

Renal regulation

The kidneys excrete hydrogen ions (H^+) and reabsorb bicarbonate, playing a long-term role in maintaining acid-base homeostasis.

Disruptions in these systems can lead to two primary conditions:

Acidosis: A state where the blood becomes too acidic ($pH < 7.35$).

Alkalosis: A condition where the blood becomes too basic ($pH > 7.45$).

Common metabolic problems.

Acidosis

Metabolic acidosis

Occurs when acid production exceeds the body's buffering capacity or when bicarbonate levels decrease. In livestock and poultry, common causes include:

Rumen acidosis (Cattle, Sheep, Goats)

Resulting from high-grain diets with inadequate fiber, this leads to rapid fermentation, excessive volatile fatty acid (VFA) production, and a drop in rumen pH.

Lactic acidosis

Prolonged hypoxia or vigorous exercise can increase lactic acid production, reducing systemic pH.

Diarrhoea

It causes bicarbonate loss exacerbating acidosis.

Symptoms of acidosis

Poor feed intake and weight loss

Diarrhoea

Lethargy

Tachycardia (in severe cases)

Respiratory acidosis

It results from impaired CO_2 elimination due to respiratory dysfunction, which increases carbonic acid levels in the blood.

Alkalosis

Metabolic alkalosis

This condition arises from excessive bicarbonate or a loss of hydrogen ions. In ruminants, it is often linked to:

Prolonged vomiting: Loss of gastric acid.

Displaced abomasum (Cattle): Reduced acid secretion.

Hypochloremia: A low chloride diet can impair the chloride-bicarbonate exchange, increasing bicarbonate.

Respiratory alkalosis

It is caused by hyperventilation, often due to stress or high environmental temperatures.

Symptoms of alkalosis

Muscle weakness and tremors

Slow growth or weight gain

Reduced feed efficiency

Species-specific metabolic problems

Livestock

Cattle

Rumen acidosis

Sub-acute rumen acidosis (SARA) is prevalent in dairy cattle on high-energy diets. It impairs digestion, milk production, and reproductive performance.

Hypocalcaemia and alkalosis

In dairy cows, alkalosis is often linked to hypocalcaemia during the transition period, predisposing animals to metabolic disorders like milk fever.

Sheep and Goats

They are susceptible to rumen acidosis due to excessive grain feeding during fattening periods. Pregnancy toxemia, associated with acid-base imbalance, affects ewes carrying multiple lambs.

Swine

High-protein diets can lead to metabolic acidosis, causing reduced growth rates.

Heat stress exacerbates respiratory alkalosis due to hyperventilation.

Poultry

Heat stress and respiratory alkalosis

Poultry, especially broilers and layers, are highly sensitive to high temperatures. Panting increases CO_2 loss, leading to respiratory alkalosis.

Consequences

Reduced egg production and shell quality in layers,

Lower feed intake and growth rates in broilers,

Acidosis in meat birds

Rapid growth rates increase the risk of metabolic acidosis due to lactate build-up during intensive activity or heat stress.

Contributing factors to acid-base disorders

Diet

High-carbohydrate or low-fibre diets lead to acidosis in ruminants.

High-protein diets predispose pigs and poultry to acidosis.

Imbalanced electrolyte levels (e.g., Na^+, K^+ and Cl^-) contribute to alkalosis.

Environmental stress

Heat stress in poultry and swine causes respiratory alkalosis.

Cold stress in livestock can lead to metabolic acidosis.

Disease

Gastrointestinal diseases causing diarrhoea lead to bicarbonate loss.

Respiratory infections impair CO_2 elimination.

Production demands

High-yielding dairy cows and fast-growing broilers have increased metabolic rates, heightening susceptibility to acid-base imbalances.

Diagnosis and monitoring

Clinical signs

Observation of symptoms like altered feed intake, lethargy, or changes in growth rates.

Blood tests

Assess pH, bicarbonate, CO_2 levels, and electrolytes.

Urine tests

Urine pH can provide insights into systemic acid-base status.

Management and Prevention Strategies

Dietary adjustments

For ruminants

Include adequate fiber to stimulate rumination and salivary bicarbonate production.

Gradually transition diets to avoid abrupt changes in rumen fermentation.

For monogastrics (swine and poultry)

Optimize dietary electrolyte balance (Dietary cation-anion difference; DCAD).

Avoid excessive protein feeding.

Environmental management

Provide shade, ventilation, and cooling systems to prevent heat stress.

Ensure access to clean drinking water to promote hydration & thermoregulation.

Supplementation

Use buffers (e.g., sodium bicarbonate) in feed to prevent acidosis.

Electrolyte supplements during heat stress can reduce alkalosis risk.

Health monitoring

Regularly monitor blood parameters in high-producing animals.

Prompt treatment of gastrointestinal or respiratory diseases.

Probiotic and enzyme additives

Probiotics help stabilize gut microflora, reducing acid build-up.

Enzymes improve feed digestion, minimizing metabolic stress.

The acid-base balance plays a pivotal role in the health and productivity of livestock and poultry. Disruptions in this balance, leading to acidosis or alkalosis, can have severe consequences, including reduced growth, poor feed efficiency, and increased susceptibility to disease. A comprehensive approach involving proper dietary management, environmental control, and regular health monitoring is essential to prevent and manage acid-base-related metabolic disorders. By addressing these challenges proactively, producers can ensure optimal animal performance and economic returns.

a) Metabolic acidosis

Metabolic acidosis is a serious disorder in ruminants, characterized by a decreased blood pH due to the accumulation of acid or the loss of bicarbonate. This condition can arise due to various factors but is often related to dietary imbalances, particularly in high-producing dairy cattle and feedlot animals, where rapid fermentation of carbohydrates results in increased acid production. In this chapter, we will explore the causes, symptoms, effects, and nutritional management of metabolic acidosis in ruminants, particularly focusing on nutritional strategies to prevent and manage this condition.

Causes of metabolic acidosis in ruminants

In ruminants, metabolic acidosis can be classified into two major types:

Acute ruminal acidosis (grain overload)

This form of acidosis results from the rapid fermentation of highly fermentable carbohydrates primarily starches from grains. When ruminants are abruptly introduced to large amounts of grains or other fermentable carbohydrates, the fermentation rates are increased in the rumen. As a result, volatile fatty acids (VFAs) and lactic acid accumulate, lowering the pH of the rumen. Lactic acid, in particular, is poorly absorbed and acts as a potent acid, exacerbating the drop in ruminal pH.

Sub-acute ruminal acidosis (SARA)

SARA is a more common and chronic form of acidosis that occurs when the rumen pH consistently drops below the optimal range (typically pH < 5.5) for extended periods. It is often due to diets that contain high amounts of fermentable carbohydrates and inadequate fiber content. Unlike acute acidosis, SARA develops gradually and can persist for weeks or months, leading to reduced feed intake, milk production, and compromised overall health.

Pathophysiology of metabolic acidosis

In ruminants, the fermentation of carbohydrates by rumen microbes produces VFAs such as acetate, propionate, and butyrate. These VFAs are crucial energy sources for ruminants. However, when excessive fermentable carbohydrates (such as grains) are introduced into the diet, the balance of microbial fermentation is disturbed. As a result, lactic acid-producing bacteria, such as *Streptococcus bovis* and *Lactobacillus species*, proliferate. This increases lactic acid levels, which leads to a dramatic drop in ruminal pH, causing acute or sub-acute acidosis.

In cases of acute acidosis, the rumen's buffering capacity is overwhelmed, and the systemic absorption of lactic acid into the bloodstream can lead to metabolic acidosis. As the blood becomes more acidic, it impairs normal physiological processes and may lead to dehydration, shock, and in severe cases, death.

Clinical signs and effects

The clinical manifestations of metabolic acidosis in ruminants can vary depending on the severity and duration of the condition. Some of the key symptoms include:

Reduced feed intake: Ruminants experiencing acidosis often reduce their intake of feed as a protective mechanism.

Diarrhoea: Loose, watery stools are common due to impaired digestion and fermentation in the rumen.

Dehydration: Excessive acid accumulation can lead to osmotic shifts, causing dehydration and electrolyte imbalances.

Laminitis: Chronic acidosis can predispose ruminants to laminitis, a painful inflammatory condition of the hooves.

Reduced milk production and fat content: Dairy cows with SARA often exhibit reduced milk yield and lower milk fat percentages due to impaired rumen function.

Weight loss and poor growth: In beef cattle, chronic acidosis can lead to poor growth rates and reduced feed efficiency.

Nutritional management

The nutritional management of metabolic acidosis in ruminants revolves around preventing the excessive fermentation of carbohydrates and ensuring the rumen's buffering capacity is maintained. Several strategies can help prevent and manage acidosis:

Balanced diets

One of the most effective ways to prevent metabolic acidosis is to provide a balanced diet that meets the energy requirements of ruminants without overloading the rumen with fermentable carbohydrates. A well-formulated diet should include:

Adequate fibre

Fibre (neutral detergent fibre or NDF) plays a crucial role in maintaining rumen health by promoting saliva production, which serves as a natural buffer. Diets with insufficient fibre can increase the risk of acidosis. Long-fibre forages, such as hay and silage, should be incorporated into the diet to stimulate rumination and saliva production.

Controlled carbohydrate intake

High-grain diets are often necessary for high-producing dairy cows and feedlot cattle, but the amount of starch and rapidly fermentable carbohydrates should be carefully managed. Grain should be introduced gradually, and highly fermentable grains (e.g., wheat, barley) should be used cautiously. Slowly fermentable grains like corn may be a better choice.

Buffer

Dietary buffers, such as sodium bicarbonate and magnesium oxide, can help neutralize excess acids in the rumen. They can be included in the ration to maintain a more stable rumen pH.

Gradual dietary transitions

Abrupt dietary changes, especially an increase in grain or other fermentable carbohydrates, can overwhelm the rumen and lead to acidosis. It is important to introduce new feeds gradually to allow rumen microbes time to adjust. For instance, transitioning from a forage-based diet to a high-grain diet should be done over 2–3 weeks to prevent a sudden drop in rumen pH.

Feed additives

Several feed additives can help manage ruminal pH and reduce the risk of acidosis:

Ionophores

Ionophores, such as monensin, can alter rumen fermentation patterns, favoring the production of propionate over lactate. This helps to stabilize rumen pH and reduce the risk of acidosis. Ionophores are commonly used in feedlot cattle diets to improve feed efficiency and reduce the incidence of acidosis.

Probiotics

Certain probiotics, such as *Saccharomyces cerevisiae* (yeast) and lactic acid-utilizing bacteria, can help stabilize rumen fermentation by promoting the growth of beneficial microbes and reducing lactic acid accumulation.

Tannins and essential oils

Some studies suggest that tannins and essential oils, when included in the diet, may reduce the production of lactic acid and stabilize ruminal fermentation. These natural additives could be explored as alternatives or supplements to traditional feed additives.

Frequent feeding

Offering smaller, more frequent meals can help prevent large fluctuations in rumen pH. This strategy ensures that fermentable carbohydrates are processed more evenly throughout the day, reducing the risk of lactic acid accumulation.

Monitoring and adjusting diets

Regular monitoring of feed intake, rumen pH, and animal performance is essential in preventing metabolic acidosis. For dairy cows, milk fat levels can be a useful indicator of rumen health, as low milk fat percentage often correlates with SARA. Similarly, fecal consistency can provide clues about the digestive health of feedlot cattle. Diets should be adjusted based on these observations to maintain rumen stability.

Metabolic acidosis in ruminants is a complex disorder primarily related to dietary imbalances, particularly the overconsumption of fermentable carbohydrates. Acute acidosis can lead to life-threatening consequences, while chronic forms like SARA reduce productivity and animal well-being. However, effective nutritional management strategies, including the provision of adequate fiber, controlled carbohydrate intake, gradual dietary transitions, the use of buffers and additives, and frequent feeding, can significantly reduce the risk of metabolic acidosis in ruminants. By maintaining proper rumen function, farmers can optimize the health and productivity of their herds, ensuring better animal welfare and improved economic outcomes.

b) Metabolic alkalosis

Metabolic alkalosis is a less common but significant metabolic disorder in ruminants, characterized by an elevated blood pH due to an excessive accumulation of bicarbonate or a loss of acid from the body. It usually arises as a consequence of disruptions in acid-base balance, often linked to dietary factors or metabolic disturbances such as displaced abomasum, excessive potassium intake, or improper feeding strategies. In this chapter, we will explore the causes, symptoms, physiological effects, and nutritional management of metabolic alkalosis in ruminants, focusing on preventive and corrective strategies.

Causes of metabolic alkalosis in ruminants

Metabolic alkalosis occurs when the body loses hydrogen ions (H+) or gains an excess of bicarbonate ions (HCO_3^-), leading to an increase in blood pH (above 7.45). In ruminants, there are several potential causes of metabolic alkalosis:

Hypochloremia and potassium imbalance

One of the most common causes of metabolic alkalosis in ruminants is hypochloremia, which refers to a reduction in chloride (Cl^-) levels in the blood. Chloride is an essential electrolyte involved in maintaining acid-base balance. The loss of chloride is frequently associated with gastrointestinal disturbances, especially abomasal disorders such as displaced abomasum, abomasal volvulus, or prolonged vomiting and diarrhoea. When chloride is lost, bicarbonate remains unbuffered, leading to alkalosis.

High dietary potassium (K□) intake

High dietary potassium (K^+) intake can exacerbate alkalosis by interfering with the kidney's ability to excrete bicarbonate. Diets rich in potassium, often from forages such as alfalfa or legumes, can promote the development of alkalosis by inducing hypochloremia and reducing renal H+ excretion.

Excessive use of sodium bicarbonate

In dairy cows, sodium bicarbonate is commonly used as a dietary buffer to prevent ruminal acidosis, especially when high-concentrate diets are fed. However, excessive use of sodium bicarbonate can lead to an accumulation of bicarbonate in the bloodstream, tipping the balance towards alkalosis. While buffering the rumen is beneficial in many cases, overuse must be avoided to prevent systemic alkalosis.

Renal dysfunction

The kidneys play a critical role in maintaining acid-base balance by excreting hydrogen ions and reabsorbing bicarbonate. When renal function is impaired, the kidney's ability to regulate this balance may be compromised, leading to the retention of bicarbonate and the development of alkalosis.

Prolonged loss of gastric secretions

Conditions that lead to prolonged loss of gastric secretions, such as abomasal torsion or volvulus, can contribute to metabolic alkalosis. Gastric secretions are rich in hydrochloric acid (HCl), and their loss can result in the retention of bicarbonate in the blood.

Pathophysiology of metabolic alkalosis

In metabolic alkalosis, the blood becomes abnormally alkaline due to the loss of hydrogen ions or the gain of bicarbonate. This disrupts the normal acid-base balance, leading to a variety of physiological effects. The alkalosis interferes with normal cellular function, as many enzymes and metabolic processes are pH-dependent. Ruminants affected by alkalosis often experience reduced feed intake, decreased growth, and impaired production due to altered metabolic processes.

Alkalosis also affects the respiratory system, as the body attempts to compensate for the increased blood pH by reducing carbon dioxide (CO_2) levels. This is achieved through decreased respiratory rate and depth, which helps retain CO_2 and increase carbonic acid (HCO_3) levels, thereby lowering blood pH. However, respiratory compensation is often incomplete and cannot fully correct metabolic alkalosis.

Additionally, metabolic alkalosis leads to shifts in electrolyte balances, particularly potassium and calcium. Hypokalemia (low blood potassium) and hypocalcemia (low blood calcium) are common secondary effects, which can cause muscle weakness, decreased gut motility, and other clinical signs.

Clinical signs and effects

The clinical signs of metabolic alkalosis in ruminants can be subtle or pronounced, depending on the severity and duration of the condition. Key clinical signs include:

Reduced feed intake

Affected animals often exhibit a reduction in feed intake, which can impact milk production in dairy cows or weight gain in beef cattle.

Dehydration

Due to electrolyte imbalances, animals may show signs of dehydration, despite adequate water intake.

Muscle weakness and tremors

Alkalosis disrupts electrolyte balances, particularly potassium and calcium levels, leading to muscle weakness, tremors, and in severe cases, tetany (muscle stiffness and spasms).

Decreased rumen motility

Reduced rumen contractions (ruminal atony) and impaired digestion may result in reduced feed efficiency and production.

Decreased milk production

In dairy cows, alkalosis is often associated with reduced milk yield and lower milk quality.

Changes in respiration

Respiratory rate may decrease as the body attempts to retain CO_2 and compensate for the alkaline state.

Nutritional management

Managing metabolic alkalosis in ruminants requires a multifaceted approach that includes dietary modification, electrolyte management, and addressing underlying causes such as abomasal disorders. Below are several key strategies for the nutritional management of metabolic alkalosis:

Correcting chloride deficiency

Chloride plays a critical role in maintaining acid-base balance. Diets deficient in chloride can predispose ruminants to alkalosis, especially in animals experiencing gastrointestinal disturbances. To prevent and correct alkalosis, it is important to:

Supplement chloride

Adding chloride to the diet, either through common salt (NaCl) or other chloride sources such as ammonium chloride (NH_4Cl), can help restore the chloride balance and promote bicarbonate excretion by the kidneys.

Increase salt intake

Ensuring adequate salt intake in the diet can help maintain electrolyte balance and promote water intake, which is important for restoring hydration and promoting normal kidney function.

Managing potassium levels

Excessive potassium intake can exacerbate alkalosis by interfering with the kidneys' ability to excrete bicarbonate. To prevent this, potassium levels in the diet should be carefully managed:

Monitor forage potassium levels

Forages, particularly legumes like alfalfa, can be high in potassium. It is important to test forages and adjust diets accordingly, especially in animals prone to alkalosis.

Limit high-potassium forages

Reducing the intake of high-potassium forages and replacing them with lower-potassium alternatives can help mitigate the risk of alkalosis.

Avoiding excessive sodium bicarbonate

While sodium bicarbonate is a common buffer used to prevent acidosis, excessive use can lead to alkalosis. It is important to:

Adjust bicarbonate supplementation

Avoid overfeeding sodium bicarbonate, especially in animals already at risk of alkalosis. Monitoring rumen pH and adjusting buffering agents accordingly can help prevent excess bicarbonate accumulation.

Use of ammonium chloride

Ammonium chloride is a dietary acidifier that can be used to correct metabolic alkalosis. It works by increasing the excretion of bicarbonate and reducing blood pH. Adding ammonium chloride to the diet can be effective in managing alkalosis, but care must be taken to avoid over-acidification.

Hydration and electrolyte support

Ensuring adequate water intake is critical in managing alkalosis, as dehydration can exacerbate electrolyte imbalances. Providing oral rehydration solutions that contain balanced electrolytes can help correct imbalances in chloride, potassium, and sodium.

Metabolic alkalosis in ruminants, though less common than acidosis, can have significant negative impacts on animal health, productivity, and well-being. It is primarily caused by electrolyte imbalances, such as hypochloremia and excessive potassium intake, as well as the overuse of sodium bicarbonate. Nutritional management of metabolic alkalosis involves correcting these imbalances through chloride supplementation, managing potassium levels in the diet, and ensuring adequate hydration. By addressing these factors, farmers and nutritionists can prevent and manage metabolic alkalosis, thereby improving the overall health and productivity of ruminants.

c) Respiratory acidosis

Respiratory acidosis is a condition in ruminants where the pH of the blood drops due to an accumulation of carbon dioxide (CO_2) caused by impaired ventilation or gas exchange. This excess CO_2, which normally should be exhaled, combines with water in the body to form carbonic acid (H_2CO_3). As this carbonic acid dissociates, it releases hydrogen ions (H^+), thereby lowering blood pH and leading to a condition of acidosis. Respiratory acidosis is primarily caused by respiratory dysfunction rather than dietary imbalances. However, nutrition can play a supportive role in managing the overall condition of affected ruminants. This chapter will cover the causes, symptoms, physiological effects, and nutritional strategies for managing respiratory acidosis in ruminants.

Causes of respiratory acidosis in ruminants

Respiratory acidosis in ruminants occurs due to conditions that impair the ability of the lungs to effectively exchange gases, specifically the failure to eliminate sufficient CO_2 from the body. Some of the main causes include:

Pneumonia

Pneumonia is a leading cause of respiratory acidosis in ruminants. It can result from bacterial, viral, or fungal infections, which cause inflammation in the lungs. This inflammation hinders proper gas exchange and leads to the accumulation of CO_2, resulting in respiratory acidosis.

Aspiration pneumonia

Aspiration pneumonia occurs when food, liquid, or ruminal contents are inhaled into the lungs, causing infection or blockage. This leads to impaired lung function and the retention of CO_2, contributing to respiratory acidosis.

Pulmonary oedema

Pulmonary oedema is the accumulation of fluid in the lungs, which can occur due to heart failure, toxins, or allergic reactions. The presence of fluid impairs oxygen and CO_2 exchange, leading to the accumulation of CO_2 and, subsequently, respiratory acidosis.

Chronic obstructive pulmonary disease (COPD)

The COPD in ruminants can result from long-term exposure to environmental pollutants, dust, or noxious gases. This condition leads to chronic inflammation of the airways, resulting in reduced lung capacity and the retention of CO_2, ultimately leading to respiratory acidosis.

Physical obstructions and trauma

Physical blockages, such as tumors or foreign objects obstructing the airways, can prevent proper ventilation. Trauma to the chest or diaphragm can also impair breathing, causing CO_2 accumulation and respiratory acidosis.

Pathophysiology of respiratory acidosis

In healthy ruminants, CO_2 is produced as a normal by-product of metabolism and is expelled through the lungs during respiration. However, when lung function is compromised, the body's ability to expel CO_2 becomes impaired. As CO_2 accumulates in the blood, it combines with water to form carbonic acid. This acid dissociates into bicarbonate (HCO_3^-) and hydrogen ions (H^+), leading to a decrease in blood pH (acidosis).

The body attempts to compensate for respiratory acidosis through renal mechanisms. The kidneys increase the excretion of hydrogen ions and reabsorb bicarbonate to buffer the excess acidity. However, renal compensation is a slow process and cannot fully offset the rapid changes caused by respiratory dysfunction.

Clinical signs of respiratory acidosis

The clinical presentation of respiratory acidosis in ruminants can vary depending on the severity of the underlying respiratory condition. Some key clinical signs include:

Laboured or difficult breathing (dyspnoea)

Affected animals often exhibit increased respiratory effort as they struggle to expel CO_2.

Coughing

In cases of pneumonia or aspiration pneumonia, coughing may be frequent and productive.

Nasal discharge

Particularly in cases of respiratory infections, nasal discharge may be observed.

Lethargy and weakness

As acidosis progresses, animals may become weak and lethargic due to the accumulation of CO_2 and decreased oxygenation.

Increased heart rate

Tachycardia may occur as the heart tries to compensate for the decreased oxygen levels.

Cyanosis

In severe cases, cyanosis (bluish discoloration of the mucous membranes) may develop due to inadequate oxygenation.

Reduced feed intake

Acidosis and respiratory distress often lead to reduced appetite and poor nutritional intake.

Nutritional management

While respiratory acidosis is primarily caused by respiratory disorders, nutritional strategies can support the management of affected ruminants by improving overall health, enhancing immune function, and preventing secondary complications such as metabolic acidosis. Nutritional management focuses on supporting lung function, optimizing metabolic balance, and providing adequate energy to meet the animal's needs during periods of respiratory stress. The following strategies are important for the nutritional management of respiratory acidosis:

Balanced energy and protein intake

Respiratory acidosis can increase an animal's energy requirements due to the increased effort required for breathing. Maintaining an adequate energy supply is crucial to prevent weight loss and ensure that the animal's metabolic demands are met.

High-quality forages

Providing high-quality, easily digestible forages helps maintain energy intake without placing excess strain on the rumen. Good-quality hay, silage or pasture should be the primary source of nutrition.

Concentrates and grains

In cases where energy requirements are elevated, controlled supplementation with grains or concentrates can provide additional energy. However, care must be taken not to overload the rumen with rapidly fermentable carbohydrates, which could lead to ruminal acidosis, a secondary condition that can exacerbate respiratory acidosis.

Adequate protein is also essential to support immune function, tissue repair and overall health. Protein sources should be of high quality to ensure that the animal's needs are met, especially during illness.

Electrolyte and acid-base balance

Supporting acid-base balance is critical in managing respiratory acidosis. Although respiratory acidosis is caused by the accumulation of CO_2, ensuring proper electrolyte balance can help the body's compensatory mechanisms.

Bicarbonate supplementation

If necessary, oral or intravenous bicarbonate supplementation can be used to buffer the blood and prevent further drops in pH. However, this must be done cautiously, as over-supplementation can lead to metabolic alkalosis.

Sodium, chloride, and potassium

Ensuring adequate intake of these electrolytes can help maintain proper acid-base balance. Chloride, in particular, plays a key role in maintaining acid-base homeostasis. For this reason, providing a balanced electrolyte solution can support animals affected by respiratory acidosis.

Antioxidants and immune support

Respiratory infections and inflammatory conditions often accompany respiratory acidosis. Therefore, supporting the immune system through nutrition is essential:

Vitamin E and selenium

These are potent antioxidants that help reduce oxidative stress and improve immune function. Adequate levels of these nutrients can help ruminants fight infections such as pneumonia.

Vitamin A and zinc

Both nutrients are critical for maintaining epithelial tissue health, including the respiratory lining. They can improve the immune response to respiratory pathogens and help repair damaged tissues.

Hydration and fluid therapy

Respiratory acidosis can lead to dehydration, either due to the animal's reduced intake of water or due to excessive respiratory effort. Dehydration further complicates acid-base imbalances and impairs the kidney's ability to excrete acid. Ensuring proper hydration is critical for recovery.

Oral or intravenous fluids

In severe cases, intravenous fluids may be necessary to restore hydration and support renal function. Oral rehydration solutions containing balanced electrolytes can also be used to prevent dehydration and support recovery.

Minimizing respiratory irritants

Nutritional management should also consider the environment in which the ruminants are housed. Dusty, poorly ventilated environments can exacerbate respiratory problems. Feeding practices that reduce dust exposure—such as moistening feed or using pelleted feeds—can reduce respiratory irritation.

Respiratory acidosis in ruminants is a condition that arises from impaired lung function and the subsequent accumulation of CO_2, leading to a decrease in blood pH. While nutritional management is not a direct treatment for the underlying respiratory disorders, it plays a supportive role in managing the condition. Providing balanced energy and protein, supporting electrolyte and acid-base balance, and ensuring proper hydration are essential for aiding recovery. In addition, antioxidants and immune-boosting nutrients can help combat the respiratory infections that often accompany respiratory acidosis. By addressing these factors, nutrition can play a key role in the overall management of respiratory acidosis in ruminants, promoting recovery and improving animal welfare.

d) Respiratory alkalosis

Respiratory alkalosis is a metabolic condition in ruminants characterized by an increased blood pH due to excessive loss of carbon dioxide (CO_2) from the bloodstream. This condition arises when hyperventilation causes excessive CO_2 to be expelled from the lungs, which leads to a reduction in carbonic acid (H_2CO_3) and, consequently, a decrease in hydrogen ions (H□). Respiratory alkalosis is relatively rare in ruminants, but it can occur under specific conditions, such as heat stress, high altitudes, or certain environmental and metabolic disturbances. In this chapter, we will explore the causes, physiological effects, clinical signs, and nutritional management strategies for respiratory alkalosis in ruminants.

Causes of respiratory alkalosis in ruminants

Respiratory alkalosis occurs when the rate of CO_2 exhalation exceeds the rate of CO_2 production, disrupting the balance between CO_2 and oxygen in the bloodstream. This hyperventilation leads to excessive loss of CO_2, reducing the amount of carbonic acid and shifting the acid-base balance toward alkalosis. Common causes of respiratory alkalosis in ruminants include:

Heat stress

Ruminants, particularly cattle and sheep, are susceptible to heat stress in hot climates or during periods of high humidity. Heat stress increases the respiratory rate as animals attempt to cool themselves by panting. This rapid breathing results in excessive loss of CO_2, leading to respiratory alkalosis. Heat stress is especially common in dairy cows, as their high metabolic rates and milk production increase their susceptibility to overheating.

High altitude

When ruminants are exposed to high altitudes, the lower oxygen levels in the atmosphere (hypoxia) stimulate an increase in breathing rate to compensate for reduced oxygen availability. This hyperventilation leads to excessive loss of CO_2, causing respiratory alkalosis. Cattle raised in mountainous regions or those transported to higher elevations are at risk of developing this condition.

Pain or anxiety

Situations that cause acute pain, fear, or anxiety in ruminants can trigger rapid, shallow breathing, which may result in hyperventilation and respiratory alkalosis. For example, handling stress during transportation or in veterinary settings can lead to an increase in breathing rate, contributing to CO_2 loss.

Acidosis compensation

In some cases, respiratory alkalosis can develop as a compensatory mechanism for metabolic acidosis. For example, animals with ruminal acidosis may hyperventilate as a way to eliminate CO_2 and elevate blood pH, inadvertently pushing the body toward respiratory alkalosis.

Pathophysiology of respiratory alkalosis

Under normal conditions, CO_2 is produced as a metabolic by-product and excreted through the lungs during respiration. When CO_2 is exhaled excessively due to hyperventilation thus the level of carbonic acid in the blood decreases. Carbonic acid dissociates into bicarbonate (HCO_3^-) and hydrogen ions (H^+), playing a key role in buffering blood pH. With less carbonic acid, there are fewer hydrogen ions available, leading to an increase in blood pH, which defines alkalosis.

Respiratory alkalosis affects the body in several ways. The reduction in blood CO_2 causes a corresponding reduction in bicarbonate levels as the body attempts to restore balance. This leads to an imbalance in electrolytes, particularly potassium (K^+), calcium (Ca^{2+}), and chloride (Cl-), which are essential for normal cellular function. Hypokalemia (low blood potassium) and hypocalcaemia (low blood calcium) are common consequences of respiratory alkalosis, leading to muscle weakness, decreased gut motility, and other metabolic disturbances.

Clinical signs

The clinical signs of respiratory alkalosis in ruminants are often subtle and may be masked by the underlying causes, such as heat stress or high-altitude hypoxia. Key clinical signs include:

Increased respiratory rate (tachypnea)

Rapid and shallow breathing is the primary sign of respiratory alkalosis, as the animal attempts to regulate its body temperature or compensate for hypoxia.

Muscle tremors and weakness

Due to electrolyte imbalances, particularly hypokalemia and hypocalcaemia, affected animals may exhibit muscle tremors, weakness, or even stiffness.

Decreased rumen motility

Electrolyte imbalances can reduce ruminal contractions, leading to poor digestion and reduced feed intake.

Reduced feed intake

Animals suffering from respiratory alkalosis may show a decreased appetite due to the stress of rapid breathing and electrolyte imbalances.

Increased thirst

Animals may drink more water in response to heat stress, which can further dilute blood electrolytes and exacerbate imbalances.

Nutritional management

Nutritional management plays a key role in both preventing and mitigating the effects of respiratory alkalosis in ruminants. The goal of nutritional management is to restore acid-base balance, correct electrolyte imbalances, and support the animal's overall health during periods of respiratory stress. Several strategies can be employed to manage respiratory alkalosis nutritionally:

Electrolyte supplementation

Respiratory alkalosis leads to significant disturbances in electrolyte balance, particularly potassium, calcium, and chloride. Restoring these electrolytes is essential for correcting the physiological effects of alkalosis.

Potassium

Supplementing potassium in the diet can help correct hypokalemia and improve muscle function. Potassium chloride (KCl) is a common supplement that can be added to feed or water to replenish potassium levels.

Calcium

Supplementing calcium is critical in preventing hypocalcemia, which can lead to muscle weakness and poor rumen motility. Calcium carbonate or calcium propionate can be used as supplements to maintain proper calcium levels.

Chloride

Chloride plays a key role in maintaining acid-base balance by promoting bicarbonate excretion. Supplementing chloride through common salt (sodium chloride, NaCl) or ammonium chloride (NH_4Cl) can help restore normal electrolyte balance and prevent further alkalosis.

Adjusting dietary cation-anion balance (DCAB)

The dietary cation-anion balance (DCAB) refers to the balance between positively charged ions (cations) such as sodium (Na^+) and potassium (K^+), and negatively charged ions (anions) such as chloride (Cl^-) and sulfur (S^{2+}). A high DCAB diet (high in cations) can predispose animals to alkalosis, while a low DCAB diet (higher in anions) can help prevent or mitigate alkalosis by promoting acid production.

Lowering DCAB

In cases where respiratory alkalosis is a concern, adjusting the DCAB of the diet by increasing anion content (e.g., chloride and sulphur) can help maintain acid-base balance. Adding ammonium chloride or calcium sulphate ($CaSO_4$) to the diet can reduce the DCAB and promote a more balanced acid-base status.

Water management

Proper hydration is critical for maintaining electrolyte balance and supporting respiratory function, especially during periods of heat stress. However, excessive water intake can dilute electrolytes and exacerbate the effects of alkalosis.

Electrolyte-rich water

Providing water with added electrolytes, particularly potassium and chloride, can help restore balance. Electrolyte solutions should be offered during periods of heat stress or high-altitude exposure to prevent electrolyte dilution and correct imbalances.

Cooling strategies

To reduce the risk of respiratory alkalosis due to heat stress, providing shade, fans, or misting systems can help cool animals and reduce their respiratory rate. Reducing panting helps prevent excessive CO_2 loss and the onset of alkalosis.

Optimizing forage quality

Forage is the primary source of nutrition for ruminants, and its quality can influence the risk of respiratory alkalosis. High-forage diets are generally more alkalogenic than concentrate-rich diets due to their higher potassium content.

Low-potassium forages

Feeding forages that are lower in potassium, such as certain grasses, can help reduce the risk of alkalosis, especially in animals that are prone to hyperventilation. Monitoring forage potassium levels through regular testing can help guide dietary adjustments.

5. Minimizing environmental stress

Nutritional management alone may not be sufficient to prevent respiratory alkalosis. Addressing environmental factors that contribute to respiratory distress, such as heat or altitude is essential.

Heat stress management

Providing adequate shade, cooling systems, and reducing handling during the hottest parts of the day can prevent respiratory alkalosis due to hyperventilation.

Altitude acclimatization

For animals transported to high-altitude regions, gradual acclimatization can help prevent hyperventilation and reduce the risk of alkalosis.

Respiratory alkalosis in ruminants, though less common than other metabolic disorders, can have significant impacts on animal health and productivity. It arises primarily from hyperventilation due to heat stress, high altitude, or other environmental or physiological factors. Nutritional management of respiratory alkalosis focuses on restoring acid-base balance, correcting electrolyte imbalances, and supporting overall health. Key strategies include electrolyte supplementation, adjusting dietary cation-anion balance, providing proper hydration, and minimizing environmental stress. By addressing these factors, respiratory alkalosis can be effectively managed, improving the well-being and productivity of ruminants.

39

Obesity in Pet Animals: Dogs, Cats and Horses

Obesity in pets is a growing concern and can lead to several health issues, including diabetes, arthritis, cardiovascular problems, and decreased lifespan. Preventing obesity in dogs, cats, and horses requires a balanced dietary regimen, regular exercise, and monitoring of body weight. Here is a comprehensive guide to recommended dietary regimens and formulated diets, along with the proportion of ingredients suitable for each species.

Prevention of obesity in dogs

Dietary regimen

Caloric control

Dogs require a controlled-calorie diet to prevent excess weight gain. The number of calories should be adjusted based on the dog's age, breed, activity level, and current body weight.

Meal frequency

Feeding dogs smaller, more frequent meals instead of free-feeding helps manage portion control and prevents overeating.

High fibre diet

Fibre promotes satiety without adding significant calories. Including adequate amounts of fibre helps prevent obesity.

Low-fat diet

High-fat diets can lead to quick weight gain. A diet lower in fat content but with sufficient essential fatty acids helps prevent obesity while maintaining overall health.

Recommended formulated diets

Protein content (20-30%)

Dogs need high-quality protein sources like chicken, turkey, or fish to maintain muscle mass while losing fat. Animal-based proteins are preferred because they contain all essential amino acids.

Carbohydrates (30-50%)

The carbohydrate content should include complex carbs like sweet potatoes, brown rice, or barley, which provide a slow release of energy without causing spikes in blood sugar.

Fat content (8-12%)

Low-fat content is essential to prevent excess weight gain. However, omega-3 and omega-6 fatty acids should be included to promote healthy skin and coat.

Fibre content (5-10%)

Adding vegetables such as carrots, peas, and green beans can increase fibre intake, promote satiety, and prevent overeating.

Ingredients proportion example (Per 1,000 grams of dry food)

Chicken or fish meal: 250-300 grams

Brown rice or sweet potatoes: 300-400 grams

Peas, carrots, and other vegetables: 50-100 grams

Fats (chicken fat or fish oil): 50-80 grams

Fibre (beet pulp or oats): 50-100 grams

Vitamin and mineral mix: 10-20 grams

Prevention of obesity in cats

Dietary regimen

Protein-rich, low-carbohydrate diet

Cats are obligate carnivores, meaning they require a diet high in animal-based proteins and low in carbohydrates. Overconsumption of carbohydrates is one of the leading causes of obesity in cats.

Portion control

Cats should be fed measured portions of food rather than being allowed to graze throughout the day, as free-feeding can lead to overeating.

Moisture content

Cats tend to be poor drinkers. A diet high in moisture (such as wet food) helps with hydration and weight control by promoting satiety without excess calorie intake.

Recommended formulated diets

Protein content (40-50%)

High-quality animal proteins like chicken, turkey, or fish are essential for muscle maintenance and overall health. Meat-based protein sources provide essential amino acids like taurine, crucial for heart health.

Carbohydrate content (5-10%)

Cats have little need for carbohydrates, and excess carbs can lead to obesity. Limiting carbohydrates by using low-glycemic ingredients like pumpkin or green peas is recommended.

Fat content (10-20%)

While fat is important for energy, it should be moderated to prevent weight gain. Healthy fats like omega-3 and omega-6 fatty acids should still be included to maintain skin and coat health.

Fibre content (5-8%)

Fibre from sources such as pumpkin or psyllium husk helps regulate digestion and prevents constipation, which is common in overweight cats.

Ingredients proportion example (Per 1,000 grams of dry food)

Chicken or fish meal: 400-500 grams

Pumpkin or green peas: 50-80 grams

Fish oil or chicken fat: 80-100 grams

Fibre (*Psyllium husk* or pumpkin): 40-60 grams

Vitamin and mineral mix: 10-20 grams

Water (for moisture in wet food diets): ~70% moisture in wet food

Prevention of obesity in horses

Dietary regimen

Forage-based diet

Horses should primarily consume forage, such as grass or hay, as it is the foundation of their diet. High-fibre diets are essential for a horse's digestive health and prevent rapid weight gain.

Restrict caloric intake

Overfeeding grains or high-calorie concentrates can lead to obesity. Restricting these foods and relying on pasture or low-calorie hay (such as timothy or orchard grass) is key to controlling a horse's weight.

Controlled grazing

Limiting grazing time on lush, calorie-dense pastures can help prevent weight gain. Grazing muzzles can also be used to limit the amount of grass consumed.

Monitor body condition score (BCS)

Regularly assess the horse's body condition using a scoring system (1-9) and adjust feed intake accordingly.

Recommended formulated diets

Forage content (70-90%)

High-quality grass hay or pasture is the bulk of a horse's diet. Alfalfa can be used for horses needing more calories, but should be limited in overweight horses.

Concentrates (0-20%)

Overweight horses should have minimal access to grain-based concentrates. For horses that require supplements (older horses), grain alternatives like beet pulp or ration balancers can be used.

Fat content (4-6%)

Horses need less dietary fat than dogs or cats, but adding small amounts of fat through vegetable oils can provide necessary fatty acids while avoiding weight gain.

Fibre content (30-35%)

Fibre is crucial for horses to promote healthy digestion. Using hay with low non-structural carbohydrate (NSC) content, such as Timothy or Orchard Grass, is important in weight management.

Ingredients proportion example (Per 1,000 grams of feed)

Timothy or orchard grass hay: 700-900 grams

Beet pulp or ration balancer: 50-150 grams

Vegetable oil: 30-60 grams (if needed for added fat)

Vitamin and mineral mix: 10-20 grams

General tips for all pets

Exercise

Regular physical activity is a key to prevent obesity in pets. Dogs benefit from daily walks and play sessions; cats need interactive toys and climbing structures, while horses need regular turnout and exercise.

Monitor weight

Regularly weighing your pet and adjusting their diet as necessary can help prevent obesity before it becomes a serious health concern.

Healthy treats

Use low-calorie treats such as vegetables for dogs and horses or small amounts of lean meat for cats. Treats should not exceed more than 10% of the daily caloric intake.

Veterinary guidance

Always consult with a veterinarian when formulating a diet plan, especially if our pet is overweight or has specific health conditions.

Preventing obesity in dogs, cats, and horses requires a tailored approach based on each species' unique dietary needs. By providing balanced nutrition, managing caloric intake, and ensuring adequate exercise, pet owners can maintain their animals at a healthy weight. A diet rich in high-quality proteins, moderate fats, low carbohydrates, and sufficient fibre, combined with regular veterinary checkups, will go a long way in keeping pets fit and healthy throughout their lives.

40

Significance of Oesophageal Groove and its Closer Time

The oesophageal groove (also known as the reticular groove) is a physiological structure and reflex in ruminants that plays a crucial role in the digestion of milk and liquids, particularly in young animals. Its significance and the timing of its closure vary across different species of animals, reflecting adaptations to their feeding behaviors, nutritional needs, and developmental stages.

Anatomy and function of the oesophageal groove

The oesophageal groove is a muscular fold located in the reticulum, the first compartment of the ruminant stomach. When activated, it forms a channel that directs ingested liquids (e.g., milk in young animals) directly to the omasum and abomasum, bypassing the rumen. This mechanism prevents milk from fermenting in the rumen, ensuring efficient digestion and nutrient absorption. The closure of the groove is triggered by specific stimuli, such as suckling, milk temperature, and sensory cues like the smell or taste of milk.

Species-specific significance and closure time

The significance and closure time of the oesophageal groove differ among species, influenced by their physiological needs and feeding practices.

1. **Cattle (*Bos taurus*)**

   Significance: In calves, the oesophageal groove is critical for survival during the milk-feeding stage. By directing milk to the abomasum (the true stomach), it avoids fermentation in the rumen, which can lead to bloating and metabolic disturbances.

   Closure time: The reflex is most active in calves and begins to diminish as they transition to a diet of solid feed, usually around 8–12 weeks of age. In adult cattle, the reflex is rarely triggered, although specific stimuli (e.g., salty solutions) can induce it.

2. **Sheep and Goats (*Ovis aries* and *Capra hircus*)**

   Significance: In lambs and kids, the oesophageal groove ensures the proper digestion of milk, which is their primary food source during early life. This mechanism is crucial for maximizing nutrient absorption and supporting rapid growth.

Closure time: Similar to cattle, the groove in lambs and kids begins to lose functionality as they start consuming more fibrous feed, typically by 8–10 weeks of age. However, it can still respond to certain stimuli in older animals, especially in goats, which are more sensitive to salty or sugary liquids.

3. **Buffaloes (*Bubalus bubalis*)**

   Significance: In buffalo calves, the oesophageal groove is vital for directing milk to the abomasum. Given their susceptibility to ruminal dysfunction during milk-feeding, the proper functioning of this reflex is critical.

   Closure time: The reflex in buffalo calves is active for a slightly longer period compared to cattle, as buffaloes are slower in transitioning to solid feed. It may remain partially functional until 4–5 months of age.

4. **Camelids (Camels, Alpacas, and Llamas)**

   Significance: Although camelids are not true ruminants, they have a three-compartment stomach, and the oesophageal groove helps direct milk to the third compartment (equivalent to the abomasum). This ensures the efficient digestion of milk and minimizes the risk of fermentation in the forestomach.

   Closure time: The reflex remains active during the early life stages and gradually diminishes as camelids transition to foraging on fibrous plant material.

5. **Deer and other wild ruminants**

   Significance: In species such as deer, the oesophageal groove is essential for the early development of fawns or calves, ensuring milk is efficiently digested without fermenting in the rumen.

   Closure time: The closure reflex is active in neonates and young animals but fades as they begin to graze and develop functional rumen flora.

6. **Non-ruminants (Pigs, Horses)**

   Significance: In non-ruminants like pigs and horses, there is no true oesophageal groove. However, the closure of the lower oesophageal sphincter in these species helps prevent reflux and ensures proper gastric processing. While not homologous to the oesophageal groove in ruminants, the function serves to optimize digestion.

Factors affecting oesophageal groove closure

1. ***Age***: The reflex is strongest in neonates and gradually diminishes as the animal matures and transitions to solid feed.
2. ***Dietary composition***: The type of feed influences the retention or decline of the reflex. Milk and milk replacers are strong triggers, while fibrous feeds reduce the need for the reflex.
3. ***Stimuli***: Suckling, temperature, and taste play key roles. For example, warm milk stimulates closure in calves. Salt solutions can trigger the groove in adult ruminants, particularly in goats and sheep.
4. ***Management practices***: Feeding methods (e.g., bottle-feeding vs. bucket-feeding) affect the activation of the oesophageal groove. Incorrect feeding practices can lead to milk entering the rumen, causing digestive disorders.

Clinical and practical implications

The oesophageal groove has several implications for animal management, nutrition, and health:

1. ***Milk replacer feeding:*** Young animals require proper stimulation of the oesophageal groove for efficient milk replacer utilization. Improper feeding techniques (e.g., feeding milk from buckets instead of allowing natural suckling) can result in fermentation in the rumen, leading to diarrhoea, bloat, or acidosis.
2. ***Rumen development:*** The decline of oesophageal groove activity is associated with the maturation of the rumen. Feeding fibrous and solid feeds encourages rumen development, preparing young ruminants for a forage-based diet.
3. ***Therapeutic applications:*** In adult ruminants, the oesophageal groove can be pharmacologically or manually stimulated to direct oral medications or fluids to the abomasum. This is particularly useful in treating conditions requiring bypass of the rumen.
4. ***Nutritional disorders***

   Failure of the oesophageal groove to close properly can result in nutritional inefficiencies or disorders. For example:

   Calf Scours: Improper digestion of milk due to groove dysfunction.

   Ruminal acidosis: Caused by fermentation of milk or liquids in the rumen.

Evolutionary and ecological significance

The oesophageal groove reflects an evolutionary adaptation to the unique dietary needs of young ruminants. By enabling direct milk digestion, it ensures that neonates rapidly obtain the energy and nutrients needed for growth and survival. Its gradual obsolescence in adulthood correlates with the transition to a herbivorous diet and the reliance on microbial fermentation in the rumen.

The oesophageal groove is a critical adaptation in young ruminants, ensuring efficient digestion of milk and supporting early growth. The timing of its closure varies among species, influenced by dietary habits, developmental stages, and ecological needs. Understanding the oesophageal groove's function and closure mechanisms has significant implications for animal husbandry, veterinary medicine, and nutritional management, highlighting its importance in both natural and managed environments.

41

Laboratory Tests for Haemostasis in Animals

Haemostasis refers to the physiological process that prevents blood loss following vascular injury while maintaining fluid blood flow within the vascular system. It involves a complex interplay between the vascular endothelium, platelets, coagulation factors, and fibrinolytic proteins. In animals, laboratory tests for haemostasis play a critical role in diagnosing bleeding disorders, monitoring treatment, and investigating hypercoagulable states. This article discusses the principles, applications, and limitations of laboratory tests used to assess haemostasis in animals.

In veterinary medicine, laboratory tests for haemostasis are essential diagnostic tools for identifying bleeding disorders, hypercoagulable states, and monitoring therapeutic interventions. They are crucial in both emergency and routine clinical settings for animals. These tests provide insights into the functionality of primary haemostasis (platelet plug formation), secondary haemostasis (coagulation cascade and fibrin clot formation), and tertiary haemostasis (fibrinolysis and clot breakdown).

The range of laboratory tests includes basic screening tools like platelet counts and prothrombin time (PT), advanced coagulation assays like thromboelastography (TEG), and specific factor assays for diagnosing inherited or acquired coagulopathies. However, the interpretation of these tests requires careful consideration of species-specific differences, pre-analytical and analytical factors, and the underlying pathophysiology of haemostatic disorders. The discussion explores the principles, applications, and limitations of haemostasis tests in animals, providing a comprehensive understanding of their role in veterinary diagnostics.

Components of haemostasis

Haemostasis is traditionally divided into three phases:

1. ***Primary haemostasis***: Involves vascular constriction and the formation of a platelet plug.

2. ***Secondary haemostasis***: Involves activation of the coagulation cascade to form a stable fibrin clot.
3. ***Fibrinolysis***: Refers to the breakdown of the clot once the vessel is repaired.

Disorders of haemostasis can affect one or more of these phases, leading to either bleeding or thrombotic conditions.

1. Evaluation of primary haemostasis

Platelet count and morphology

Methods

Platelet count is commonly determined using automated hematologyanalyzers or manually via a blood smear.

Interpretation

Thrombocytopenia (low platelet count) is a common cause of bleeding in animals. Examination of platelet morphology can provide clues to platelet function disorders.

Species variation

Certain species, such as cats, may show platelet clumping, which can lead to underestimation of platelet count.

Buccal mucosal bleeding time (BMBT)

Principle: Measures the time taken for a standardized cut on the buccal mucosa to stop bleeding.

Clinical use: Primarily used to evaluate platelet function and vascular integrity.

Limitations: Results are affected by technical variability and are not reliable in animals with thrombocytopenia.

Von Willebrand factor (vWF) assay

Importance: Von Willebrand disease (vWD) is a common hereditary disorder in dogs, particularly in breeds like Doberman Pinschers.

Testing: vWF antigen levels are measured using enzyme-linked immunosorbent assays (ELISA).

Interpretation: Low vWF levels indicate vWD, which impairs platelet adhesion to the sub-endothelium.

2. Evaluation of secondary haemostasis

Prothrombin time (PT)

Principle: Assesses the extrinsic and common coagulation pathways, including factors VII, X, V, II, and fibrinogen.

Method: Citrated plasma is mixed with tissue factor and calcium, and the clotting time is measured.

Clinical use: Prolonged PT indicates deficiencies in these pathways, often seen in conditions like vitamin K antagonism (e.g., rodenticide toxicity).

Activated partial thromboplastin time (aPTT)

Principle: Evaluates the intrinsic and common pathways by measuring clotting time after the addition of an activator, phospholipid, and calcium to plasma.

Clinical use: Prolonged aPTT suggests deficiencies or inhibitors affecting factors VIII, IX, XI, XII, X, V, II, or fibrinogen.

Activated clotting time (ACT)

Principle: Similar to aPTT but performed at the point of care using whole blood.

Clinical use: Screening test for coagulation disorders, though less sensitive and specific than aPTT.

Species consideration: ACT reference intervals vary between species.

Fibrinogen quantification

Methods: Determined by heat precipitation, thrombin time assay, or automated analyzers.

Clinical relevance: Hyperfibrinogenemia is associated with inflammation, while hypofibrinogenemia may occur in disseminated intravascular coagulation (DIC) or liver failure.

Thrombin time (TT)

Principle: Measures the conversion of fibrinogen to fibrin in plasma.

Clinical use: Prolonged TT indicates hypofibrinogenemia, dysfibrinogenemia, or the presence of fibrinogen inhibitors (e.g., heparin).

3. Evaluation of fibrinolysis

D-Dimer and fibrin degradation products (FDPs)

Principle: Detect breakdown products of fibrinogen and fibrin, indicating ongoing fibrinolysis or thrombosis.

Clinical use: Elevated levels are associated with DIC, thromboembolic diseases, and severe inflammation.

Species specificity: D-dimer assays are validated for dogs and horses but may require species-specific reference intervals.

Clot retraction

Principle: Assesses platelet function and fibrin polymerization by evaluating the retraction of a clot over time.

Limitations: Rarely used in routine practice due to technical challenges.

4. Global haemostasis tests

Thromboelastography (TEG) and rotational thromboelastometry (ROTEM)

Principle: Provide a dynamic assessment of the entire coagulation process, including clot initiation, formation, strength, and breakdown.

Applications

Monitoring anticoagulant therapy,

Guiding transfusion therapy in coagulopathies,

Assessing hypercoagulability in diseases such as protein-losing nephropathy and/or hyperadrenocorticism

Advantages: Provide a comprehensive assessment of haemostasis.

Limitations: Require specialized equipment and expertise.

5. Disorders and diagnostic approaches

Bleeding disorders

Inherited disorders: like Von willebrand disease, Hemophilia A (factor VIII deficiency) and Hemophilia B (factor IX deficiency).

Acquired disorders: Thrombocytopenia from immune-mediated diseases, infections or drugs,

Coagulation factor deficiencies due to liver disease or vitamin K antagonists,

Diagnostic strategy: Begin with platelet counts and PT/aPTT, followed by specialized assays (e.g., vWF assay, specific factor assays).

Hypercoagulable states

Conditions

Protein-losing nephropathy or enteropathy,

Hyperadrenocorticism,

Neoplasia,

Testing: TEG/ROTEM and D-dimer/FDPs are key tools for detecting hypercoagulability.

Disseminated intravascular coagulation (DIC)

Pathophysiology: A systemic, uncontrolled activation of coagulation leading to microthrombosis and secondary fibrinolysis.

Diagnostic criteria

Prolonged PT/aPTT,

Thrombocytopenia,

Elevated D-dimers/FDPs,

Decreased fibrinogen levels,

6. Preanalytical and analytical considerations

Sample collection: Use citrate-anticoagulated blood, ensuring a proper blood-to-anticoagulant ratio (9:1).

Handling and processing: Avoid hemolysis and process samples promptly to minimize artefacts.

Species variations: Reference intervals vary widely among species, necessitating species-specific interpretation.

Limitations: Test results can be influenced by medications (e.g., heparin, anticoagulants), concurrent diseases, and preanalytical errors.

Laboratory tests for haemostasis are indispensable tools for diagnosing and managing bleeding and thrombotic disorders in animals. Routine tests like platelet counts, PT, and aPTT are complemented by advanced assays such as TEG and vWF testing to provide a comprehensive evaluation of haemostasis. Understanding the principles, applications, and limitations of these tests is crucial for their effective interpretation in clinical practice.

Accurate interpretation of haemostasis tests requires a clear understanding of the principles behind each assay and recognition of species-specific variations in haemostatic mechanisms. Additionally, the reliability of these tests depends on proper sample collection, handling, and processing to avoid pre-analytical errors. Awareness of these factors is essential for obtaining meaningful results that can be applied effectively in clinical practice.

Haemostatic disorders in animals can result from congenital defects, acquired diseases, or external factors such as toxins and medications. Laboratory testing aids in identifying the underlying cause, whether it is thrombocytopenia, coagulation factor deficiencies, or fibrinolytic abnormalities. These insights allow for targeted interventions, such as factor replacement therapy, anticoagulant management, or supportive care in cases like disseminated intravascular coagulation (DIC).

Laboratory tests for haemostasis are vital tools in veterinary medicine, offering precise and actionable insights into the complex mechanisms of blood clotting and fibrinolysis. Their proper application and interpretation not only enhance diagnostic accuracy but also improve patient outcomes by facilitating timely and effective treatment strategies.

a) Blood groups in humans: An in-depth exploration

Blood groups are essential classifications of blood based on the presence or absence of specific antigens on red blood cells (RBCs) and corresponding antibodies in plasma. These antigens, which are inherited, play a significant role in transfusion medicine, organ transplantation, and maternal-foetal health. Human blood groups are determined by various systems, with the ABO and RH systems being the most clinically significant.

ABO blood group system

The ABO system, discovered by Karl Landsteiner in 1900, is the most important blood group system in humans.

ABO antigens

The ABO system is defined by the presence of A and B antigens on the surface of RBCs. The antigens are glycoproteins or glycolipids resulting from specific enzymatic activity encoded by the ABO gene on chromosome 9.

Four main blood types exist:

Type A: Only A antigen is present on RBCs.

Type B: Only B antigen is present on RBCs.

Type AB: Both A and B antigens are present on RBCs.

Type O: Neither A nor B antigen is present on RBCs.

ABO antibodies

Individuals produce antibodies against the antigens they lack:

Type A individuals have anti-B antibodies.

Type B individuals have anti-A antibodies.

Type AB individuals have no anti-A or anti-B antibodies making them universal recipients.

Type O individuals have both anti-A and anti-B antibodies, making them universal donors (for RBC transfusions).

Clinical relevance

Blood transfusions: ABO incompatibility can lead to hemolytic transfusion reactions (HTRs), which are life-threatening.

Organ transplantation: ABO matching is critical to avoid rejection.

Hemolytic disease of foetus and newborn (HDFN): Maternal antibodies against foetal ABO antigens can cause mild to moderate hemolysis in the foetus.

RH blood group system

The RH blood group system, discovered by Landsteiner and Wiener in 1940, is the second most important blood group system. It includes several antigens, the most significant being D antigen.

RH positive and RH negative

RH Positive: Individuals with the D antigen on their RBCs.

RH Negative: Individuals without the D antigen.

RH antibodies

Unlike ABO antibodies, RH antibodies (e.g., anti-D) are not naturally occurring. They form after exposure to RH-positive blood through transfusion or pregnancy.

Clinical relevance

Transfusions: RH incompatibility can lead to delayed hemolytic reactions in sensitized individuals.

Pregnancy: If an RH-negative mother carries an RH-positive foetus, she may develop anti-D antibodies. These antibodies can cross the placenta in subsequent pregnancies, causing severe HDFN.

Prevention of RH sensitization

RH immunoglobulin (RHIg) is administered to RH-negative mothers during and after pregnancy to prevent sensitization.

Other blood group systems

In addition to ABO and RH, over 40 other blood group systems exist. While less commonly encountered, some have significant clinical implications.

1. **Kell blood group system**

   Contains several antigens, with K (Kell) being the most immunogenic.

   Anti-K antibodies can cause HTRs and severe HDFN.

2. **Duffy blood group system**

   Includes antigens like Fy^a and Fy^b.

   Duffy-negative individuals (common in people of African descent) are resistant to Plasmodium vivax malaria, as the parasite uses Duffy antigens to invade RBCs.

3. **Kidd blood group system:** Includes antigens like Jk^a and Jk^b. Anti-Kidd antibodies are associated with delayed HTRs.

4. **MNS blood group system:** This blood group system includes numerous antigens like M, N, S, and s. It is rarely implicated in HTRs or HDFN.

5. **Lutheran blood group system**

   Includes antigens like Lu^a and Lu^b.

   Rarely clinically significant but may cause mild HTRs.

Genetics of blood groups

Blood group antigens are inherited in a Mendelian fashion, with codominance playing a major role.

ABO inheritance

The ABO gene has three alleles: A, B, and O.

A and B are codominant, while O is recessive.

An individual inherits one allele from each parent, leading to the four possible blood types.

RH inheritance

The RH system is more complex, involving the RHD and RHCE genes.

The presence or absence of the RHD gene determines RH positivity or negativity.

Blood typing and testing

Blood typing

Blood typing identifies an individual's blood group and is crucial before transfusion or transplantation. The two main methods are:

Forward typing

Determines which antigens are present on RBCs by reacting them with anti-A and anti-B sera.

Reverse typing

Determines which antibodies are present in plasma by reacting it with A and B RBCs.

Cross-matching

Cross-matching ensures donor-recipient compatibility by testing the recipient's plasma against the donor's RBCs.

Clinical significance of blood groups

Transfusion medicine

Blood group compatibility is essential to prevent hemolytic reactions, which can result in fever, kidney failure, shock, and even death.

Universal donors (Type O negative) and recipients (Type AB positive) simplify emergency transfusions.

Hemolytic disease of the foetus and newborn (HDFN)

HDFN occurs when maternal antibodies attack foetal RBCs. It is most severe with RH incompatibility but can also occur in other systems like ABO or Kell.

Disease susceptibility and resistance

Certain blood groups are associated with disease susceptibility:

Duffy-negative individuals resist *P. vivax* malaria.

Type O individuals have reduced risk of severe malaria and thrombosis but may have increased susceptibility to gastrointestinal infections.

Blood group distribution

The distribution of blood groups varies worldwide due to genetic and environmental factors. For example:

Type O is the most common globally, particularly in South America and Africa.

Type AB is the rarest globally.

Advancements and challenges

Advances

Molecular typing: DNA-based methods improve accuracy in identifying rare blood groups.

Artificial blood: Research into synthetic or universal donor blood substitutes aims to address shortages.

Challenges

Blood shortages remain a global issue, especially for rare blood types.

Managing sensitized patients with multiple antibodies requires specialized resources.

The study of human blood groups is a cornerstone of transfusion medicine, transplantation, and maternal-fetal health. The ABO and Rh systems dominate clinical relevance, but other systems like Kell, Duffy, and Kidd also have significant implications. Understanding blood group genetics, compatibility testing, and their relationship with disease is crucial for ensuring safe and effective medical care. As science advances, novel technologies like molecular typing and artificial blood promise to revolutionize the field, addressing current limitations and ensuring the availability of lifesaving treatments for diverse populations.

b) Blood groups in animals

Blood groups are classifications based on the presence or absence of specific antigens on the surface of red blood cells (RBCs) and corresponding antibodies in the plasma. These antigens and antibodies are crucial for ensuring compatibility in blood transfusions, organ transplantation, and determining maternal-foetal compatibility. In animals, as in humans, the identification of blood groups is vital for veterinary medicine, especially for transfusion practices and preventing hemolytic reactions.

The blood group systems in animals differ significantly between species. They are determined by genetic inheritance, and their clinical importance varies depending on the species. Below is an outline of blood group systems in commonly studied animals.

Canine blood groups

Dogs have a complex blood group system, primarily defined by the dog erythrocyte antigen (DEA) system. The DEA system includes over 12 blood group types, with the most clinically significant being DEA 1.

DEA 1

Subtypes include DEA 1.1 and DEA 1.2.

DEA 1.1 is highly antigenic and can trigger severe hemolytic reactions if mismatched during transfusions.

Dogs do not have naturally occurring antibodies against DEA 1, making a first incompatible transfusion less likely to cause a reaction. However, sensitization occurs, and subsequent mismatched transfusions may lead to life-threatening reactions.

Blood typing in dogs

This is performed using blood typing cards, gel column tests, or molecular methods. Cross-matching (major and minor) is crucial to identify compatibility before transfusion.

Feline blood groups

Cats have a simpler blood group system, classified into three groups: A, B, and AB.

Group A

Group A is the most common blood type in domestic cats. Cats with type-A blood may have naturally occurring anti-B antibodies, which can cause mild to moderate transfusion reactions.

Group B

Group B found in certain purebred cats, such as British Shorthairs and Devon Rex. Cats with type-B blood have strong anti-A antibodies, leading to severe hemolytic reactions if transfused with type A blood.

Group AB

Rare in cats and considered the universal recipient. No naturally occurring antibodies against A or B antigens.

Clinical relevance

Neonatal isoerythrolysis (NI) is a significant concern in cats. It occurs when kittens with type A or AB blood inherit anti-A antibodies from a type B queen's colostrum, leading to hemolysis.

Equine blood groups

Horses have eight major blood group systems: A, C, D, K, P, Q, U, and T. Among these, the A, C, and Q systems are most significant in clinical practice.

Unique features

Horses can express multiple blood group antigens simultaneously. Naturally occurring antibodies are rare, but sensitization can occur due to previous transfusions or pregnancy.

Clinical implications

Neonatal isoerythrolysis is a concern in foals born to mares sensitized to incompatible blood groups. Blood typing and cross-matching are critical for equine transfusion therapy.

Bovine blood groups

Cattle possess 12 blood group systems, with the B and J systems being the most notable.

B system

B system is highly polymorphic, with over 60 antigens. This diversity complicates transfusion compatibility testing.

J system

The J system is the only blood group system in cattle with naturally occurring antibodies. Anti-J antibodies can cause hemolytic reactions if mismatched.

Clinical relevance

Blood typing is rarely performed in cattle due to the complexity of their blood group systems. Instead, emergency transfusions often use cross-matching or whole blood from the same individual.

Blood groups in other species

Sheep and goats: Have multiple blood group systems, but their clinical importance is limited. Cross-matching is preferred over extensive blood typing.

Pigs: Possess multiple blood group systems (e.g., A, E), mainly studied in research and agriculture.

Rabbits and rodents: Blood groups are of research interest but rarely used clinically.

Birds: Avian blood groups are not well-defined. Cross-matching is usually sufficient for transfusions.

Blood group testing and techniques

Blood group testing determines the compatibility of donor and recipient blood. The methods vary from simple field tests to sophisticated molecular techniques.

Blood typing methods

1. ***Agglutination tests:*** Use species-specific reagents to detect antigen-antibody interactions. It is simple and widely used in veterinary clinics.
2. ***Gel column testing***: More precise, separating RBCs based on size and density during antigen-antibody reactions.
3. ***Immunochromatographic tests***: This test is common for quick blood typing, especially in dogs and cats.
4. ***Molecular testing***: Molecular testing is used to detect blood group genotypes using DNA analysis. It is useful for complex or rare blood group systems.

Cross-matching

Cross-matching identifies potential incompatibilities between donor and recipient blood, regardless of their blood types.

Major cross-match: Tests recipient serum against donor RBCs.

Minor cross-match: Tests donor serum against recipient RBCs.

Importance

Prevents hemolytic transfusion reactions,

Essential in species with complex or poorly understood blood group systems.

Clinical importance of blood groups

Understanding blood groups is essential in veterinary medicine for the following reasons:

1. ***Blood transfusions***

   Compatibility testing minimizes transfusion reactions.

   In dogs and cats, even the first transfusion can cause reactions if mismatched in sensitized animals.

2. ***Neonatal isoerythrolysis (NI)***

   Occurs in species like horses and cats,

   Blood typing and matching the mare or queen with the sire prevent this condition.

3. ***Disease diagnosis and management:*** Blood group studies can assist in diagnosing immune-mediated haemolytic anaemia (IMHA) or autoimmune reactions.

4. ***Veterinary research:*** Blood groups are valuable in studying population genetics and disease resistance in various species.

Limitations and challenges

Species-specific complexity: The diversity and polymorphism of blood group systems make standardized testing challenging.

Availability: Specialized tests and reagents are not universally available, especially for less commonly studied species.

Cost: Blood typing and cross-matching can be expensive, limiting their routine use in some veterinary practices.

Sensitivity: Some methods lack the sensitivity required for detecting rare antigens or low titres of antibodies.

Future directions

Advancements in molecular diagnostics and genomics hold promise for improving blood group testing in animals. These technologies can:

Enhance the accuracy of blood typing.

Identify rare or novel blood group antigens.

Facilitate universal donor identification programs in veterinary medicine.

Blood groups are fundamental to understanding compatibility in blood transfusions, maternal-foetal interactions, and immune responses. Each species has unique blood group systems with varying clinical significance. The ability to accurately type and cross-match blood is crucial for preventing adverse transfusion reactions and managing conditions like neonatal isoerythrolysis. While challenges such as species diversity and limited resources exist, ongoing research and advancements in diagnostic techniques continue to improve the understanding and application of blood group knowledge in veterinary medicine.

c) Blood transfusion: A lifesaving medical procedure

Blood transfusion is the process of transferring blood or blood components from a donor to a recipient. It is a critical medical intervention that saves millions of lives worldwide each year. Blood transfusions are used to replace lost components in cases of severe blood loss, manage certain medical conditions, and support patients undergoing surgery. This comprehensive discussion explores the principles, types, indications, procedures, and challenges of blood transfusion.

Principles of blood transfusion

The success of a blood transfusion depends on compatibility between the donor and recipient. Blood groups, determined by specific antigens on red blood cells (RBCs), play a key role in ensuring safety. The most critical blood group systems in transfusion medicine are the ABO system and the RH system.

ABO compatibility

Type O blood is the universal donor for RBCs, while type AB is the universal recipient. Plasma compatibility works oppositely, with type AB plasma being the universal donor.

RH factor

RH-negative recipients should receive RH-negative blood to avoid sensitization.

Blood typing and cross-matching are performed before transfusion to prevent incompatibility reactions.

Types of blood transfusions

Blood transfusions can involve whole blood or specific components, depending on the patient's needs.

Whole blood transfusion

Whole blood contains all blood components (RBCs, plasma, platelets, and white blood cells). Rarely used today except in cases of massive blood loss or in emergencies where component separation is not feasible.

Component therapy

Modern transfusion practices favour component therapy, which involves transfusing only the needed parts of blood:

1. ***Red blood cell (RBC) transfusions***: RBCs transfusion used to treat anaemia and blood loss. Packed RBCs (PRBCs) are concentrated and reduce the risk of volume overload.
2. ***Platelet transfusions:*** Platelet transfusion indicated for thrombocytopenia or platelet dysfunction. It can prevent or treat bleeding in conditions like leukemia or during chemotherapy.
3. ***Plasma transfusions***: Plasma contains clotting factors, antibodies, and other proteins. Its transfusion used for coagulation disorders, liver disease, or massive transfusion protocols.
4. ***Cryoprecipitate***: The cryoprecipitate derived from plasma which is rich in fibrinogen, factor VIII, and von Willebrand factor. It is used for conditions like hemophilia A or hypofibrinogenemia.
5. ***Granulocyte transfusions***: Rarely used but may be indicated in severe neutropenia with life-threatening infections.

Indications for blood transfusion

Blood transfusions are indicated in various clinical scenarios:

1. ***Acute blood loss:*** Acute blood losses occur in conditions such as severe trauma, surgery, or gastrointestinal bleeding. It is indicated to restore oxygen-carrying capacity and blood volume.
2. ***Anaemia:*** Blood transfusion indicated in chronic anaemia from conditions like kidney disease or cancer as well as in symptomatic anaemia unresponsive to other treatments.
3. ***Coagulation disorders:*** Blood transfusion suggested in coagulation disorders such as haemophilia, disseminated intravascular coagulation (DIC), or liver dysfunction. Though, plasma or cryoprecipitate transfusions provide necessary clotting factors.

4. ***Thrombocytopenia:*** Low platelet count due to chemotherapy, bone marrow disorders, or immune conditions. Platelet transfusions reduce the risk of spontaneous bleeding.
5. ***Neonatal Transfusion:*** Exchange transfusion for severe neonatal jaundice or RH incompatibility. Replaces damaged or incompatible RBCs.
6. ***Other indications:*** Blood transfusion as supportive care needed during organ transplantation or stem cell transplantation as well as treatment of autoimmune haemolytic anaemia.

Blood transfusion procedure

1. ***Pre-transfusion testing***
   1. *Blood typing*: To determine the ABO and RH group of both donor and recipient.
   2. *Cross-matching*: Cross-matching ensures compatibility by mixing donor RBCs with recipient plasma and then identifies potential incompatibility reactions.
2. ***Donor selection***

   Donors are screened for infectious diseases (HIV, hepatitis, syphilis, etc.).

   Healthy individuals aged 18–65 years are typically eligible.
3. ***Administration:*** Blood products are administered intravenously under sterile conditions.

   Dosage and rate depend on the patient's condition:

   Rapid transfusion for massive haemorrhage,

   Slower rates for patients at risk of volume overload.

Monitoring

Vital signs are closely monitored during and after transfusion.

Patients are observed for signs of adverse reactions, such as fever, chills, or hypotension.

Risks and complications of blood transfusion

Although blood transfusions are generally safe, complications can occur:

Immediate reactions

1. *Allergic reactions*:

   Mild reactions to plasma proteins,

   Symptoms include itching, hives, and rash.

2. *Febrile non-haemolytic reactions*:

   Caused by recipient antibodies reacting to donor white blood cells,

   Characterized by fever and chills,

3. *Hemolytic reactions*:

   Occur due to ABO or RH incompatibility.

   Symptoms include fever, chills, hemoglobinuria, and shock.

4. *Infections*:

   Rare due to rigorous donor screening,

   Risks include bacterial contamination or viral infections.

5. *Iron overload*: Multiple transfusions can lead to excess iron deposition in organs, requiring chelation therapy.

Delayed reactions

1. *Delayed haemolytic reaction*:

   Caused by antibodies formed after prior sensitization,

   Symptoms include mild anaemia and jaundice.

2. *Graft-versus-host disease (GVHD)*:

   Rare but fatal condition caused by transfused white cells attacking the recipient,

   Prevented by irradiating blood products for at-risk individuals,

Advancements in blood transfusion

Blood component technology

Advances in separating blood into components have revolutionized transfusion medicine, improving efficiency and reducing risks.

Synthetic blood substitutes

Research into artificial blood products, such as hemoglobin-based oxygen carriers (HBOCs), aims to address shortages and improve safety.

Pathogen reduction technology

Techniques to inactivate pathogens in blood products enhance safety.

Genomic matching

Molecular testing ensures precise compatibility in patients with rare blood types or multiple antibodies.

Ethical and logistical challenges

Blood supply shortages

Maintaining an adequate blood supply is challenging, especially during disasters or pandemics.

Cultural and religious barriers

Some individuals may refuse transfusions due to personal beliefs, necessitating alternatives like erythropoietin therapy.

Costs

Blood transfusions can be expensive, particularly in low-resource settings.

Future directions

Personalized transfusion medicine: Incorporating individual genetic profiles for precise matching.

Artificial blood development: Creating universally compatible substitutes to reduce reliance on donors.

Improved donation practices: Encouraging voluntary donations and enhancing donor recruitment strategies.

Blood transfusion is a cornerstone of modern medicine, saving lives in emergencies, chronic illnesses, and surgical interventions. The practice has evolved significantly, with component therapy, stringent testing protocols, and advanced technology ensuring safety and efficacy. Despite its remarkable benefits, challenges such as shortages, complications, and logistical barriers remain. Research continued into artificial blood products, molecular matching, and improved donation practices holds promise for addressing these challenges, ensuring that blood transfusions remain a reliable and lifesaving intervention for patients worldwide.

d) Blood transfusions in animals: A comparative perspective

Blood transfusion is an essential veterinary medical procedure used to treat a variety of conditions in animals, including blood loss, anemia, and clotting disorders. While the principles of blood transfusion are similar across species, the specifics differ due to variations in blood group systems, immunological reactions, and clinical applications. This discussion explores blood transfusion practices in different animal species, focusing on their unique challenges and solutions.

Principles of blood transfusion in animals

Indications for transfusion

Animal blood transfusions are performed for:

Acute hemorrhage due to trauma, surgery, or poisoning,

Chronic anemia from conditions like parasitic infections or autoimmune diseases,

Coagulopathies or clotting disorders,

Shock or hypovolemia caused by severe blood loss.

Key considerations

Blood compatibility: Understanding species-specific blood group systems is crucial to minimize transfusion reactions.

Cross-matching: Performed to ensure compatibility, especially in species with known blood group complexities.

Volume and rate: Adjusted based on the animal's size and condition.

Blood group systems in animals

Animals, like humans, have blood groups determined by surface antigens on red blood cells. However, the complexity and clinical relevance of these systems vary by species.

Canines (dogs)

Blood group system: Dog erythrocyte antigen (DEA) system includes over a dozen identified groups. DEA 1 is the most clinically significant. DEA1.1-positive dogs can cause severe hemolytic reactions if transfused into DEA 1.1-negative recipients.

Universal donors: Dogs negative for DEA 1, 3, 4, 5, and 7 are considered universal donors.

First transfusion exception: Dogs often tolerate their first transfusion without prior cross-matching due to a lack of naturally occurring antibodies.

Felines (cats)

Blood group system: The AB system includes three blood types: A, B and AB. Type A is most common, while type B is found in certain breeds like British Shorthairs. Type AB is rare.

Naturally occurring antibodies: Cats have strong pre-existing antibodies against incompatible blood types, making cross-matching essential even for a first transfusion.

Mik antigen: A newly recognized feline antigen that can cause incompatibility.

Equines (horses)

Blood group system: Horses have more than 30 blood groups classified into 8 major systems (A, C, D, K, P, Q, T, U). The Aa and Qa antigens are the most immunogenic.

Cross-matching: Critical due to the high diversity in equine blood groups.

Neonatal Isoerythrolysis: Foals can suffer from hemolysis if they inherit antigens from the sire that are not present in the dam.

Bovines (cattle)

Blood group system: Cattle have 12 blood group systems, with B and J being the most significant. The B system has extensive polymorphism, complicating compatibility. The J antigen can lead to antibody formation in transfusion recipients.

Other species

Goats and sheep: Have multiple blood groups but rarely require transfusions due to the absence of naturally occurring antibodies.

Swine (pigs): Have a complex system but limited use of transfusions in practice.

Birds: Lack distinct blood group systems; cross-matching is less critical, but donor-recipient compatibility is still important.

Exotic pets: Blood transfusions in species like rabbits, ferrets, and reptiles require tailored approaches due to limited knowledge of their blood group systems.

Blood transfusion procedure in animals

1. Donor selection

Criteria: Healthy, disease-free animals of appropriate size and weight.

Screening: Donors are tested for infectious diseases and compatibility.

Frequency: Donors can typically donate 10–20% of their blood volume safely every 4–6 weeks.

2. Collection and storage

Blood is collected using sterile techniques into anticoagulant-containing bags.

Storage varies: Fresh whole blood is used immediately for clotting factor replacement. Packed RBCs and plasma can be stored under refrigeration or frozen, depending on the component.

3. Administration

Blood is transfused intravenously using a calculated dose based on the recipient's size and need:

Dogs: 20 mL/kg can raise the PCV (packed cell volume) by ~10%.

Cats: 10 mL/kg can raise PCV by ~10%.

Rate of administration starts slowly to monitor for reactions, then increases if tolerated.

4. Complications of blood transfusion in animals

Despite its lifesaving benefits, transfusion carries risks:

Immediate reactions

Hemolytic reactions: Occur if incompatible blood is transfused.

Symptoms: Fever, jaundice, hemoglobinuria, and shock.

Non-hemolytic reactions: Include fever, itching, or urticaria due to immune response to white cells or plasma proteins.

Volume Overload: Particularly in smaller animals like cats or those with pre-existing heart conditions.

Delayed reactions

Delayed hemolysis: Occurs days to weeks later due to newly formed antibodies.

Iron overload: Can develop with repeated transfusions, particularly in chronic conditions.

Infectious diseases

Transmission of infections like Babesia, Ehrlichia, or FeLV is possible if donor screening is inadequate.

Special considerations by species

Dogs

Donor dogs should ideally weigh over 25 kg and be negative for DEA 1 and other key antigens. The absence of naturally occurring antibodies simplifies the first transfusion.

Cats

Cross-matching is mandatory for all transfusions due to naturally occurring antibodies. Donor cats should ideally weigh at least 4.5 kg and be screened for FeLV and FIV.

Horses

Blood typing is not commonly performed in emergencies; cross-matching is the standard. Commercial availability of equine-specific blood products is limited.

Livestock

Blood transfusions are rare but may be performed in valuable animals during emergencies like hemorrhage or neonatal isoerythrolysis.

Exotic Species

Limited research and resources make transfusions challenging. Autotransfusion (recycling the animal's lost blood) is sometimes used as an alternative.

Advancements in veterinary transfusion medicine

Blood banks: Animal blood banks now supply screened and typed blood products for dogs, cats, and even horses.

Synthetic blood substitutes: Hemoglobin-based oxygen carriers (HBOCs) are under research for use in animals.

Point-of-care cross-matching kits: Simplify compatibility testing in emergencies.

Ethical and logistical challenges

Donor welfare: Ensuring the health and safety of donor animals is a top priority.

Cost and accessibility: Blood transfusions are expensive and may not be readily available in remote areas.

Limited research: Knowledge gaps exist for exotic and less commonly treated species.

Blood transfusions in animals are lifesaving interventions with diverse applications across species. While the principles remain consistent, species-specific differences in blood group systems, compatibility testing and clinical needs require tailored approaches. Advancement in veterinary transfusion medicine including animal blood banks and synthetic blood substitutes, are addressing challenges and expanding the scope of care. Continued research and innovation will further improve outcomes and accessibility, ensuring better health and welfare for animals in need of transfusion support.

e) Blood volume and blood pressure: Understanding their relationship and importance

Blood volume and blood pressure are two critical physiological parameters in the cardiovascular system. These elements are intimately connected, and their regulation is essential for maintaining homeostasis, ensuring adequate tissue perfusion, and supporting overall health. Understanding blood volume and blood pressure and how they influence each other is fundamental for diagnosing and managing various cardiovascular and systemic conditions. This discussion explores the nature of blood volume, its regulation, how it influences blood pressure, and the mechanisms that balance these two critical factors.

Blood volume

Blood volume refers to the total amount of blood circulating within an organism's body. In humans, blood volume typically makes up about 7-8% of total body weight, with an average volume of approximately 5 litres in an adult. The blood volume of different species of animals and poultry varies depending on species, size, and physiological characteristics. Typically, blood volume is expressed as a percentage of body weight and serves as a critical component in delivering oxygen, nutrients, and waste products throughout the body. In mammals and birds, blood volume generally makes up 6-10% of body weight. However, this range can vary based on the species, age, health condition, and environmental factors.

Large mammals

For other large mammals like cows, horses, and sheep, blood volume remains within this range, typically about 7-8% of body weight.

Small mammals

Blood volume in small mammals, such as dogs and cats, tends to be slightly higher as a percentage of their body weight, usually around 8-10%. For instance, a cat weighing 4 kg has an approximate blood volume of 0.32-0.4 litres.

Poultry and birds

Birds, including poultry species like chickens, have a blood volume closer to 6-7% of their body weight. This is slightly lower compared to mammals. A chicken weighing 2 kg, for example, will have a blood volume of about 0.12-0.14 litres.

Livestock and farm animals

In cattle and pigs, blood volume makes up approximately 7% of their body weight.

For sheep and goats, this figure is closer to 6-7%.

Average blood volume in litres

Below are specific examples of blood volume in animals:

Cattle: A cow weighing 500 kg has a blood volume of approximately 35-40 litres (7-8% of body weight).

Horses: A horse weighing 500 kg has a blood volume of about 35-40 litres.

Dogs: A dog weighing 20 kg will have a blood volume of around 1.6-2.0 litres (8-10% of body weight).

Pigs: A pig weighing 100 kg will have a blood volume of around 6-7 litres (6-7%).

Chickens: A chicken weighing 2 kg has a blood volume of approximately 0.12-0.14 litres.

Sheep: A sheep weighing 50 kg has a blood volume of about 3.0-3.5 litres.

Factors influencing blood volume

1. ***Species***: Mammals tend to have higher blood volume percentages compared to birds.
2. ***Body size***: Larger animals have greater total blood volume, but the percentage decreases slightly compared to smaller animals.
3. ***Health condition***: Dehydration or anaemia can reduce blood volume, while pregnancy can slightly increase it.
4. ***Activity levels***: Active animals, such as racing horses, may have more blood volume to meet oxygen demands.

Blood volume is a critical physiological parameter in animals and poultry. While it averages around 6-10% of body weight depending on the species, the exact value in litres can be calculated using body weight. For example, in large animals like cattle or horses, total blood volume can reach 35-40 litres, whereas in smaller species such as poultry, it is significantly lower, with a 2 kg chicken having about 0.12-0.14 litres of blood. This understanding is essential in veterinary medicine, livestock management, and poultry farming for monitoring health, fluid therapy, and overall well-being.

Component of blood volume

The components of blood volume include:

Plasma: The liquid portion of blood, consisting mostly of water (around 90%), electrolytes, proteins (such as albumin, globulins, and fibrinogen), hormones, and waste products. Plasma constitutes about 55% of total blood volume.

Red Blood Cells (RBCs): These cells are responsible for oxygen and carbon dioxide transport. They make up approximately 40-45% of blood volume, a proportion known as the hematocrit.

White Blood Cells (WBCs): Although they are important for immune function, WBCs make up only a small fraction of blood volume (less than 1%).

Platelets: These are involved in blood clotting and also contribute minimally to blood volume.

Blood Pressure

Blood pressure is the force exerted by circulating blood on the walls of blood vessels, primarily arteries. Blood pressure is essential for driving blood through the circulatory system and ensuring adequate perfusion of organs and tissues. Blood pressure is typically measured in milli meters of mercury (mmHg) and is given as two values: systolic and diastolic.

Systolic pressure (peak pressure during heart contraction): The pressure exerted on blood vessel walls during the contraction of the heart's ventricles, as blood is pumped into the arteries.

Diastolic pressure (pressure during heart relaxation): The pressure in the arteries when the heart is at rest between beats (diastole), reflecting the recoil of the arterial walls.

The mean arterial pressure (MAP) is often used to describe the overall blood pressure and is calculated using the formula:

The regulation of blood pressure involves multiple mechanisms, including:

Cardiac output (CO): The volume of blood pumped by the heart per minute (typically 4-6 liters per minute in an adult at rest). Cardiac output is influenced by heart rate and stroke volume.

Total peripheral resistance (TPR): The resistance to blood flow offered by the vasculature which is determined by the size and tone of blood vessels.

Blood volume: The total volume of blood in the body, which directly impacts venous return to the heart and, consequently, cardiac output.

The formula or methods for determining blood pressure are broadly similar across species but can vary based on size, anatomy, and physiology. Here is an overview for humans and animals

General formula for blood pressure (humans and animals)

Blood pressure is typically expressed as

Blood pressure (BP) = Cardiac output (CO) x Total peripheral resistance (TPR)

Cardiac output (CO) = Heart rate (HR) × Stroke volume (SV)

Total peripheral resistance (TPR) is the resistance to blood flow in the vessels.

This formula applies broadly to humans and animals because blood pressure depends on similar physiological processes.

Human blood pressure

Normal range: Approximately 120/80 mmHg (systolic/diastolic).

In human being BP is measured using sphygmomanometers (manual or digital) on the upper arm.

Blood pressure in animals

Blood pressure varies widely among species due to differences in cardiovascular anatomy, metabolic rate, and size. Here's an outline:

Large mammals (e.g., horses, cattle)

Normal BP: 140/90 mmHg (varies depending on activity and size).

Small mammals (e.g., cats, dogs)

Dogs: 130-160/80-100 mmHg (breed-dependent).

Cats: 120-160/80-100 mmHg.

Rodents (e.g., rats, mice):

Higher relative pressures: ~120-140/90 mmHg.

Measurement often uses tail-cuff or invasive techniques.

Birds (including Poultry)

Birds have a high metabolic rate and relatively higher blood pressures:

Poultry (chickens)

Normal BP: ~ 140-160/90-110 mmHg.

Measurement is invasive due to avian anatomy. Birds require higher pressures to support oxygenation during flight.

Reptiles (snakes, lizards, turtles): Lower BP due to slower metabolism:

Typical range: 30-50/20-30 mmHg.

Fish

Fish have very low blood pressure since they have a single-circuit circulatory system:

Average BP: ~20-30 mmHg.

Measuring blood pressure in animals

Indirect methods

Oscillometric or Doppler techniques are used in small animals (dogs, cats).

Tail-cuff methods are common in rodents.

Direct (invasive) methods

Arterial catheterization is used for birds, reptiles, or larger mammals.

Factors affecting blood pressure across species

Body size: Larger animals generally have higher pressures.

Metabolic rate: Higher metabolism (birds, rodents) correlates with higher blood pressure.

Environment: Aquatic vs. terrestrial species has different circulatory demands.

Relationship between blood volume and blood pressure

Blood volume plays a critical role in determining blood pressure when blood volume increases or decreases; the pressure exerted by the blood on the arterial walls also changes.

The relationship can be explained through basic cardiovascular physiology:

Increased blood volume

When blood volume increases, the heart has to pump more blood, which increases cardiac output (CO). The increased volume of blood in the circulatory system also causes more pressure to be exerted on the walls of the arteries, leading to increased blood pressure. Volume overload (as seen in conditions like hypervolemia) can result from excessive sodium intake, kidney dysfunction, or excessive intravenous fluid administration, causing an increase in both systolic and diastolic blood pressure. Conditions like heart failure can exacerbate the effects of increased blood volume, as the heart's ability to pump blood efficiently is impaired, causing a backup of blood in the vessels and contributing to elevated pressure.

Decreased blood volume

Hypovolemia (reduced blood volume), caused by blood loss, dehydration, or fluid loss through vomiting or diarrhoea, leads to a decrease in cardiac output, as less blood returns to the heart. This results in lower blood pressure. The body compensates for reduced blood volume by activating several mechanisms, including vasoconstriction (narrowing of blood vessels) and increasing heart rate, to maintain blood pressure. However, in severe cases of hypovolemia, blood pressure can fall dangerously low, leading to hypotension and impaired tissue perfusion.

Regulation of blood volume and blood pressure

Several regulatory systems ensure the balance of blood volume and pressure, maintaining them within optimal ranges for health.

Renin-angiotensin-aldosterone system (RAAS)

The RAAS is one of the most important mechanisms for regulating both blood volume and blood pressure.

When blood volume or blood pressure drops, the kidneys release renin, which activates the conversion of angiotensinogen to angiotensin I, then to angiotensin II. Angiotensin II has multiple effects, including:

Vasoconstriction: Narrowing of blood vessels to increase blood pressure.

Aldosterone release: Stimulates the kidneys to retain sodium and water, increasing blood volume and, consequently, blood pressure.

Antidiuretic hormone (ADH)

The ADH, also known as vasopressin, is produced by the hypothalamus and released by the pituitary gland in response to low blood volume or high blood osmolality. ADH acts on the kidneys to promote water reabsorption, which increases blood volume and helps restore normal blood pressure.

Baroreceptor reflex

Baroreceptors are stretch-sensitive sensors located in the aorta and carotid arteries. They continuously monitor blood pressure and send signals to the brain. When blood pressure falls, baroreceptors trigger compensatory mechanisms, such as increased heart rate (tachycardia) and vasoconstriction, to raise blood pressure. Conversely, when blood pressure is elevated, the body responds by slowing the heart rate and dilating blood vessels.

Fluid shifting mechanism

Fluid can shift between compartments in the body, including the vascular, interstitial, and cellular spaces. In response to changes in blood volume, the body can shift fluid into or out of the bloodstream to help regulate pressure. For instance, in dehydration, interstitial fluid may enter the bloodstream to maintain blood volume.

Disorders of blood volume and blood pressure

Several medical conditions are associated with abnormalities in blood volume and blood pressure regulation:

Hypertension (high blood pressure)

Chronic hypertension is a condition in which the blood pressure is consistently elevated. This can be caused by various factors, including increased blood volume, excessive sodium intake, kidney dysfunction, or hormonal imbalances. Long-term hypertension can damage blood vessels and organs, increasing the risk of stroke, heart disease, and kidney failure.

Hypotension (low blood pressure)

Hypotension occurs when blood pressure falls too low to adequately perfuse organs and tissues. This can be caused by blood loss, dehydration, heart problems, or severe infection. Symptoms may include dizziness, fainting, and shock, and it requires prompt medical intervention to restore blood volume and pressure.

Hypervolemia and hypovolemia

Hypervolemia (excess blood volume) and hypovolemia (insufficient blood volume) are conditions that can lead to either elevated or low blood pressure, respectively. Both require careful management to avoid complications like organ damage, tissue hypoxia, and circulatory collapse.

Blood volume and blood pressure are vital physiological parameters that must be carefully regulated to ensure proper circulatory function. Blood volume directly influences cardiac output and blood pressure, while mechanisms like the renin-angiotensin-aldosterone system, ADH, and baroreceptor reflexes help maintain balance. Disruptions in blood volume or blood pressure can lead to serious health conditions such as hypertension, hypotension, and fluid imbalances. Understanding the interplay between blood volume and blood pressure is crucial for diagnosing, treating, and managing cardiovascular diseases and ensuring the body's overall health and homeostasis.

42

Heart Muscles: Hypertrophy, Atrophy and Pacemaking

The heart, composed primarily of specialized muscle tissue known as cardiac muscle, is responsible for maintaining the circulatory system's function by pumping blood throughout the body. The unique characteristics of heart muscle allow it to respond dynamically to various physiological demands. Among these adaptive responses, hypertrophy, atrophy, and the process of pacemaking are fundamental to heart function and health. Understanding these processes is crucial for comprehending how the heart responds to different stressors, such as physical activity, disease, or aging, and how its electrical and mechanical functions are integrated.

Nature of cardiac muscle

Cardiac muscle is a specialized type of muscle tissue found only in the heart. It is unique because it combines features of both skeletal and smooth muscle, providing both strength and rhythm in its contractions. Cardiac muscle fibres are striated and connected by intercalated discs, which allow rapid transmission of electrical impulses between cells, ensuring synchronized contraction. Unlike skeletal muscle, cardiac muscle has a relatively high density of mitochondria, providing the energy necessary for continuous activity.

a) Hypertrophy of heart muscle

Hypertrophy refers to the enlargement or thickening of the heart muscle cells (cardiomyocytes) in response to various physiological or pathological stimuli. Unlike hyperplasia (an increase in cell number), hypertrophy involves an increase in the size of existing cells.

There are two main types of cardiac hypertrophy

Physiological hypertrophy

This occurs as a normal response to increased workload, such as from regular exercise or pregnancy. It is usually characterized by an increase in both the size of the heart muscle fibres and their functional capacity. This type of hypertrophy is typically well-regulated and does not impair heart function.

Pathological hypertrophy

This occurs in response to chronic stress, such as high blood pressure (hypertension), valve disease, or heart failure. Pathological hypertrophy is often maladaptive and can lead to a reduction in the heart's efficiency and ultimately contribute to the development of heart failure.

Causes of hypertrophy

Increased workload

Conditions that increase the heart's workload, such as hypertension (increased after load) or valve disease (e.g. aortic stenosis), can trigger hypertrophy. The heart compensates for the increased resistance by enlarging the myocardium to generate greater force for blood ejection.

Hormonal influences

Various hormones and signalling molecules, including angiotensin-II, aldosterone, and catecholamines, contribute to the development of hypertrophy. These molecules increase the expression of genes involved in cell growth and protein synthesis in cardiomyocytes.

Exercise-induced hypertrophy

Endurance athletes, particularly those involved in activities like long-distance running or cycling may experience physiological hypertrophy. This adaptation is typically benign and supports the increased cardiovascular demand from sustained aerobic activity.

Effects and consequences

Functional adaptation

In cases of physiological hypertrophy, the heart muscle becomes more efficient, enabling better oxygen delivery and performance.

Maladaptive effects

In pathological hypertrophy, the heart may undergo fibrosis (scarring) and alterations in the heart's structure, leading to a decrease in the heart's ability to pump efficiently. This can eventually lead to diastolic dysfunction (impaired filling of the heart) and systolic dysfunction (impaired ejection of blood), both of which contribute to heart failure.

b) Atrophy of heart muscle

Atrophy refers to the reduction in the size or mass of the heart muscle, typically as a result of reduced workload or prolonged lack of use. Just as hypertrophy is a response to increased demand, atrophy is a response to decreased demand or insufficient blood supply.

Causes of atrophy

Disuse

Prolonged periods of inactivity or immobility, such as bed rest after surgery, can lead to a reduction in heart muscle mass. This is often observed in patients with severe chronic illness or those with extended periods of immobilization.

Nutritional deficiency

A lack of essential nutrients, particularly those involved in cellular energy production (like ATP), can lead to the weakening and shrinking of the heart muscle.

Aging

As individuals age, there is a gradual decrease in the size of heart muscle fibers. This process, combined with a decline in the regenerative capacity of the heart, can contribute to a weakened heart over time.

Ischemia

Reduced blood flow due to blockages in coronary arteries (coronary artery disease) can lead to myocardial atrophy. In severe cases, prolonged ischemia can result in necrosis (cell death) and irreversible loss of heart tissue.

Effects and consequences

Decreased cardiac output

Atrophy can impair the ability of the heart to generate sufficient force, leading to a reduction in cardiac output, which in turn can affect tissue perfusion and oxygen delivery.

Risk of heart failure

When atrophy is associated with coronary artery disease or a myocardial infarction (heart attack), it can result in the thinning of the myocardium and compromised function, potentially leading to heart failure.

Reduced heart size

In extreme cases, such as severe disuse or chronic malnutrition, the heart may shrink significantly, weakening its capacity to perform the vital function of pumping blood.

c) Pacemaking in the heart

Cardiac pacemaker mechanism

The heart possesses its own intrinsic pacemaker mechanism that controls the rhythm of the heartbeat. The pacemaker activity is generated by specialized cells in the sinoatrial (SA) node, located in the right atrium of the heart. These pacemaker cells have the unique ability to depolarize spontaneously, setting the rhythm for the heart's contractions. The action potential generated by the SA node spreads through the atria, causing them to contract, and then travels to the atrioventricular (AV) node. From the AV node, the signal passes through the bundle of His, dividing into the right and left bundle branches, and finally to the Purkinje fibres, stimulating the ventricles to contract.

Pacemaker Potential

Cardiac pacemaker cells exhibit a phenomenon known as pacemaker potential or prepotential. Unlike most cells, pacemaker cells do not have a stable resting membrane potential. Instead, their membrane potential gradually depolarizes, reaching a threshold that triggers an action potential. This gradual depolarization is due to the influx of sodium (Na□) and calcium (Ca^{2}□) ions through specialized ion channels, particularly the HCN (hyperpolarization-activated cyclic nucleotide-gated) channels.

Heart rate regulation

While the SA node sets the baseline heart rate (often referred to as the intrinsic rate), it is influenced by various external factors:

Autonomic nervous system

The sympathetic nervous system releases nor-epinephrine, increasing the heart rate by enhancing pacemaker activity. Conversely, the parasympathetic nervous system, primarily through the vagus nerve, releases acetylcholine, which slows the heart rate.

Hormonal influence

Hormones like adrenaline (epinephrine) and thyroid hormones can also increase heart rate by enhancing pacemaker activity and the conduction of electrical impulses through the heart.

Disorders of pacemaking

Disruptions in pacemaker function can result in various arrhythmias (irregular heartbeats), including:

Bradycardia

Abnormally slow heart rate, which can occur if the SA node fails to generate impulses at the appropriate frequency, or if the AV node takes over as the primary pacemaker.

Tachycardia

It is abnormally fast heart rate which can occur due to increased sympathetic activity, electrolyte imbalances or abnormal pacemaker activity elsewhere in the heart.

Atrial fibrillation

It is a condition where multiple ectopic pacemaker sites in the atria generate rapid, disorganized electrical impulses leading to ineffective atrial contraction and a risk of clot formation.

The heart muscle, through processes like hypertrophy, atrophy, and its intrinsic pacemaking ability, plays a central role in adapting to both internal and external changes. Hypertrophy allows the heart to meet increased demands, while atrophy may reflect pathological conditions or aging. The pacemaker activity of the heart ensures the rhythmic contraction necessary for effective blood circulation, and any disturbances in this process can lead to serious arrhythmias. A thorough understanding of these mechanisms is essential for diagnosing and managing heart diseases, as they directly affect the efficiency of the cardiovascular system and overall health.

43

Advanced and Sensor Based Technologies for Detection of Metabolic Disorders in Animals

Advances in technology are revolutionizing the detection of metabolic diseases in animals, making diagnostics more precise, rapid, and non-invasive. From wearable sensors and metabolomics to AI and nanotechnology, these tools are enhancing animal health and productivity. As these technologies become more accessible, they promise to transform veterinary medicine and animal husbandry, ensuring early detection, effective management, and improved outcomes for animals.

Sensor-based technologies have emerged as a game-changer in the detection and management of metabolic diseases in dairy animals. These tools provide continuous monitoring, early detection, and real-time insights, enabling better health management and increased productivity. This chapter explores the most promising sensor-based technologies used for diagnosing metabolic diseases in dairy animals, their applications, and the future of sensor-driven animal health management.

Metabolic diseases like ketosis, milk fever (hypocalcemia), acidosis, and displaced abomasum significantly impact dairy productivity and animal welfare. Early detection is critical for reducing treatment costs, improving recovery rates, and minimizing production losses. Traditional diagnostic methods are accurate and often invasive, time-consuming, and not feasible for large-scale monitoring. Therefore, sensor-based technologies offer a solution by enabling continuous and non-invasive monitoring of physiological and biochemical parameters.

Detection of metabolic disorders in animals is an area of growing importance in veterinary and animal sciences. Advances in technology have enabled the development of more accurate, rapid, and non-invasive diagnostic tools. This chapter explores the latest technologies particularly sensor based technologies for detecting metabolic disorders in animals, focusing on methods like wearable biosensors, metabolomics, imaging technologies, point-of-care testing, and artificial intelligence, etc.

I. Wearable biosensors and monitoring devices

1. Wearable biosensors

Wearable biosensors are small, portable devices attached to the animal's body to measure various physiological metrics. These sensors play a pivotal role in detecting metabolic diseases by providing real-time data.

Key features

Physiological monitoring: Sensors measure parameters like heart rate, body temperature, rumination, and activity levels.

Biochemical analysis: Advanced wearable biosensors can measure metabolites like β-hydroxybutyrate (BHB) and glucose levels.

Applications

a) ***Ketosis detection***: Ketosis, caused by a negative energy balance, leads to elevated BHB levels. Wearable sensors capable of detecting changes in BHB concentrations provide early warnings of subclinical ketosis.

b) ***Milk fever monitoring***: Wearable sensors that monitor body temperature and calcium levels help detect hypocalcemia, a common metabolic disorder during the peri-parturient period.

c) ***Ruminal acidosis***: Accelerometers measure feeding and chewing activity, while pH sensors implanted in the rumen detect changes in acidity, a key indicator of acidosis.

Examples

Rumen boluses: Devices like Smart bow's boluses are placed in the rumen to continuously measure pH, temperature, and motility.

Collar sensors: These track rumination, feeding, and activity patterns to detect anomalies linked to metabolic diseases.

Advantages

Non-invasive and stress-free monitoring,

Early detection and timely intervention,

Reduces dependence on manual testing,

2. Automated milk analyzers

Milk analysis provides valuable insights into the metabolic health of dairy animals. Automated milk sensors are integrated into milking systems to assess milk composition in real-time.

Key metrics monitored

Milk fat-to-protein ratio: A high fat-to-protein ratio indicates ketosis, while a low ratio suggests acidosis.

Electrical conductivity: Changes in conductivity can signal mastitis or metabolic imbalances.

Milk ketones: Sensors detect ketone bodies in milk as an early marker of ketosis.

Applications

a) ***Subclinical ketosis detection***: Automated milk testing systems identify changes in milk ketone levels before clinical signs appear.

b) ***Nutritional monitoring***: Milk composition analysis helps optimize feed rations to prevent metabolic disorders.

Examples

Lely astronaut: A robotic milking system with integrated sensors for milk composition and quality analysis.

DeLaval herd navigator: Provides real-time data on ketosis, urea levels, and overall metabolic health.

Advantages

Real-time feedback during milking

Non-invasive and suitable for herd-wide monitoring

Enhances precision feeding strategies

3. Ruminal pH sensors

Rumen health is critical to preventing metabolic diseases like acidosis and ketosis. Rumen pH sensors provide continuous monitoring of acidity levels, offering insights into rumen function.

Technology

Implantable sensors: Devices are inserted into the rumen and transmit pH data wirelessly.

Telemetry systems: Data is sent to cloud-based platforms for analysis and alerts.

Applications

a) ***Acidosis detection***: Persistent low pH levels indicate sub-acute or acute ruminal acidosis.

b) ***Diet management***: The pH sensors help monitor the impact of dietary changes on rumen health, preventing imbalances.

Examples

SmaXtec rumen bolus: Monitors rumen pH, temperature, and drinking behaviour to detect acidosis and other disorders.

Advantages

Continuous, real-time monitoring

Provides early warnings for dietary adjustments

Helps maintain optimal rumen function

4. Infrared and thermal sensors

Infrared (IR) and thermal imaging sensors are increasingly used to detect physiological changes associated with metabolic diseases.

Technology

Infrared thermography (IRT): Measures surface temperature variations to identify inflammation or metabolic stress.

Infrared spectroscopy: Analyzes blood, milk, or saliva for biochemical markers like glucose and BHB.

Applications

a) ***Fever monitoring***: Elevated body temperatures detected by thermal sensors can indicate infections or metabolic disorders.

b) ***Mastitis detection***: Thermal imaging identifies udder inflammation linked to metabolic stress.

c) ***Energy balance assessment***: IR spectroscopy detects changes in metabolites indicative of energy imbalances.

Advantages

Non-invasive and stress-free

Ideal for early disease detection

Useful for monitoring large herds

5. Motion and activity sensors

Changes in movement and activity patterns often precede clinical signs of metabolic diseases. Motion sensors, typically integrated into collars, pedometers, or leg tags, provide valuable behavioural data.

Key metrics monitored

Step counts: Reduced mobility may indicate lameness or metabolic stress.

Rumination time: Decreased chewing activity can signal digestive or metabolic issues.

Resting behaviour: Abnormal lying times are linked to discomfort caused by disorders like milk fever or ketosis.

Applications

a) ***Lameness detection***: Reduced activity levels combined with other parameters help diagnose lameness linked to metabolic conditions.

b) ***Heat stress monitoring***: Activity sensors detect heat stress, which exacerbates metabolic issues.

c) ***General health monitoring***: Early identification of stress-related behavioral changes.

Examples

Cow manager sensors: Track activity, rumination, and temperature to detect health anomalies.

Fitbit-like wearables: Devices adapted for livestock to monitor health and productivity.

Advantages

Continuous behavioral monitoring

Early detection of subclinical diseases

Easy integration into herd management systems

6. Real-time data integration and analytics

The value of sensor-based technologies is amplified when integrated with advanced data analytics systems. Real-time data collected by sensors is analyzed using algorithms to generate actionable insights.

Features

Cloud-based platforms: Data from sensors is transmitted to cloud systems for storage and analysis.

Artificial intelligence (AI): Machine learning algorithms predict the onset of metabolic diseases based on historical and real-time data.

Dashboards: User-friendly interfaces allow farmers and veterinarians to monitor herd health.

Examples

Precision livestock farming (PLF) systems: Combine data from multiple sensors for comprehensive herd monitoring.

AI-powered alerts: Provide early warnings and recommendations for disease management.

Advantages

Enhanced decision-making

Early intervention minimizes economic losses

Scalable for large herds

II. Metabolomics

Metabolomics is the comprehensive study of metabolites in biological samples. It provides a holistic view of metabolic processes and reveals biomarkers of disease.

Techniques such as nuclear magnetic resonance (NMR) spectroscopy, mass spectrometry (MS), and chromatography are used to identify and quantify metabolites in blood, urine, or tissue samples.

Applications

Metabolic profiling: Metabolomics helps detect subclinical conditions like ketosis and fatty liver in livestock.

Biomarker discovery: Identification of novel biomarkers for diseases like equine metabolic syndrome (EMS) and bovine ketosis.

Dietary impact assessment: Evaluates how nutrition influences metabolic health.

Advantages

Provides a detailed metabolic snapshot,

Aids in understanding disease mechanisms and pathways,

Facilitates precision medicine for individual animals,

III. Advanced imaging technologies

Imaging technologies provide non-invasive methods to visualize internal metabolic changes and organ function.

Magnetic resonance imaging (MRI)

Used to assess liver and pancreatic function,

Identifies changes in fat distribution related to metabolic disorders,

Positron emission tomography (PET)

Tracks metabolic activity in tissues,

Commonly used in research to study metabolic disorders in large animals,

Infrared thermography (IRT)

Detects temperature variations that may indicate metabolic stress or inflammation,

Useful in identifying early signs of mastitis or laminitis, conditions linked to metabolic imbalances.

Ultrasound imaging

Non-invasively detects fatty liver syndrome and other organ-related metabolic issues.

IV. Point-of-care testing (POCT) devices

POCT devices allow on-site, rapid, and accurate testing of biomarkers relevant to metabolic diseases. Portable diagnostic kits or handheld devices analyze blood, urine, or milk samples for critical parameters like glucose, ketones, and electrolytes.

Applications

Dairy industry: Milk ketone testers detect subclinical ketosis.

Equine practice: Handheld glucose and insulin meters for EMS detection.

Small animal clinics: Portable analyzers for quick diagnosis of diabetes or thyroid disorders.

Advantages

Reduces turnaround time compared to laboratory testing,

Easy to use in field conditions,

Cost-effective for routine screenings,

V. Artificial intelligence and machine learning

Artificial intelligence (AI) has become integral to improving the detection and management of metabolic diseases. AI uses algorithms to analyze large datasets from sensors, imaging, and biochemical tests to identify patterns associated with diseases.

Applications

Predictive analytics: Machine learning models predict the likelihood of metabolic disorders based on historical and real-time data.

Pattern recognition: AI helps identify subtle patterns in imaging or metabolomic data that might be missed by the human eye.

Precision livestock farming: AI-powered systems integrate data from wearables, environmental sensors, and feed analysis to optimize herd health.

Advantages

Improves diagnostic accuracy,

Enables early detection and personalized treatment,

Reduces reliance on invasive procedures,

VI. Genomics and transcriptomics

Genomic and transcriptomic analyses are transforming how metabolic diseases are diagnosed and understood at the molecular level. Genomics involves studying an animal's entire DNA to identify genetic predispositions to metabolic diseases. Transcriptomic sanalyzes RNA to understand gene expression changes during metabolic stress.

Applications

Identifying genetic markers for insulin resistance or obesity in horses and companion animals,

Screening livestock for traits associated with higher metabolic efficiency,

Studying the effects of stress on gene expression in dairy cattle,

Advantages

Enables selective breeding for disease resistance,

Provides insights into molecular mechanisms underlying diseases,

Facilitates targeted nutritional and pharmacological interventions,

VII. Non-invasive sampling techniques

Non-invasive methods minimize stress and discomfort for animals during the diagnostic process. Techniques include saliva, sweat, hair, or fecal sampling to measure metabolic markers.

Applications

Saliva cortisol tests for stress-related metabolic disorders,

Hair analysis for long-term assessment of mineral imbalances,

Sweat biomarkers in horses for dehydration and electrolyte monitoring.

Advantages

Stress-free and suitable for repeated testing,

Ideal for monitoring wildlife or zoo animals,

Reduces the risk of sample contamination,

VIII. Nanotechnology in diagnostics

Nanotechnology has opened up new possibilities for the early detection of metabolic diseases. Nanoparticles are used to enhance the sensitivity and specificity of diagnostic assays.

Applications

Development of nanosensors for glucose and lactate detection,

Nano-enabled assays for detecting early-stage metabolic acidosis,

Delivery of targeted therapies for metabolic conditions,

Advantages

High sensitivity and rapid response,

Potential for multiplexed testing,

Miniaturization enables portable devices.

Challenges and future directions

While these technologies offer significant advantages, challenges remain:

Cost and accessibility: Many advanced technologies are expensive and not widely available, especially in rural or resource-limited settings.

Training requirements: Adoption of sophisticated tools like AI and metabolomics requires specialized training for veterinarians and technicians.

Data management: Handling and interpreting large datasets generated by wearables, imaging, and genomics is complex.

Future research should focus on

Integration: Combining data from multiple sources (wearables, imaging, metabolomics) for comprehensive diagnostics.

Miniaturization: Developing more affordable and portable devices.

Standardization: Establishing global standards for diagnostic markers and protocols.

Advanced and sensor-based technologies have revolutionized the detection and management of metabolic disorders in dairy animals. From wearable biosensors and milk analyzers to ruminal pH sensors and AI-driven analytics, metabolomics, point of care testing devices, genomics and transcriptomics, non-invasive sampling techniques and nanotechnology in diagnostics, etc., these tools enable early detection, real-time monitoring, and improved animal welfare. As advancements continue, these technologies promise to transform dairy farming, making it more sustainable and productive.

44

Omega-3 Fatty Acids as Functional Feed Supplements

Omega-3 fatty acids, primarily eicosapentaenoic acid (EPA) and docosahexaenoic acid (DHA), are renowned for their anti-inflammatory properties. They modulate inflammation by competing with arachidonic acid, reducing pro-inflammatory eicosanoids, and enhancing anti-inflammatory resolvins and protectins. The optimal supplementation dose varies among animal species due to differences in metabolism, physiology, and dietary requirements.

Dogs

For dogs, omega-3 fatty acids are often used to manage inflammatory conditions like arthritis and dermatitis. Doses of 50-100 mg/kg body weight of combined EPA and DHA are commonly recommended, with variations based on the severity of inflammation and the dog's health status.

Cats

In cats, omega-3 supplementation is used for similar conditions as in dogs. Typical dose ranges from 30-50 mg/kg body weight of EPA and DHA, taking into account their strict carnivorous diet and specific fatty acid metabolism.

Horses

Horses benefit from omega-3 supplementation in conditions such as osteoarthritis and inflammatory airway disease. The recommended dose is about 4-6 g of combined EPA and DHA per day for an average-sized horse (~500 kg). Flaxseed oil, rich in alpha-linolenic acid (ALA), is also commonly used, although its conversion to EPA and DHA is limited.

Cattle and Sheep

In ruminants, omega-3 fatty acids are used to improve immune function and reduce inflammation, particularly during stress periods. Doses range from 0.5% to 1% of the total dry matter intake, typically through supplementation with fish oil or flaxseed.

Pigs

In pigs, omega-3 fatty acids improve immune response and reduce inflammatory diseases. A supplementation dose of 1-2% of the diet, providing 0.5-1 g EPA+DHA per kg of feed, is recommended.

The exact dose should be tailored to the animal's health, diet, and intended therapeutic goals, ensuring it does not disrupt the omega-6: omega-3 balance.

45

Important Vitamins and Minerals as Immune Modulating Supplements

Vitamins and minerals play crucial roles in supporting immune function across various animal species. Their supplementation can enhance immune responses, reduce disease susceptibility, and improve overall health. The appropriate doses vary depending on species, physiological state, and environmental conditions.

Cattle

1. ***Vitamins***

   Vitamin A: Enhances mucosal immunity; 30,000–50,000 IU/day.

   Vitamin D: Supports macrophage function; 10,000–15,000 IU/day.

   Vitamin E: Antioxidant, improves phagocyte function; 500–1,000 IU/day.

2. ***Minerals***

   Zinc: Promotes wound healing and immune cell activity; 40–60 mg/kg feed.

   Enhances neutrophil and antibody responses; 0.3 mg/kg feed.

Sheep and Goats

1. ***Vitamins***

   Vitamin A: 5,000–10,000 IU/day.

   Vitamin E: 200–400 IU/day to enhance antioxidant defenses.

   Vitamin *D*: 2,000–3,000 IU/day to maintain calcium balance and immunity.

2. ***Minerals***

   Zinc: 25–50 mg/kg feed to support skin and immune health.

   Selenium: 0.2–0.3 mg/kg feed for antioxidant protection.

Poultry

1. ***Vitamins***

   Vitamin A: 8,000–10,000 IU/kg feed for mucosal barrier support.

   Vitamin E: 150–300 IU/kg feed to enhance humoral immunity.

   Vitamin C: 200–500 mg/kg feed, especially during heat stress.

2. ***Minerals***

   Zinc: 40–80 mg/kg feed to support feather and skin health.

   Selenium: 0.3 mg/kg feed to reduce oxidative stress.

Swine

1. ***Vitamins***

   Vitamin A: 3,000–5,000 IU/kg feed for epithelial integrity.

   Vitamin E: 50–200 IU/kg feed for anti-inflammatory effects.

   Vitamin D: 1,000–2,000 IU/kg feed to modulate immune cells.

2. ***Minerals***

   Zinc: 100–250 mg/kg feed to enhance barrier immunity.

   Selenium: 0.2–0.3 mg/kg feed for antioxidant support.

Dogs and Cats

1. ***Vitamins***:

   Vitamin A: 100–200 IU/kg body weight/day.

   Vitamin E: 10–20 IU/kg body weight/day to combat oxidative stress.

   Vitamin C: 5–10 mg/kg body weight/day for enhanced immunity (dogs).

2. ***Minerals***:

   Zinc: 25–50 ppm in diet to support skin and immune health.

   Selenium: 0.2–0.5ppm diet to reduce inflammation.

Horses

1. ***Vitamins***

   Vitamin A: 15,000–30,000 IU/day for epithelial health.

   Vitamin E: 500–2,000 IU/day for immune modulation.

   Vitamin C: 1–3 g/day (especially for older horses or under stress).

2. ***Minerals***

   Zinc: 40–60 mg/kg feed to enhance wound healing.

   Selenium: 1–3 mg/day to support antioxidant activity.

Aquatic species

1. ***Vitamins***

   Vitamin C: 100–1,000 mg/kg feed to enhance disease resistance.

   Vitamin E: 50–200 mg/kg feed for antioxidant protection.

2. ***Minerals***

   Zinc: 20–50 mg/kg feed for growth and immunity.

   Selenium: 0.2–0.5 mg/kg feed for oxidative stress reduction.

Supplementation of vitamins and minerals is vital for supporting immune health in animals. However, doses should be tailored to species, physiological demands, and environmental stressors. Excessive supplementation can be harmful, emphasizing the importance of balanced dietary management.

46

Probiotics and Prebiotics as Feed Additives for Livestock and Poultry

Probiotics and prebiotics are widely used in animal nutrition to promote gut health, enhance immune function, and improve production performance. Probiotics are live microorganisms (e.g., *Lactobacillus, Bifidobacterium, Bacillus*, and *Saccharomyces*) that confer health benefits when administered in adequate amounts. Prebiotics, such as fructooligosaccharides (FOS) and mannanoligosaccharides (MOS), are non-digestible compounds that stimulate the growth of beneficial gut microbes. The optimal supplementation levels vary among species based on their physiology, diet, and production goals.

Cattle

Probiotics

Lactobacillus acidophilus, Bifidobacterium bifidum, Enterococcus faecium, and *Saccharomyces cerevisiae* are commonly used.

Dose

Lactobacillus spp.: 1×10^9– 1×10^{10} CFU/head/day.

Saccharomyces cerevisiae: 5–10 g/head/day or 10^8– 10^9 CFU/g feed.

Prebiotics

MOS and FOS are used to improve gut integrity and nutrient absorption.

Dose

MOS: 5–10 g/head/day.

FOS: 2–4 g/head/day.

Benefits

Enhanced milk production and quality in dairy cattle

Improved feed efficiency and weight gain in beef cattle

Sheep and Goats

Probiotics

Common strains include Lactobacillus plantarum and Saccharomyces cerevisiae.

Dose

Lactobacillus spp.: 1×10^8–1×10^9 CFU/head/day.

Saccharomyces cerevisiae: 2–4 g/head/day.

Prebiotics

MOS and inulin are widely used.

Dose

MOS: 2–5 g/head/day.

Inulin: 1–3 g/head/day.

Benefits

Enhanced fiber digestion and rumen fermentation,

Improved immunity and growth performance,

Poultry

Probiotics

Lactobacillus spp., *Bacillus subtilis*, *Bifidobacterium spp.*, and *Saccharomyces cerevisiae* are common.

Dose

Lactobacillus spp.: 1×10^6–1×10^8 CFU/g feed.

Bacillus subtilis: 1×10^6–1×10^7 CFU/g feed.

Prebiotics

MOS, FOS, and beta-glucans are used to enhance gut health and pathogen resistance.

Dose

MOS: 0.5–1 g/kg feed.

FOS: 0.5–2 g/kg feed.

Benefits

Improved growth rate and feed conversion ratio (FCR)

Reduced incidence of gut pathogens like *Salmonella* and *E. coli*

Swine

Probiotics

Lactobacillus rhamnosus, *Bacillus subtilis*, and *Saccharomyces boulardii* are widely used.

Dose

Lactobacillus spp.: 1×10^7– 1×10^8 CFU/g feed.

Bacillus subtilis: 1×10^7 CFU/g feed.

Prebiotics

MOS, FOS, and chicory-derived inulin are used for promoting gut health.

Dose

MOS: 1–2 g/kg feed.

Inulin: 2–4 g/kg feed.

Benefits

Enhanced weight gain and FCR

Reduced post-weaning diarrhea and improved immune response

Dogs and Cats

Probiotics

Lactobacillus acidophilus, *Bifidobacterium animalis* and *Enterococcus faecium* are common.

Dose

Lactobacillus spp.: 1×10^6– 1×10^8 CFU/g feed.

Prebiotics

FOS, inulin, and galactooligosaccharides (GOS) are popular.

Dose

FOS: 0.5–1 g/kg feed.

Inulin: 1–2 g/kg feed.

Benefits

Improved gut microbiota balance and digestion

Enhanced skin and coat health

Horses

Probiotics

Saccharomyces cerevisiae and *Lactobacillus spp.* are used to support hindgut fermentation.

Dose

Lactobacillus spp.: 1 × 10^8 CFU/day.

Saccharomyces cerevisiae: 10–20 g/day.

Prebiotics

MOS, FOS, and pectin are widely used.

Dose:

MOS: 5–10 g/day.

FOS: 2–5 g/day.

Benefits

Enhanced nutrient absorption and reduced risk of colic.

Improved performance in athletic horses.

Aquatic species

Probiotics

Bacillus subtilis, *Lactobacillus spp.*, and *Pediococcus spp.* are commonly used to improve water quality and gut health.

Dose

Bacillus subtilis: 1 × 10^6– 1 × 10^7 CFU/g feed.

Prebiotics

MOS, FOS, and beta-glucans are effective.

Dose

MOS: 0.5–1 g/kg feed.

Benefits

Improved growth performance, disease resistance, and water quality

Rabbits

Probiotics

Lactobacillus acidophilus, *Bacillus subtilis*, and *Enterococcus faecium* are common.

Dose

Lactobacillus spp.: 1 × 10^7 CFU/g feed.

Prebiotics

FOS and MOS are used to prevent digestive disorders.

Dose

FOS: 1–2 g/kg feed.

Benefits

Reduced incidence of diarrhoea and improved growth rates.

Camelids

Probiotics

Saccharomyces cerevisiae and *Lactobacillus spp.* are used to enhance fibre digestion.

Dose

Saccharomyces cerevisiae: 5–10 g/day.

Prebiotics

MOS and FOS are effective for improving gut health.

Dose

MOS: 3–5 g/day.

Benefits

Improved nutrient utilization and immune response

Wildlife and zoo animals

Probiotics

Species-specific strains like *Lactobacillus spp.* are used.

Dose

Adjusted based on the species and body weight (e.g., 1×10^6 CFU/g feed).

Prebiotics

MOS, FOS, and pectin are used.

Dose: 0.5–2 g/kg feed.

Benefits

Enhanced health and adaptation to captivity

Probiotics and prebiotics are valuable tools for improving gut health and production performance in animals. The choice of strain, type, and dose depends on species-specific requirements and health goals. Balancing these supplements in diets is critical to achieving optimal benefits while avoiding over-supplementation.

47

Important Fluids and Electrolytes Therapy for Animals

Fluids and electrolytes are critical in managing dehydration, electrolyte imbalances, and shock in animals. These life-saving interventions restore homeostasis in various disease conditions. Fluid therapy is tailored based on species, body weight, disease condition, and hydration status, ensuring proper restoration of blood volume, electrolyte balance, and acid-base homeostasis.

Principles of fluid therapy

1. **Types of fluids**

   Crystalloids: Normal saline, Ringer's solution, and lactated Ringer's solution (LRS).

   Colloids: Plasma, hydroxyethyl starches, and dextrans.

   Dextrose solutions: For energy supply and hypoglycemia.

2. **Routes of administration**:

   Intravenous (IV): For severe dehydration and shock.

   Subcutaneous (SC): For mild dehydration.

   Oral (PO): When the gastrointestinal tract is functional.

3. **Calculation of fluid needs**:

   Maintenance rate: 50–60 ml/kg/day.

   Deficit Replacement: % Dehydration × Body weight (kg) × 10 = Fluid deficit (ml).

   Ongoing losses: Added based on clinical estimates.

Cattle

1. **Dehydration and diarrhoea**: Use isotonic saline or LRS to correct dehydration and electrolyte imbalances.

   Dose:

   Initial bolus: 40–80 ml/kg IV over 1–2 hours for severe dehydration.

   Maintenance: 5% dextrose in saline at 50 mL/kg/day.

2. **Acidosis (e.g., ruminal acidosis)**: Administer bicarbonate-rich fluids.

   Dose: Sodium bicarbonate (1.3% solution) at 5–10 mL/kg/hour IV.

3. **Shock**: Hypertonic saline (7.2%) is used for rapid volume expansion.

   Dose: 4–5 ml/kg IV over 15 minutes, followed by isotonic fluids.

Sheep and Goats

1. **Pregnancy toxemia**: Treat with 5% dextrose or LRS.

   Dose: 100–200 ml of 5% dextrose IV or SC twice daily.

2. **Dehydration**

   Oral rehydration solutions (ORS) with electrolytes for mild cases

   Severe dehydration requires IV fluids.

   Dose: LRS or isotonic saline at 50–60 ml/kg/day.

3. **Hypocalcemia (Milk fever):** Administer calcium borogluconate.

   Dose: 50–100 ml IV slowly over 5–10 minutes.

Horses

1. **Colic and dehydration**: Use isotonic crystalloids or LRS.

   Dose

   Initial bolus: 20 ml/kg IV over 30 minutes.

   Maintenance: 50–60 ml/kg/day IV.

2. **Electrolyte imbalance**: Administer electrolyte solutions with sodium, potassium, and chloride.

   Dose: Potassium chloride at 20–40 mEq/L in fluids IV slowly.

3. **Endotoxemia and shock**: Colloids like plasma or hetastarch are used.

   Dose: Hetastarch at 10 ml/kg IV bolus.

Dogs

1. **Acute vomiting and diarrhoea**: Balanced crystalloids like LRS or Normosol-R.

   Dose

   Initial bolus: 20–30 mL/kg IV over 15–30 minutes.

   Maintenance: 2–4 mL/kg/hour.

2. **Hypoglycemia**: Administer dextrose solutions.

   Dose: 1–5 mL/kg of 5% dextrose IV bolus, followed by a continuous infusion.

3. **Shock (Hypovolemic)**: Administer colloids or hypertonic saline.

   Dose

   Hypertonic saline: 4–6 ml/kg IV over 15 minutes.

   Colloids: 10–20 ml/kg/day IV.

Cats

1. **Chronic Kidney Disease (CKD)**: Subcutaneous administration of isotonic fluids,

   Dose: 75–100 mL/day SC (divided into 2–3 doses).

2. **Dehydration**: Use LRS or Normosol-R.

   Dose: 20–30 mL/kg IV over 1–2 hours, followed by maintenance at 2–4 mL/kg/hour.

3. **Electrolyte imbalances (e.g., hypokalemia)**: Administer potassium chloride.

   Dose: 0.1–0.5 mEq/kg/hour IV, not exceeding 0.5 mEq /kg /hour

Swine

1. **Dehydration and scours**: ORS or IV fluids with electrolytes

   Dose:

   ORS: 50–100 ml/kg/day PO.

   IV fluids: 20–30 ml/kg over 1–2 hours.

2. **Shock**: Use hypertonic saline or colloids.

   Dose: 4–5 ml/kg IV hypertonic saline over 10–15 minutes.

3. **Hypoglycemia (Piglets)**: Administer dextrose.

   Dose: 1–2 mL of 5% dextrose SC or PO every 2–4 hours.

Poultry

1. **Dehydration**: Provide ORS through drinking water.

   Dose: 1 g of ORS per liter of water.

2. **Electrolyte Imbalances (Heat stress)**: Sodium bicarbonate or potassium chloride in drinking water,

   Dose: 1–2 g/L of water.

3. **Shock**: Administer glucose-electrolyte solutions.

   Dose: 5% glucose with electrolytes at 1–2 g/L water.

Aquatic Species (Fish)

1. **Osmotic imbalance**: Use salt baths.

   Dose: 1–3% saline solution (10–30 g/L water) for 10–15 minutes.

2. **Dehydration and stress**: Administer ORS in water.

   Dose: 1–2 g/L of water.

Rabbits

1. **Dehydration**: Subcutaneous or oral administration of fluids,

   Dose: 50–60 mL/kg/day SC in divided doses,

2. **Electrolyte imbalances**: Provide LRS or saline with dextrose.

   Dose: 10–20 mL/kg IV over 2–4 hours.

Wildlife and exotic animals

1. **Dehydration**: SC or IV administration of isotonic fluids,

   Dose: 50–100 mL/kg/day SC or IV.

2. **Electrolyte imbalances**: ORS or IV fluids based on species-specific needs.

   Dose: Adjust to weight and clinical status.

Fluids and electrolytes are essential for treating life-threatening conditions across animal species. Accurate assessment of hydration, electrolyte balance, and disease conditions is crucial for determining the type, dose, and route of fluid administration. Close monitoring and tailored therapy maximize the effectiveness of these interventions.

48

Whole Blood, Plasma/Serum and Blood Components Transfusion in Animals

Whole blood, plasma, serum, and blood transfusions are critical life-saving therapies in veterinary medicine. They are used to restore blood volume, correct anemia, provide clotting factors, and address protein deficiencies in various disease conditions. This chapter discusses their administration doses, species-specific considerations, and disease conditions where these fluids are essential.

Key components of blood therapy

1. ***Whole blood***: Contains red blood cells (RBCs), plasma, clotting factors, and platelets.
2. ***Plasma***: The liquid portion of blood with clotting factors, albumin, and immunoglobulins but without RBCs and platelets.
3. ***Serum***: Plasma without clotting factors.
4. ***Packed red blood cells (pRBCs)***: Concentrated RBCs used to treat anemia without volume overload.

General principles of blood transfusion

1. ***Indications***: Severe anemia (PCV <15–20%), hypoproteinemia, coagulopathies, or shock.
2. ***Compatibility testing***: Blood typing and crossmatching are essential in species prone to transfusion reactions.
3. ***Volume calculation***:

   Formula for volume required:

   Blood volume (L) = Body weight (kg) x Blood volume factor (ml/ kg)

Approximate blood volume factors by species

Human: 70-80 ml/ kg

Dogs: 80–90 ml/kg.

Cats: 60–70 ml/kg.

Horses: 70–100 ml/kg.

Cattle: 60–70 ml/kg.

Sheep and Goats: 60-70 ml/kg.

Pigs: 60-80 ml/kg.

Birds: 60-120 ml/kg.

Reptiles: 50-70 ml/kg.

Mice/Rats: 50-80 ml/kg.

Example calculations

1. For a 10 kg dog: Blood volume = 10 kg x 90 ml/kg = 900 ml/kg (or 0.9 L)
2. For a 5 kg cat: Blood volume = 5 kg x 70 ml/kg = 350 ml/kg (or 0.35 L)

NB

a) Above factors are estimates and can vary based on the animal's age, health and hydration status.

b) Blood volume is typically expressed as a percentage of body weight: 6-10 % for most of mammals, higher for birds and some other species.

4. ***Rate of administration***

   Initial: 0.5–1 mL/kg/hour for the first 15–30 minutes.

   Increase to 5–10 mL/kg/hour if no adverse reactions occur.

Cattle

1. **Hemorrhagic Diseases (e.g., trauma, surgery)**

   Whole blood

   Dose: 10–20 mL/kg IV. Use when PCV <15% or acute hemorrhage occurs.

   Plasma

   Dose: 10–15 mL/kg IV for clotting disorders or hypoproteinemia.

2. **Hypoproteinemia**

   Plasma or serum

   Dose: 10–15 mL/kg IV or SC for chronic protein loss.

3. **Clotting Disorders (e.g., DIC)**:

   Fresh frozen plasma is preferred.

   Dose: 10–20 mL/kg IV over 4 hours.

Horses

1. **Severe anaemia**

   Whole Blood

   Dose: 20–30 mL/kg IV to raise PCV by 5–10%.

   Example: A 500-kg horse requires ~5–10 liters for severe anemia.

2. **Clotting disorders (e.g., rodenticide toxicity):**

   Plasma:

   Dose: 10–20 mL/kg IV.

3. **Failure of Passive Transfer (Foals)**: Administer plasma rich in immunoglobulins.

   Dose: 1–2 liters IV (10–20 mL/kg).

Dogs

1. **Anemia** (e.g., IMHA, hemorrhage):

   Whole blood

   Dose: 10–20 mL/kg IV. Aim to raise PCV by ~10%.

   pRBCs

   Dose: 6–10 mL/kg IV for anemia without volume overload.

2. **Coagulopathies (e.g., hemophilia)**: Fresh frozen plasma:

   Dose: 10–20 mL/kg IV for clotting factor replacement.

3. **Hypoalbuminemia**

   Plasma

   Dose: 20–30 mL/kg IV.

4. **Shock**

   Whole blood

   Dose: 20–30 mL/kg IV rapidly to restore blood volume.

Cats

1. **Severe anaemia**

   Whole blood

   Dose: 10–15 mL/kg IV.

   Example: A 5-kg cat requires ~50–75 mL to raise PCV significantly.

   pRBCs

   Dose: 6–8 mL/kg IV.

2. **Coagulopathies (e.g., hepatic failure)**
   Plasma
   Dose: 10–15 mL/kg IV.
3. **Hypoproteinemia**
   Plasma or serum
   Dose: 10–20 mL/kg IV.

Sheep and goats

1. **Anaemia** (e.g., parasitic infestations)
   Whole blood
   Dose: 15–20 mL/kg IV.
2. **Hypoproteinemia**
   Plasma
   Dose: 10–15 mL/kg IV.
3. **Clotting disorders**
   Plasma
   Dose: 10–20 mL/kg IV.

Swine

1. **Neonatal anaemia (iron deficiency)**
   Whole blood
   Dose: 10–20 mL/kg IV in severe cases.
2. **Shock**
   Whole blood or plasma
   Dose: 15–20 mL/kg IV.
3. **Hypoproteinemia**
   Plasma
   Dose: 10–15 mL/kg IV.

Poultry

1. **Severe blood loss**
   Whole blood
   Dose: 10–15 mL/kg IV or SC in small birds.
2. **Coagulopathies**
   Plasma
   Dose: 5–10 mL/kg IV.

Wildlife and exotic animals

1. **Severe anaemia or Bblood loss**

   Whole blood

   Dose: 10–20 mL/kg IV based on species and size.

2. **Hypoproteinemia**

   Plasma or serum

   Dose: 10–15 mL/kg IV.

3. **Immunodeficiencies**

   Plasma

   Dose: 10–20 mL/kg IV.

Monitoring during blood transfusion

1. ***Adverse reactions***

   Signs: Tachycardia, fever, vomiting, restlessness, or urticaria.

   Management: Stop transfusion and administer antihistamines or corticosteroids.

2. ***Rate adjustments:***

   Slow the rate during the first 15–30 minutes.

   Increase cautiously based on clinical response.

3. ***Blood parameters***: Monitor PCV, total protein, and coagulation profiles.

Blood products such as whole blood, plasma, and serum are invaluable in managing life-threatening conditions in animals. Tailored administration based on species, disease conditions, and individual requirements ensures optimal therapeutic outcomes. Proper monitoring and adherence to dosage guidelines are essential to minimize risks and maximize the benefits of blood transfusions.

49

Special Antidotes to Counteract on Adverse Effects of Blood Transfusion in Animals

Blood transfusions, while life-saving, can sometimes result in adverse reactions. These reactions range from mild allergic responses to severe hemolysis or anaphylaxis. Prompt recognition and administration of appropriate antidotes are crucial to mitigate these effects and ensure patient safety. This chapter discusses the antidotes used in different species of animals for blood transfusion reactions, their doses, and species-specific considerations.

Types of transfusion reactions

I. ***Febrile non-hemolytic reaction***: Caused by leukocyte antigens or cytokines in the donor blood.

II. ***Allergic reaction***: Due to hypersensitivity to plasma proteins.

III. ***Hemolytic reaction***: Immune-mediated destruction of transfused red blood cells (RBCs).

IV. ***Volume overload***: Excessive fluid administration leading to pulmonary edema.

V. ***Anaphylaxis***: Severe, life-threatening allergic response.

Antidotes and management of blood transfusion reactions

I. Febrile non-hemolytic reaction

Signs: Fever, tachycardia, shivering, and mild discomfort.

Antidote

1. **Paracetamol (Acetaminophen)**: Reduces fever.

   Dose

   Dogs and cats: 10–15 mg/kg orally or per rectally every 8–12 hours.

   Horses: 10–15 mg/kg orally every 12 hours.

   Cattle: 10 mg/kg orally every 12 hours.

2. **Diphenhydramine (if mild allergic reaction is suspected)**:

   Dose: 1–2 mg/kg IM or IV in dogs and cats.

II. Allergic reaction

Signs: Urticaria, pruritus, swelling, and vomiting.

Antidote

1. **Antihistamines (e.g., diphenhydramine)**

   Dogs and Cats: 1–2 mg/kg IM or IV every 8 hours.

   Horses: 0.5–1 mg/kg IV.

   Cattle: 0.5 mg/kg IM or IV.

2. **Hydrocortisone: Used in severe cases to suppress immune response**

   Dose

   Dogs and Cats: 2–4 mg/kg IV.

   Horses and Cattle: 0.5–1 mg/kg IV.

III. Hemolytic Reaction

Signs: Hemoglobinuria, jaundice, tachycardia, fever, and hypotension.

Antidote

Stop the transfusion immediately.

1. **Intravenous fluids (IV fluids)**: To maintain blood pressure and prevent kidney damage.

   Dose

   i. ***Isotonic crystalloids*** (e.g., **Lactated Ringer's solution** or **Normal saline**): 10–20 mL/kg/hour IV in all species.

   ii. **Corticosteroids (e.g., dexamethasone)**: Suppress immune-mediated hemolysis.

   Dose

   Dogs and Cats: 0.1–0.2 mg/kg IV or IM.

   Horses and Cattle: 0.05–0.1 mg/kg IV or IM.

   iii. **Furosemide**: To promote diuresis and protect kidneys.

   Dose

   Dogs and Cats: 2–4 mg/kg IV every 8 hours.

   Horses and Cattle: 0.5–1 mg/kg IV.

IV. Volume overload

Signs: Dyspnea, coughing, pulmonary edema, and jugular vein distension.

Antidote

1. **Furosemide**: A loop diuretic to reduce fluid overload.

   Dose:

   Dogs and Cats: 2–4 mg/kg IV or IM every 8 hours.

   Horses and Cattle: 0.5–1 mg/kg IV.

2. **Oxygen therapy**: Supplemental oxygen at 2–5 L/min via mask or nasal cannula.

V. Anaphylaxis

Signs: Hypotension, respiratory distress, cyanosis, and collapse.

Antidote

1. **Epinephrine**: The primary treatment for anaphylaxis.

   Dose

   Dogs and Cats: 0.01 mg/kg IM or IV every 10–15 minutes as needed.

   Horses and Cattle: 0.005–0.01 mg/kg IV or IM.

2. **Corticosteroids**: To reduce inflammation.
3. **Dexamethasone**: 0.1–0.2 mg/kg IV in all species.
4. **Antihistamines**: For ongoing allergic response.
5. **Diphenhydramine**: As above.
6. **Intravenous fluid (IV fluids)**: Rapid administration of isotonic crystalloids to counteract shock.

   Dose: 20–40 mL/kg IV bolus, repeated as needed.

Species-specific considerations

Cattle

Common reactions: Volume overload and febrile reactions.

Preferred antidotes: Furosemide for overload and paracetamol for fever.

Horses

Common reactions: Anaphylaxis and allergic reactions.

Preferred antidotes: Epinephrine for anaphylaxis and diphenhydramine for mild reactions.

Dogs

Common reactions: Hemolytic and allergic reactions.

Preferred Antidotes: Antihistamines, corticosteroids, and IV fluids for hemolysis,

Cats

Common reactions: Anaphylaxis and volume overload.

Preferred Antidotes: Epinephrine and furosemide

Sheep and Goats

Common reactions: Allergic reactions.

Preferred antidotes: Diphenhydramine and hydrocortisone.

Poultry

Common reactions: Rare but may experience anaphylaxis.

Preferred antidotes: Epinephrine and oxygen therapy.

Prevention of transfusion reactions

1. ***Compatibility testing***: Perform blood typing and crossmatching, especially in species prone to hemolytic reactions (e.g., cats and horses).
2. ***Slow initial infusion***: Start transfusions at 0.5–1 mL/kg/hour for the first 15–30 minutes, monitoring for reactions.
3. ***Filter usage***: Use blood filters to remove clots or debris.
4. ***Pre-medication***: Administer antihistamines (e.g., diphenhydramine) or corticosteroids prophylactically in at-risk patients.

Monitoring and follow-up

1. ***Vital Signs***: Monitor temperature, heart rate, respiratory rate, and blood pressure every 15 minutes during transfusion.
2. ***Clinical signs***: Watch for vomiting, urticaria, dyspnea, or hemoglobinuria.
3. ***Laboratory tests***: Post-transfusion PCV, total protein, and renal function tests.

Adverse reactions to blood transfusions can be life-threatening if not promptly managed. Understanding the appropriate antidotes, their doses, and species-specific protocols is critical to ensure effective treatment. Prevention through proper pre-transfusion testing and close monitoring minimizes the risk of complications and improves outcomes.

50

Prokinetic Drugs for Stimulating Gastrointestinal Motility

Prokinetic drugs are used to stimulate gastrointestinal motility, including the abomasum, in animals like cattle, especially in cases of abomasal disorders such as displaced abomasum or atony. Below are commonly used prokinetic drugs and their doses:

1. **Metoclopramide**

   Mechanism: Dopamine antagonist; enhances gastric emptying and increases motility.

   Dose

   Cattle: 0.1–0.3 mg/kg subcutaneously or intravenously every 8–12 hours.

   Horses: 0.04–0.08 mg/kg intravenously as a constant rate infusion (CRI).

2. **Erythromycin** (Macrolide antibiotic with prokinetic activity)

   Mechanism: Mimics motilin, stimulating smooth muscle contractions.

   Dose

   Cattle: 8.6–10 mg/kg orally every 8–12 hours.

   Horses: 0.5–1 mg/kg intravenously every 6–8 hours.

3. **Neostigmine**

   Mechanism: Acetylcholinesterase inhibitor; increases acetylcholine at neuromuscular junctions.

   Dose

   Cattle: 0.02–0.04 mg/kg subcutaneously or intramuscularly.

   Horses: 0.01–0.02 mg/kg intravenously or intramuscularly.

4. **Domperidone**

   Mechanism: Dopamine antagonist; enhances motility by increasing acetylcholine release.

Dose

Cattle and Horses: 0.1–0.2 mg/kg orally every 12–24 hours.

5. **Bethanechol**

   Mechanism: Cholinergic agonist; directly stimulates muscarinic receptors to enhance motility.

   Dose

   Cattle: 0.02–0.03 mg/kg subcutaneously every 4–6 hours.

   Horses: 0.025–0.05 mg/kg orally or subcutaneously.

6. **Lidocaine**

   Mechanism: Reduces inhibitory sympathetic input and promotes motility by its anti-inflammatory effects on the gut.

   Dose

   Cattle: 1.3 mg/kg as a continuous intravenous infusion over 24 hours.

   Horses: 1.3–1.5 mg/kg loading dose intravenously, followed by 0.05 mg/kg/min as a CRI

Considerations

Species-specific responses: Always monitor for side effects, as not all species tolerate these drugs similarly.

Underlying cause: Address primary conditions such as electrolyte imbalances, hypocalcemia, or inflammation, as these can exacerbate motility issues.

Veterinary supervision: Prokinetic drug use should always be under the guidance of a veterinarian, especially for off-label uses.

References

Aetz, A.L. and Pearson, J. E. 1972. Amer. J. Vet. Res., 33: 1941.

Afsah Hejri, S.J., Badiei, K., Pourjafar, M., Oryan, A., Keshavarzi, B., Chalmeh, A. and Mesbah, G. 2019. Iran. J. Vet. Med., 13: 163–173.

Aghion, J.E. 1953. Can. J. comp. Med., 15: 10.

Ahn, S., Shin, W., Han, Y, Bae S., Cho, C., Choi S., Oh Y and Kim, J. 2023. Korean J. Vet. Serv., 46: 161-165.

Anstee, D.J. 2010. Blood, 115: 4635–43.

Aravindh, S. and Jacob, N. 2021. J. Entomol. Zool. Studies, 9 (5): 357-361.

Arjona, A. A., D. M. Denbow, and W. D. Weaver, Jr. 1990. Comp. Bioch. Physiol. 95 A: 393–399.

Arora, S. P., Harfield, E. D., Hinds, F. C. and Garrigus, U. S. 1968. J. Anim. Sci., 27: 1765.

Ashby, S. L. and King, L.L. and Parkey, D. V. 1981. Environ. Res. 26: 95-104.

Bacic G, Karadjole T, Macesic N and Karadjole M, 2007. Vet Arhiv, 77(6): 567-577.

Bains, B. S. 1996. World Poultry-Misset 12(4): 38–41.

Banga, CB; Neser, FWC and Garrick, DJ. 2014. South Afric. J. Anim. Sci., 44: 173-177.

Berenjian, A., Sharifi, S.D., Mohammadi-Sangcheshmeh, A., Bakhtiarizadeh, M.B., 2021. Poult. Sci., 100: 100889.

Berndt, J., Kovacs, P., Ruschke, K, Kloting, N., Fasshauer, M., Schon, M. R., Korner, A., Stumvoll, M. and Bluher, M. 2007. Diabetologia., 50: 1472-1480

Black H. 1988. N Z Vet. J., 36:15-17.

Blanco-Penedo, I., Cruz, J.M., López-Alonso, M., Miranda, M., Castillo, C., Hernández, J. and Benedito, J.L. 2006. Env. Int., 32: 901–906.

Blumberg, J. 1994. In: Diet, Nutrition, and Immunity. R. A. Forse, ed. Baca Raton, FL: CRC Press., pp. 234-247.

Bortolotti, M., Mari, C., Brunelli, F., Sarti, P., and Miglioli, M. 1999. Dig.Dis. Sci. 44 (12), 2439–2442.

Bradley, A.S., Leavitt, W.D. and Johnston, D.T. 2011. Geobiol., 9: 446–457.

Braun, U., Fluckiger M and Nageli F 1993. Vet. Rec., 132(5): 103-109.

Braun U, Gansohr B and Haessiget M, 2002. J. Vet. Med. Physiol, Pathol. and Clin. Med., 49: 299-302.

Bremner, I. 1970. British J. Nutr. 24: 769–783.

Brent, B E. 1976. J Anim Sci., 43: 930-935.

Brodeur, A., Wright, A. and Cortes, Y. 2017. J. Vet. Eme. Critic Care, 27 (2): 151-163.

Burke, J. M., Whitley, N. C., Pollard, D. A., Miller, J. E., Terrill, T. H., Moulton, K. E., and Mosjidis, J. A. 2011. Vet. Parasitol., 27 (2-4): 345-9.

Busato A, Faissle D, Küpfer U. and Blum JW. 2002. J. Vet Med A, Physiol Pathol Clin Med., 9:455–60.

Camacho, P. Fan, H., Liu, Z. and He, Jia-Qiang 2016. Am. J. Cardiovasc. Dis. 6 (3): 70-80.

Camilleri, M. and Atieh, J. 2021. Front. Pharmacol., 12: 711500.

Cardoso FC, Kalscheur KF and Drackley JK, 2020. J Dairy Sci., 103(6): 5684-5693.

Carruthers, V.R., Phipps, D.E. and Bakker, R.J. 1994. Proc. N Z Soc. Anim. Prod., 54: 23-25.

Caspary, W.F., 1992. Physiology and Pathophysiology of Intestinal Absorption. Oxford University Press.

Castellini, C., Mugnai, C., Dal Bosco, A. 2002. Meat Sci., 60: 219–225.

Cheeke, P.R. and Dierenfeld E. 2010. Comparative Animal Nutrition and Metabolism. CAB International.

Chegini, A; Ghavi Hossein-Zadeh, N; Hosseini-Moghadam, H. and Shadparvar, AA. 2016. Rev. Méd. Vét., 167: 145-153.

Chen, L.E., Wu, F., Xin, Y., Zhao, A., Sun, X., 2013. J. Physiol. Sci., 63: 457–464.

Chhabra, A. and Arora, S. P. 1987. Indian J. Dairy Sci., 40: 322.

Chmelíková, E., Bolechová, P., Chaloupková, H., Svobodová, I., Jovičić, M. and Sedmíková, M. 2020. Dom. Anim. Endocrinol., 72:106428.

Christensen, J.O., Grummer, R.R., Rasmussen, F.E., Bertics, S.J., 1997. J. Dairy Sci., 80, 563-568.

Christian, Parul and West, Jr K. P. 1998. Am. J. Clin. Nutr., 68 (suppl): 435S-441S.

Chytil, R., Omori, M., Liau, G. and Ong, D. E. 1983. Fed. Proc. 42:2744–2753.

Cooke, R.R., Silva Del Rio, N., Caraviello, D.Z., Bertics, S.J., Ramos, M.H., Grummer, R.R., 2007. J. Dairy Sci., 90, 2413–2418.

Cooling, L. 2015. Clin Microbiol Infect., 28: 801–70.

Corsetti, S., Ferrara, M. and Natoli, E. 2019. Animals, 9(10): 833.

Costable, P. 2003. Vet. Clin. North Am. Food Anim Pract., 19 (3): 557-597.

Crane, S. W. 1991. J. Samll Anim. Prac., 32: 275-282.

Crane, S. W. 1991. J. Samll Anim. Prac., 32: 275-282.

Darwin and Blignaut, D. 2019. J. Adv. Vet. Anim. Res., 6(1):100–107.

Darya, G., Mohammed, H., Dehghan, Z. Nakhaei, A. and Derakhshanfar, A. 2005. Lab. Anim. Res., 41 (5): 1-14.

Davidow, E. B., Spada, E. and Proverbio, D. 2021. Front. Vet. Sci., 81792720.

Davidow, E.B., Blois, S.L., Goy-Thollot, I., Harris, L., Humm, K., and Musulin, S. 2021. J Vet Emerg Crit Care, 31:167–88.

De Schepper, S., Verheijden, S., Aguilera-Lizarraga, J., Viola, M. F., Boesmans,W., Stakenborg, N. 2018. Cell, 175 (2): 400–e13.

DeGaris PJ and Lean IJ, 2008. Vet J, 176(1): 58-69.

DeSilvestro, R. A. and Marten, J. T. 1990. J. Nutr. 120:1223–1227.

Devendra C. 1985. Non-conventional Feed Resources in Asia and the Pacific, 2nd ed. APHCA.

Dirksen, G., Liebich, H.G., Mayer, E., 1985. Bov Pract., 20, 116–120.

Djokovic R., Jozek J., Cincovic M., Nemec M., Belic B., Klinkom M. and Staric J. 2018. The 30th World Buiatrics Congress, August 28 to September 1, Sapporo, Japan. p. 127.

Doepel, L., Lapierre, H., Kennelly, J.J., 2002. J. Dairy Sci., 85: 2315–2334.

Donker, R. A., M. G. B. Nieuwland, and A. J. Van Der Zijpp. 1990. Poult. Sci. 69:599–607.

Douglas, G.N., Overton, T.R., Bateman II, H.G., Drackley, J.K., 2004. J. Dairy Sci., 87: 4210–4220.

Duffield, T., 2000. Vet Clin North Am Food Pract., 16 (2): 231-253.

Duffield, T., Kelton D.F., Lleslie K.E., Lissemore K.D. and Lumsden J.H. 1997. Can. Vet. J., 38, 713-718.

Duffield, T., Leblanc S., R. Bagg R., Leslie K., TenHag J. and Dick P. 2003: J. Dairy Sci., 86: 1171-1176.

Duffield, T., Sandals, D., Leslie, K.E., Lissemore, K., McBride, B.W., Lumsden, J.H., Dick, P. and Bagg, R., 1998. J. Dairy Sci., 81, 2354–2361.

Efron, D. T. and A. Barbul. 1998. Curr. Opin. Clin. Nutr. Metab. Care 1:531–538.

Elangovan, A. V., Sahoo, A., Agarwal, N. and Pathak, N. N. 1998. Indian, J. Anim. Sci., 68: 1093.

Elek, P., Husveth, F., Gaal, T., Newbold, J.R., 2004. J. Dairy Sci., 87 (Suppl. 1): 345.

Elitok, B., Solak, M., Kabu, M., Elitok, O.M., Soylemez Z. and Fistik T. 2010. Pak. Vet. J., 30: 150-154.

Enderli, T. A., Burtch, Stephanie, R., Templet, J. N. and Carriero, A. 2016. Orthopedic Res. Rev., 8: 41-55.

Enjalbert, F., Nicot, M.C., Bayourthe, C. and Monculon R. 2001. J. Dairy Sci., 84: 583-589.

Erdman R.A., Botts R.L., Hemken R.W. and Bull L.S. 1980. J Dairy Sci., 1980; 63:923–30.

Evoy, D., M. D. Lieberman, T. J. Fahey, III, and J. M Daly. 1998. Nutr., 14:611–617.

Fagnani, R., Belote, V. and Battaglini, A. P.P. 2014. Pesq. Bras., 34 (5): 398-402.

FAO. 2004. Food and Agriculture Organization of the United Nations, Rome.

Fettman, M. J. 2007. In: Thrall MA (edn.), Hematology and Veterinary Clinical Biochemistry, Roca Sao Paulo. pp. 311-334.

Finch, V. A. 1986. J. Anim. Sci., 62: 531.

Fratzl, P., Paris, O., Klaushofer, K. and Landis W. 1996. J Clin Invest., 97(2): 396–402.

Freeman, B. A. and J. D. Crapo. 1982. Lab. Invest. 47:412–426.

Friend, T. H. and G. R. Dellmeier. 1986. Prof. Anim. Sci., pp. 1–6.

Gantnerv, Potocnik, K. and Jovanovic S. 2009. Acta Vet. Beograd., 59: 185-191.

Garbiec, A., Karpiński, M., Wojtaś, J. and Czyżowski, P. 2024. J. Appl. Anim Welfare Sci., 27(2): 294-303.

Garry, F. B. 2002. Indigestion in ruminants. In: Large animal internal Medicine. Smith, B.P. (eds.), Mosby, St Louis, Missour. Med., 42 (1-10): 229-251.

Geishauser, T. 1995. J. Vet. Med., 42 (1-10): 229-251.

Ghanem, M, M. 2010. Turki. J. Vet. Anim Sci., 34(2): 143-153.

Ghavi Hossein-Zadeh, N. 2014. J. Agr. Sci. Tech., 16: 69-78.

Gillund, P., Reksen O, Gröhn YT and Karlberg K. 2001. J. Dairy Sci., 84:1390–6.

Goff, J.P. 2018. J. Dairy Sci., 101: 2763–2813.

Go, J.P. and Horst, R.L. 1997. J. Dairy Sci., 80:1260–1268.

Golfman, L.S. 1988. The effects of molybdenum and sulfur on the flow and solubility of various minerals along the digestive tract of steers (MSc thesis). University of Manitoba.

Golub, M. S. and M. E. Gershwin. 1985. In: Animal Stress. Gary P. Moberg (ed.) Am. J. Physiol. Soc., Bethesda, MD, pp. 177–192.

Gonzalez F.D., Muinor, R., Pereira V. and Campos R. 2011: J. Vet. Sci., 12: 251–255.

Gould, L. and Kendall, N. R. 2011. Nutr. Res. Rev., 24: 176–182.

Grace, N.D. 1975. Br. J. Nutr., 34: 73–82.

Grace, N.D. and Lee, J. 1992. N. Z. J. Agric. Res., 35: 367– 377.

Grace, N.D. and Suttle, N.F. 1979. Br. J. Nutr., 41: 125–136.

Grummer, R. R. 2008. Vet. J., 176: 10–20.

Gruner, T.M. 1999. Proc. 29th annual seminar Soc. sheep and beef cattle veterinarians, N Z Vet. Assoc., 68-78.

Guarner, F. and Schaafama, G. J. 1998.Probiotics. Int. J. Food Micro., 39: 237-238.

Guliñski, P, 2021. Vet World, 14(6), 1492.

Hansen, K. 2019. Vet Folio, March 8 2019.

HariBabu, Y. 2003. Immunology: Basic Concepts and Applications. https://www.research.uky.edu/ division-laboratory-animal-esources/guidelines-blood-collection-laboratory-animals.

Harvey, S., J. G. Phillips, A. Rees, and T. R. Hall. 1984. J. Exp. Zoo 232:633–645.

Herdt, T. H. 1991. Bovine Pract. 26:92–95.

Herdt, T. H. 2000. Vet. Clin. North Am. Food Anim. Pract., 16:215–230.

Herdt, T. H., J. Liesman, B. Gerloff, and R. Emery. 1983. Am. J. Vet. Res. 44:293–296.

Higgins, R. J., and W. S. Anderson. 1983. Vet. Rec. 113:461–463.

Hilal, E.Y., Elkhairey, M.A. and Osman, A.O. 2016. Open J. Anim. Sci., 6: 304-324.

Hintz, H.F. and Hogue, D.E. 1964. J. Nutr., 82: 495--498.

Hjerpe, C.A. 1961. J Am Vet Med Assoc., 139: 230-232.

Hoffsis GF, 1980. Traumatic pericarditis. In: Bovine medicine and surgery, ed. Amstutz HE, 2nd ed., pp. 753-756. American Veterinary Publishers, Santa Barbara, CA.

Huisingh, J. and Matrone, G. 1972. Proc. Soc. Exp. Biol. Med., 139: 518--521.

Huisingh, J., McNeill,J.J. and Matrone, G., 1974. Appl Microbiol., 28: 489--497.

ICAR, 2013. Nutrient Requirement of Animals-Poultry (ICAR-NIANP), Indian Council of Agricultural Research, New Delhi.

Inoguchi T, Li P, Umeda F, Yu HY, Kakimoto M, Imamura M, et al. 2000. Diabetes. 49(11): 1939–45.

Isolauri, E., Sutas, Y., Kankaanpaa, P., Arvilommi, H. and Salminen, S. 2001. Am. J. Clin. Nutr., 73: 444-450.

Jaramillo-López E, Itza-Ortiz MF, Peraza-Mercado G and Carrera-Chávez JM, 2017. Austral J Vet Sci., 49(3): 139-148.

Johnson, R. B. 1954. Cornell Vet. 44:6–21.

Jolly, S. S., Sharma, O.P., Garg, G. and Sharma, R. 1980. Fluoride, 13: 10-16.

Josef, F. and Claudia S. 2008. Scan. Cardiovas. J., 42 (1): 9-24.

Joshi, B. R., Kommuru, D. S., Terrill, T. H., Mosjidis, J. A., Burke, J. M., Shakya, K. P., and Miller, J. E. 2011. Vet. Parasitol., 178 (1-2): 192-7.

Julian, R. J. , Ditchfield, J. and Coleman, L. G.1959. Can. J. Comp. Med. Vet. Sci., Xxiii (2)

Kaba, T., Abera, B.and Kassa, T. 2018. BMC Veterinary Research, 14:276.

Kailarpathy, K. and Chin, J. 2000. Imm. Cell. Biol., 78: 80-88.

Kaneko, J. J. 1997. Clinical Biochemistry of Domestic Animals. Fifth ed., Academic Press, New York, USA, pp. 885-905.

Kang D, Lungu SE, Danso F, Dzou CF, Chen Y, Zheng X, Nie F, Lin H, Chen J and Zhou G. 2025. Front. Vet. Sci., 12:1470391.

Khub, S. and Bhttacharya, N. K. 1991. Anim. Prod., 52: 57.

Kim, Y. and Je, Y. 2018. Eur. J. Clin. Nutr., 1682–1689.

Knight, A.P. and Walter, R. 2001. A Guide to Plant Poisoning of Animals in North America. Teton NewMedia.

Kour, H, Grewal RS, Singh P, Lamba JS, Kaur J. and Nayyar S. 2024. J Bu□alo Sci., 13:10–23.

Krook, L. and Whalen, J. P. 2010. Clin. Imag., 34 (6): 458-461.

Kumar R. 2017. Hematol. Transfu. Int. J., 4(4):116-122.

Kumar, G. S., Kumar, T. A. and Dharmaceelan, S. 2021. Indian J. Vet. Med., 41 (1): 89-90.

Kumar, P., Charaya, G., Sindhu, N., Kumar, T. and Rana, Y.S. 2018. Indian, J, Vet. Med., 38 (1 and 2): 113-114.

Kurtz, I., Kraut, J., Ornekian, V. and Nguyen, M.K. 2008. Am J Physiol Renal Physiol, 294:F1009–F1031.

Lacasse, P., Vanacker, N., Ollier, S. and Ster, C. 2018. Res Vet Sci., 116: 40-46.

Langova, L., Novotna, I., Nemcova, P., Machacek, M., Havlicek, Z. et al., 2020. Animals, 10: 1824.

Lean, I.J., Saun, R.V. and DeGaris, P.J. 2013. Vet Clin. North AM., 29:367–86.

Ledney, G.D. and Elliott, T.B. 2010. Health Phys., 98: 145–152.

Lee, M., Jeong, S., Seo, J. and Seo, S. 2019. Asian-Australas J. Anim. Sci., 32(1): 92.

Leshchinsky, T. V. and K. C. Klasing. 2001. Poult. Sci. 80:1590–1599.

Lewis, L.D. 1995. Water, energy, protein, carbohydrates, and fats for horses. In: Lewis LD, editor. Equine clinical nutrition: feeding and care. Media (PA): Williams & Wilkins; p. 3–22.

Li, P., Yin, Y., Li, D. et al. 2007. Br. J. Nutr., 98: 237–52.

Liner, I.E. 1980. Toxic Constituents of Animal Food Stuffs, 2nd ed. Academic Press.

Lipsett, M. B., Beede, D. L. and Thon, N.A. 1961. Hormonal control of sodium, potassium, chloride and water metabolism. In: C.L. Comar and F. Bonner (edn.) Mineral metabolism vol. IB. Academic press, New York.

Lois, R. and King, J. M. 1991. J. Vet. Diagn Invest. 3:52-54.

Maas, J. 2007. Vet Clin North Am Food Anim Pract., 23:527–39.

Machado, M., Castro, M. B., Gimeno, E., Barros, S. S. and Reit-Correa, F. 2020. Toxicon., 187: 1-9.

Mailonee, P. G., Beede, D. K., Collier, R. J. and Wilcox. 1985. J. Dairy Sci., 68:1476.

Mann, S., McArt, J. and Abuelo, A. 2019. Pract. 41:205–19.

Martín-Tereso, J. and Martens, H. 2014. Vet C lin N AM., 30 (3): 643-670,

Maynard, L.A., Loosli J.K., Hintz H.F. and Warner R.G. 1987. Animal Nutrition. Tata McGraw-Hill.

McDonald P, Edwards RA, Greenhalgh JFD, Morgan CA, Sinclair LA and Wilkinson RG. 2011. Animal Nutrition, 7th ed. Benjamin Cummings.

McDowell RL. 1989. Vitamins in Animal Nutrition. Academic Press.

McDowell RL. 2003. Minerals in Animal and Human Nutrition, 2nd ed. Elsevier Science.

McDowell RL. 2012. Nutrition of Grazing Ruminants in Warm Climates. Academic Press. Select articles from Journals

McGuirk, S. M. and Butler, D. G. 1980. J .A. V. M.A., 177: 551-555.

Melendez, P., Lopez, F., Lama, J., Leon, B. and Pinedo, P. 2023. Vet Animal Sci., 19:100277.

Merck Veterinary Manual. 2005. 9thedn. Published by Merck & Co. INC. Whitehouse Station, NJ. USA.

Mischke, R. 2017. J. Hellenic Vet. Med. Soc., 65 (s): 165-180.

Mitchell, K. E. and Rossow, H. A. 2020. EC Vet Sci, 5: 1-13

Monteiro, H. F. and Faciola, A. P. 2020. J. Anim. Sci., 98(8): 248.

Mudge, M.C. 2010. In: Weiss DJ, Wardrop KJ, editors. Schalm's Veterinary Hematology. 6thedn. Philadelphia: Lippincott Williams and Wilkns, 757-62.

Neal, P. A. and Pinsent, P. J. N. 1960. Vet. Rec., 72: 175,

Nelson, D.L. and Cox, M.M. 2017. Lehninger Principles of Biochemistry, 7th ed. Macmillan Learning.

Niki, E. 1987. N. Y. Acad. Sci. 498:186–198.

Nikvand, AA, Rashnavadi, M, and Tabandeh, M R. 2018. J. Vet. Behav., 23: 15-18.

Nockels, C. F. 1986. Fed. Proc. 45:353–360.

Nockels, C. F. 1989. Poultry Digest, pp. 218–226.

Nockels, C. F. 1990. In: Proceedings 51st Annual Conference for Veterinarians. January 6-9, Ft. Collins, CO: Colorado State University.

NRC, 2001. National Research Council, Nutrient Requirements of Dairy Cattle: Seventh Revised Edition, 2001. Washington, DC: The National Academies Press.

Orskov, E. R., Benzie, D., Kay, R.N.B. 1970. Br. J. Nutr., 24:785–95.
Osborne, C. A., Clinton, C.W. and Banman, L. K. 1986. Vet. Clin.North Am.Sm. Anim. Pract., 16: 27-44.
Overton, T.R. and Waldron, M.R. 2004. J Dairy Sci., 87: E105-E119.
Pacheco, R. D. L. and Cruz, G.D. 2015. In: Puniya A., Singh R., Kamra D. (eds). Rumen Microbiology: From Evolution to Revolution. Springer, New Delhi, Cap. 21, p. 315-327.
Packer, L. and Y. Suzuki. 1993. Mol. Asp. Med. 14:229–239.
Panciera, D. L. 1994. J. Am Vet Med Assoc., 204 (5): 761-767.
Pellmar, T.C. and Ledney, G. 2005. Proc. Human Factors and Medicine Panel Research Task Group 099 Radiation Bio-effects and Countermeasures, Bethesda, MD, USA, 21–23.
Phillips, R.W. and Case, G. L. 1980. A.J.V.R. 41:1039-1044.
Pirzada, W.H. and Hussain, S.Z. 1998. Trop. Anim. Health Prod. 30: 209-15.
Polledo, L., Grinwis, CCM, Graham, P., Dunning, M. and Balker, K. 2018. Vet Pathol., 55 (6): 880-888.
Pond, W.G., Church, D.B., Pond, K.R. and Schoknecht, P.A. 2004. Basic Animal Nutrition and Feeding, 5th edn., Wiley.
Prasad, C. S. and Arora, S. P. 1979. Indian J. Dairy Sci., 32: 375.
Prasse, Keith W. 1983. The Bovine Proc., N0. 15: 79-83.
Priyanka, M. and Dey, S. 2018. Vet. World, 11(9): 1307-1315.
Qi, K., Owens, F. N. and Lu, C. D. 1994. Small Rum. Res., 14: 115-126.
Radostits, O.M., Gay, C.C., Hinchcliff, K.W. and Constable, P.D. 2010. Veterinary Medicine – A Textbook of the Diseases of Cattle, Horses, Sheep, Pigs and Goats. Saunders Elsevier.
Rahbari, A., Salehi, A.q, Sharifi, S. D. and Honarbakhsh, S. 2025. Poult. Sci., 104: 104468.
Rosanthal, G., Russel, J.B. and Wilson, D. B. 1988. J. Nutr., 118: 271-279.
Round, J. L. 2001. Eur. J. Nutr., 40 (5): 245-54.
Rukkwamsuk, T., Kruip, T.A.M. and Wensing, T. 1999. Vet. Quart., 21:71–7.
Santos, M.M., Bregadioli, G.C., Santos, L.G., Curti, J.M., Duarte, C.A. et al., 2020. Res Vet Sci, 128: 24-34.
Santra, A. and Pathak, N. N. 1999. Asian Aus. J. Anim. Sci., 12: 1285.
Schnieder, P.L., Beede, D. K., Wilcox, C. J. and Collier, R. J. 1984. J. Dairy Sci., 67: 2546.
Schultz, L.H. 1970. J. Dairy Sci., 54: 962-973.
Seifi, HA and Kia, S. 2018. Iran. J. Vet. Sci. Technol., 9: 1-15.
Seifter, J. L. and Chang, Hsin-Yun. 2016. Kidney Dis., 2: 170-186.
Selye, H. 1973. Am. Scientist, 61: 692–699.
Senthilkumar, V., Mohamed S. A., Kathiravan, G., Subramanian, M. and Mani, K. 2013. Int. J. Adv. Vet. Sci. Technol., 2: 64-71.
Senthilkumar, V., Mohamed S. A., Kathiravan, G., Subramanian, M. and Mani, K. 2015. Ind. J. Vet. Anim. Sci. Res., 44: 102-104.
Seth, M., Jackson,. KV. and Giger, U. 2011. Am J Vet Res., 72:203–209.
Shaver, R.D., 1997. J. Dairy Sci., 80(10): 2449-2453.
Siegel, H. S. 1995. Br. Poult. Sci. 36:003–22.
Silva, D.C.D., Fernandes B.D., Lima, J.M.D.S., Silva, B.A.D., Rodrigues, G.P. et al., 2020. Rev Bras Saúde Prod Anim., 21: 2020.
Singh, U.B. 1987. Advanced Animal Nutrition for Developing Countries. Indo-Vision.
Sivertsen, T. and Plassen, C. 2004. Acta vet. Scandi., 45:69-77.
Skillman, J.J., Hedley-Whyte J. and Pallotta, J.A. 1971. Anim. Surg., 174(6): 911–22.
Smith, B. 1993. NewZealand Vet. J., 21: 40.
Smith, G. S. 1992. J. Range Mgmt., 45: 25-30.

Smith, T. C., McDowell, E. G., Found, F. F. and Holstead, I. A. 1975. Sci., 8.

Stauffacher, W., Orci, L., Cameron, D. P., Burr, I.M. and Renold, A. 1971. Proc.1970 Laurention Hormone Conf., pp. 41-95.

Stone, W.C. 2004. J. Dairy Sci., 87: E13-E26.

Suber, R. L., Hentges, J. R., Gudat, J. C. and Edds, G. T. 1979. A.J .Y.R., 40: 1005-1008.

Sun, J., Zhan, Y., Mariosa, D., Larsson, H., Almqvist, C., Ingre, C., Zagai, U., Pawitan, Y. and Fang, F. 2019. Eur. J. Neurol., 26 (11): 1355.

Sun, X., Wang, Y., Wang, E., Zhang, S., Wang, Q. et al., 2021. Anim., 11(8): 2401.

Sundrum, A. 2015. Anim., 5(4): 978-1020.

Suttle, N. F. 2010. Mineral Nutrition of Livestock, 4th ed. CAB International.

Tello, L. and Perez-Freytes R. 2017. Vet Clin. North Am. Small Animal Pract., 47 (2): 505-519.

Thirunavukkarasu, M; Kathiravan, G; Kalaikannan, A and Jebarani, W. 2010. Agric. Econom. Res. Rev., 23: 77-81.

Thisson, Jean-Paul, Underwood, Louis E. and Ketsisegas, Jean-Marie. 1999. Nutr. Rev., 57: 167.

Thrall, M. A. 2007. Hematology and Veterinary Clinical Biochemistry, Roca Sao Paulo.

Tosti, V., Bertozzi, B. and Fontana, L. 2018. J. Gerontol A Biol Sci Med Sci., 73: 318–326.

Usman, M., Noor, Ahmad, U., Khan, A., Rahman, A., Kazmi, Suef, Arshad, U. and Abdullah, O. B. 2025. Indus J. Biosci. Res., 3 (2): 748.

Veenhuizen, J. J., Drackley, J. K., Richard, M. J., Sanderson, T. P., Miller, L.D. and Young, J. W. 1991. J. Dairy Sci., 74: 4238-4253.

Wang, L., Lin, B., Zhai, M., Cui, W., Hull, L., Zizzo, A., Li, X., Kiang, J.G. and Xiao, M. 2022. Toxics, 10: 785.

Wardrop, K. J. 2007. J. Vet. Intern Med. 21: 205-206.

Wensvoort, P. and van der Velden, M. A. 1980. Vet. Quat., 2:3, 125-135.

Wise, G. H. 1939. J. Dairy Sci., 22: 465.

Wise, G. H. and Anderson, G. W. 1930. J. Dairy Sci., 22: 697.

Wise, G. H., Anderson, G. W. and Miller, P. G. 1942. J. Dairy Sci., 25: 529.

Wu, G. 2020. In: Animal Agriculture Bazer, Academic Press, Cambridge, MA, USA, pp 471-490.

Xiao, Q., Garaulet, M. and Scheer, F. 2019. Int. J. Obes., 43:1701–11.

Yazer, M.H., Judd, W.J., Davenport, R.D., Dake, L.R., Lomas-Francis, C. and Hue-Roye, K. 2006. Transfus., 46:1537–42.

Zhang, F., Nan, X., Wang, H., Zhao, Y. and Guo, Y. 2020. Anim., 10 (9): 1526.

Zhao, C., Liu, G., Li, X., Guan, Y., Wang, Y., Yuan, X., Sun, G., Wang, Z. and Li, X. 2018. BMC Vet. Res., 14:135.

Index

F

G

H

I

K

L

M

N

R

W

Z